Lukas Kölsch

Dynamic Incentives for Optimal Control of Competitive Power Systems

Karlsruher Beiträge zur
Regelungs- und Steuerungstechnik
Karlsruher Institut für Technologie

Band 17

Dynamic Incentives for Optimal Control of Competitive Power Systems

by
Lukas Kölsch

KIT Scientific Publishing

Karlsruher Institut für Technologie
Institut für Regelungs- und Steuerungssysteme

Dynamic Incentives for Optimal Control of Competitive Power Systems

Zur Erlangung des akademischen Grades eines Doktor-Ingenieurs von der KIT-Fakultät für Elektrotechnik und Informationstechnik des Karlsruher Instituts für Technologie (KIT) genehmigte Dissertation

von Lukas Kölsch, M.Sc.

Tag der mündlichen Prüfung: 10. Dezember 2021
Hauptreferent: Prof. Dr.-Ing. Sören Hohmann
Korreferent: Prof. Dr.-Ing. Andreas Rehkopf

Impressum

KIT Scientific Publishing

Karlsruher Institut für Technologie (KIT)
KIT Scientific Publishing
Straße am Forum 2
D-76131 Karlsruhe

KIT Scientific Publishing is a registered trademark of Karlsruhe Institute of Technology.
Reprint using the book cover is not allowed.

www.ksp.kit.edu

Print on Demand 2022 – Gedruckt auf FSC-zertifiziertem Papier

ISSN 2511-6312
ISBN 978-3-7315-1209-7
DOI 10.5445/KSP/1000147941

Preface

This dissertation has been written during my work as research assistant at the Institute of Control Systems (IRS) at Karlsruhe Institute of Technology (KIT).

First and foremost, I would like to express my deepest gratitude to my doctoral supervisor Prof. Dr.-Ing. Sören Hohmann for giving me the opportunity of pursuing my PhD under his chair, his excellent scientific guidance throughout my time at the IRS and his permanent openness to explore and find new research fields. I also owe sincere gratitude to Prof. Dr.-Ing. Andreas Rehkopf for acting as a second examiner and for his personal encouragement during the final phase of my work.

I am highly indepted to my former master's students (and now colleagues) Felix Strehle, Pol Jané Soneira, and Lukas Rausche, whose exciting scientific acumen in general and careful proofreading of this dissertation in particular have made a crucial impact on shaping the manuscript in its present form. Likewise, I would like to thank Dr.-Ing. Stefan Krebs and Dr.-Ing. Martin Pfeifer for their ongoing interest and motivation, which always pushed me to find the right path. I am very thankful for the best office colleagues one could imagine, namely Florian Köpf, Simon Rothfuß, and Manuel Schwartz, with whom I was allowed to experience many prosperous discussions of both scientific and non-scientific nature, discovering the spontaneous self-therapeutic effect of collaborative whiteboard discussions. I would also like to thank Rishabh Vyas, who helped me numerous times with special linguistic questions about the final manuscript even at the most pathological times of the night, as well as all my other undergraduate and master's students, whose willingness to write their own thesis under my supervision was instrumental in generating numerous new ideas.

Furthermore, I am very grateful for the financial support provided by the German Research Foundation as part of the DFG Priority Programme "Hybrid and Multimodal Energy Systems".

Finally, my heartfelt and very special thanks go to my parents Viola and Roland for their unconditional mental support during the writing process and throughout my whole life.

Karlsruhe, December 2021 Lukas Kölsch

Abstract

Incentive mechanisms can play a pivotal role in accelerating technologically challenging transformation processes such as the energy transition. The aim of those incentives is to provide an environment where the interplay of each competitor's individually optimal actions is also globally optimal in the sense of an overall objective. The dissertation at hand presents a feedback control framework towards optimal incentive mechanisms for today's and future power systems in the triad of system stability, economic efficiency, and grid support. A key innovation of the developed control scheme is the incorporation of temporally and spatially differentiated real-time price signals that result from the solution of static and dynamic optimization problems. Inclusion of locally available measurement information, consistent co-modeling of the underlying physical network including resistive losses, and continuous-time formulation of all subsystems pave the way towards a real *closed-loop* incentive control framework. Particular importance is attached to a rigorous separation between market and network participants, respecting the principle of unbundling. After comprehensive analysis of the resulting closed-loop system, the control scheme is further adapted to a novel real-time congestion management system. Two case studies illustrate practical advantages of the developed approach compared to existing concepts. The port-based system modeling, the absence of any centralized control authority, and the ability for automatic, decentralized regulation of all prices across the entire network finally enable seamless extensibility by additional incentive components.

Kurzfassung

Technologisch herausfordernde Transformationsprozesse wie die Energiewende können durch passende Anreizsysteme entscheidend beschleunigt werden. Ziel solcher Anreize ist es hierbei, ein Umfeld idealerweise so zu schaffen, dass das Zusammenspiel aller aus Sicht der beteiligten Wettbewerber individuell optimalen Einzelhandlungen auch global optimal im Sinne eines übergeordneten Großziels ist. Die vorliegende Dissertation schafft einen regelungstechnischen Zugang zur Frage optimaler Anreizsysteme für heutige und zukünftige Stromnetze im Zieldreieck aus Systemstabilität, ökonomischer Effizienz und Netzdienlichkeit. Entscheidende Neuheit des entwickelten Ansatzes ist die Einführung zeitlich wie örtlich differenzierter Echtzeit-Preissignale, die sich aus der Lösung statischer und dynamischer Optimierungsprobleme ergeben. Der Miteinbezug lokal verfügbarer Messinformationen, die konsequente Mitmodellierung des unterlagerten physikalischen Netzes inklusive resistiver Verluste und die durchgängig zeitkontinuierliche Formulierung aller Teilsysteme ebnen den Weg von einer reinen *Anreiz-Steuerung* hin zu einer echten *Anreiz-Regelung*. Besonderes Augenmerk der Arbeit liegt in einer durch das allgemeine Unbundling-Gebot bedingten rigorosen Trennung zwischen Markt- und Netzakteuren. Nach umfangreicher Analyse des hierbei entstehenden geschlossenen Regelkreises erfolgt die beispielhafte Anwendung der Regelungsarchitektur für den Aufbau eines neuartigen Echtzeit-Engpassmanagementsystems. Weitere praktische Vorteile des entwickelten Ansatzes im Vergleich zu bestehenden Konzepten werden anhand zweier Fallstudien deutlich. Die port-basierte Systemmodellierung, der Verzicht auf zentralisierte Regeleingriffe und nicht zuletzt die Möglichkeit zur automatischen, dezentralen Selbstregulation aller Preise über das Gesamtnetz hinweg stellen schließlich die problemlose Erweiterbarkeit um zusätzliche optionale Anreizkomponenten sicher.

Contents

List of Figures

List of Tables

Abbreviations and Symbols

Abbreviations

Abbreviation	Description
AC	alternating current
ADP	adaptive dynamic programming
aFRR	automatic frequency restoration reserve
BG	balancing group
BM	balancing market
BoP	balance of payments
BRP	balance responsible party
CC	cell coordinator
CLF	control-Lyapunov function
CPP	conventional power plant
CVPP	commercial virtual power plant
DAPI	distributed averaging proportional-integral
DC	direct current
DER	distributed energy resource
DMPC	distributed model predictive control
DSO	distribution system operator
EEX	European Energy Exchange
EI	energy internet
ENTSO-E	European Network of Transmission System Operators for Electricity
EPEX	European Power Exchange
ES	electricity supplier
ETS	emissions trading scheme
FACTS	flexible AC transmission system
FCR	frequency containment reserve
HJB	Hamilton-Jacobi-Bellman
IDA	interconnection and damping assignment
IEEE	Institute of Electrical and Electronics Engineers
ISO	independent system operator
LCL	filter composed of series inductor, parallel capacitor, and series inductor
LQ	linear-quadratic

Abbreviation	Description
LQR	linear-quadratic regulator
mFRR	manual frequency restoration reserve
MG	microgrid
MOC	modified optimal control
MPC	model predictive control
ODE	ordinary differential equation
OPF	optimal power flow
PBC	passivity-based control
PDE	partial differential equation
PES	Power & Energy Society
PHS	port-Hamiltonian system
PMSM	permanent magnet synchronous motor
PMU	phasor measurement unit
PPO	power plant operator
RES	renewable energy source
RP	retail prosumer
SG	synchronous generator
SLP	standard load profile
SM	synchronous motor
TSO	transmission system operator
TVPP	technical virtual power plant
VAR	Volt-Ampere reactive
VPP	virtual power plant
WEM	wholesale electricity market
WoC	web-of-cells
WP	wholesale prosumer

Latin Symbols

Symbol	Description
A_i	positive damping coefficient
B_{ii}	negative of self-susceptance
B_{ij}	negative of line susceptance
B'_{ij}	specific parallel susceptance
$b_{\mathrm{sh},i}$	shunt susceptance
C^1	set of n-times continuously differentiable functions
$C_{\mathrm{DC},i}$	DC-side capacitance
$\boldsymbol{d}$	disturbance input vector
$\boldsymbol{D}_{\mathrm{c}}$	incidence matrix of overall communication
$\hat{\boldsymbol{D}}_{\mathrm{c}}$	incidence matrix of inter-cell communication

Symbol	Description
$\boldsymbol{D}_{\mathrm{c},k}$	incidence matrix of communication in cell k
$D_{\mathrm{d},i}$	damping-torque coefficient
$d_{\mathrm{d},i}$	damping coefficient
$\boldsymbol{D}_{\mathrm{p}}$	plant incidence matrix
$\hat{\boldsymbol{D}}_{\mathrm{p}}$	incidence matrix of cell interconnection
$\mathbf{e}_k$	unit vector
f	mechanic friction constant
f_i	nodal frequency
$\boldsymbol{G}$	input matrix of port-Hamiltonian system
$G_{\mathrm{DC},i}$	DC-side conductance
G_{ij}	negative of line conductance
G'_{ij}	specific parallel conductance
$g_{\mathrm{sh},i}$	shunt conductance
H	Hamiltonian of port-Hamiltonian system
$\boldsymbol{I}$	identity matrix
$I_{\mathrm{DC},i}$	DC-side current
$\vec{I}_i$	nodal current
$\boldsymbol{J}$	interconnection matrix of port-Hamiltonian system
J_i	moment of inertia
$\boldsymbol{K}$	diagonal matrix of $\boldsymbol{\kappa}$ for each node
L_i	deviation of angular momentum from nominal value $\Gamma_i\Omega_{\mathrm{m}}$
L_{s}	stator inductance
m	number of edges
M_i	torque
m_{p}	number of power lines
N	number of players
n	number of states
n_{g}	number of $\mathcal{S}$ and $\mathcal{I}$ nodes
$n_{\mathcal{I}}$	number of $\mathcal{I}$ nodes
$n_{\mathcal{L}}$	number of $\mathcal{L}$ nodes
n_{p}	number of buses
n_{pp}	number of pole pairs
$n_{\mathcal{S}}$	number of $\mathcal{S}$ nodes
$n_{\mathcal{W}}$	number of wholesale prosumers
$n_{\mathcal{Z}}$	number of cells
$p_{\mathrm{DC},i}$	DC-side active power injection
$p_{\mathrm{g},i}$	active power generation
$\hat{p}_{\mathrm{g},k}$	aggregated active power generation in cell k
$\boldsymbol{p}_{\mathrm{g},k}$	vector of active power generations in cell k
$\boldsymbol{p}_{\mathrm{g},\pi}$	vector of active power generations of wholesale prosumer π
p_i	active power flow from node i to neighboring nodes
P_{ij}	(sending-end) active power flow from node i to j

Symbol	Description
$p_{\text{inj},i}$	active power injection
$p_{\ell,i}$	active power demand
$\hat{p}_{\ell,k}$	aggregated active power demand in cell k
$\boldsymbol{Q}$	state weight matrix
$q_{\text{g},i}$	reactive power generation
q_i	reactive power flow from node i to neighboring nodes
Q_{ij}	(sending-end) reactive power flow from node i to j
$q_{\text{inj},i}$	reactive power injection
$q_{\ell,i}$	reactive power demand
$\boldsymbol{R}$	dissipation matrix of port-Hamiltonian system
r	number of weights
$R_{\text{DC},i}$	DC-side resistance
R'_{ij}	specific series resistance
$\boldsymbol{R}_{\text{p}}$	plant dissipation matrix
$\boldsymbol{r}_{\text{p}}$	vector of nonlinear resistances
$R_{\text{s},i}$	stator resistance
$\vec{\boldsymbol{S}}$	vector of inter-cell complex power flows
$\boldsymbol{S}$	control weight matrix
$\vec{s}_i$	complex power injection
$\vec{S}_{ij}$	complex power flow
S_m	apparent power flow over line m
t_{f}	end time
t	time
$T'_{\text{d0},i}$	d-axis open-circuit transient time constant
$T''_{\text{d0},i}$	d-axis open-circuit subtransient time constant
$T'_{\text{q0},i}$	q-axis open circuit transient time constant
$T''_{\text{q0},i}$	q-axis open-circuit subtransient time constant
$\boldsymbol{u}$	input vector
$U_{\text{AC},i}$	AC-side voltage magnitude of inverter
$U_{\text{DC},i}$	DC-side voltage of inverter
$U_{\text{f},i}$	magnitude of excitation voltage
$U_{\text{g},i}$	terminal (internal) voltage magnitude
$\vec{U}_i$	nodal voltage phasor
U_i	nodal (external) voltage magnitude
U'_i	transient internal voltage magnitude
U''_i	subtransient internal voltage magnitude
U^{nom}	nominal voltage
U^{set}	voltage setpoint
$\overline{U}^{\Delta}$	maximum deviation of voltage magnitude
$\boldsymbol{v}$	eigenvector
$V(\boldsymbol{x})$	(control-)Lyapunov function

Symbol	Description
$V(\boldsymbol{x}, \boldsymbol{w})$	extended control-Lyapunov function
$V^{\star}(\boldsymbol{x})$	value function
$\boldsymbol{w}$	vector of weights for extended control-Lyapunov function
$\boldsymbol{x}$	state vector (of closed-loop system)
$X_{\mathrm{d},i}$	d-axis synchronous reactance
$X'_{\mathrm{d},i}$	d-axis transient reactance
$X''_{\mathrm{d},i}$	d-axis subtransient reactance
X'_{ij}	specific series reactance
$X_{\mathrm{n},i}$	reactance of the mains connection
$\boldsymbol{x}_{\mathrm{p}}$	plant state vector
$X_{\mathrm{q},i}$	q-axis synchronous reactance
$X'_{\mathrm{q},i}$	q-axis transient reactance
$X''_{\mathrm{q},i}$	q-axis subtransient reactance
$X_{\mathrm{T},i}$	transformer reactance
$\vec{\boldsymbol{Y}}$	nodal admittance matrix
$\boldsymbol{y}$	output variable
$\vec{Y}_{\mathrm{L},ij}$	parallel admittance of power line
$y_{\mathrm{sh},i}$	shunt admittance
$\boldsymbol{z}$	co-state
$\vec{Z}_{\mathrm{L},ij}$	series resistance of power line

Greek Symbols

Symbol	Description
α	learning rate
Γ_i	moment of inertia of swing equation
γ_i	rotor angle in static reference frame
δ_i	absolute rotor angle in synchronous rotating reference frame
$\delta(t)$	Dirac delta function
ϵ^{κ}	restoration signal
$\boldsymbol{\zeta}$	controller co-state
η_{ij}	multiplier for neighboring nodal prices
Θ	magnetic flux
θ_i	bus voltage phase angle
$\vartheta_{\mathrm{g},i}$	rotor angle in synchronous rotating reference frame
$\vartheta_{\mathrm{n},i}$	pseudo rotor angle in synchronous rotating reference frame
ϑ_{ij}	bus voltage angle difference
ι	communality factor
$\hat{\boldsymbol{\kappa}}$	diagonal matrix of participation factors
κ_k	participation factor for cell k

Symbol	Description
κ'_k	participation factor for cell k without correction
κ''_k	participation factor for cell k without regularization and correction
Λ_k	wholesale price in cell k
$\boldsymbol{\lambda}$	Lagrange multipliers for equality constraints
λ_i	price at node i
Λ^0	uniform price
Λ_{R}	retail price
$\boldsymbol{\mu}$	Lagrange multipliers for inequality constraints
$\boldsymbol{\nu}$	vector of virtual power flows for frequency control
$\boldsymbol{\nu}_a$	vector of virtual power flows for balance of payments containment
Ξ	vector of basis functions
$\boldsymbol{\xi}$	controller state
π	circle number
ϖ	scaling factor for cell-specific cost function
ρ_i	active loss of node i
ϱ_i	reactive loss of node i
σ	balance of payments error
$\boldsymbol{\tau}$	diagonal matrix of time constants
Υ	Lagrange multiplier of modified optimal control
Φ	overall active transmission loss
$\hat{\Phi}_k$	overall active transmission loss in cell k
ϕ_k	generalized effort for zonal pricing
φ_i	load angle
χ	scaling factor for reactive power sharing
Ψ	auxiliary variable for synchronous machine limits
$\boldsymbol{\psi}$	generalized flow vector for zonal pricing
ω_i	deviation of nodal frequency from nominal value
$\Omega_{\mathrm{m},i}$	angular velocity of mechanical rotor
$\omega_{\mathrm{m},i}$	deviation of angular velocity of mechanical rotor
$\Omega_{\mathrm{n},i}$	nominal value of angular velocity

Calligraphic Symbols

Symbol	Description
$\boldsymbol{\mathcal{A}}$	Laplacian matrix
$\mathcal{B}(\boldsymbol{x})$	open ball around $\boldsymbol{x}$
$\mathcal{C}_m$	congestion factor of line m
$\mathcal{E}_{\mathrm{c}}$	set of communication edges
$\hat{\mathcal{E}}_{\mathrm{c}}$	set of inter-cell communication edges
$\mathcal{E}_{\mathrm{c},k}$	set of communication edges of cell coordinator k

Symbol	Description
$\mathcal{E}_{\mathrm{p}}$	set of plant edges
$\hat{\mathcal{E}}_{\mathrm{p}}$	set of inter-cell edges
$\mathcal{E}_{\mathrm{p}}^{\mathrm{b}}$	set of boundary edges
$\mathcal{E}_{\mathrm{p}}^{\mathrm{nb}}$	set of non-boundary edges
$\mathcal{I}$	inverter-type
$\mathcal{J}_{\mathrm{w}}$	objective function for adaptation
$\mathcal{L}$	load-type
$\mathcal{M}$	neighborhood
$\mathcal{N}_i$	neighbors of i in $\mathscr{G}$ (undirected)
$\mathcal{N}_{\mathrm{p},i}$	neighbors of i in $\mathscr{G}_{\mathrm{p}}$ (undirected)
$\mathcal{Q}$	quadric
$\boldsymbol{\mathcal{R}}$	nonlinear resistive structure
$\mathcal{S}$	synchronous machine-type
$\mathcal{T}$	tangent cone
$\mathcal{U}$	input space
$\mathcal{V}_{\mathrm{g}}$	set of synchronous machine- and inverter-type nodes
$\mathcal{V}_{\mathrm{g},k}$	set of synchronous machine- and inverter-type nodes in cell k
$\mathcal{V}_{\mathcal{I}}$	set of inverter-type nodes
$\mathcal{V}_{\ell}$	set of synchronous machine- and load-type nodes
$\mathcal{V}_{\mathcal{L}}$	set of load-type nodes
$\mathcal{V}_{\mathrm{p}}$	set of nodes of plant system
$\mathcal{V}_{\mathcal{S}}$	set of synchronous machine-type nodes
$\mathcal{V}_{\mathcal{W}}$	set of wholesale prosumer nodes
$\mathcal{V}_{\mathcal{W},\pi}$	set of nodes of wholesale prosumers π
$\mathcal{V}_{\mathcal{Z},k}$	set of nodes in cell k
$\mathcal{W}$	set of wholesale prosumers
$\mathcal{X}$	state space
$\mathcal{Z}$	set of cells

Script Symbols

Symbol	Description
$\mathscr{G}_{\mathrm{c}}$	overall communication graph
$\hat{\mathscr{G}}_{\mathrm{c}}$	aggregated cell communication graph
$\mathscr{G}_{\mathrm{c},k}$	communication graph of cell coordinator k
$\mathscr{G}_{\mathrm{p}}$	graph of physical system
$\hat{\mathscr{G}}_{\mathrm{p}}$	aggregated cell graph
$\mathscr{H}$	Hamiltonian (Pontryagin function)
$\Im$	imaginary part
$\mathscr{J}$	objective function

Symbol	Description
$\mathscr{L}$	Lagrangian
ℓ	state-dependent integrand of objective function
$\Re$	real part

Doublestruck Symbols

Symbol	Description
$\mathbb{A}_i$	set of actions of player i
$\mathbb{L}$	solution set
$\mathbb{N}$	set of natural numbers
$\mathbb{P}$	set of players
$\mathbb{R}$	set of real numbers
$\mathbb{W}^+$	set of weights leading to positive control-Lyapunov function candidate
$\mathbb{Y}^+$	set of states and weights leading to positive control-Lyapunov function candidate
$\mathbb{Y}^+_{\mathrm{cyl}}$	cylindric set
$\hat{\mathbb{Y}}^+_{\mathrm{cyl}}$	maximum cylindric set

Typewriter Symbols

Symbol	Description
$\mathtt{C}$	overall cost function
$\mathtt{C}_{\mathrm{g},i}$	cost for power generation at node i
$\breve{\mathtt{C}}_{\mathrm{g},i}$	convexified cost function for power generation at node i
$\mathtt{C}^{\ell}_k$	cost of retail prosumer in cell k
$\mathtt{C}_\pi$	cost of wholesale prosumer π
$\mathtt{J}$	objective function
$\mathtt{L}$	Lagrangian of Lagrange-type performance index
$\mathtt{P}_{\mathrm{g},i}$	profit at node i
$\mathtt{P}_k$	profit of cell coordinator k
$\mathtt{P}_\pi$	profit of wholesale prosumer π
$\mathtt{Q}$	quadratic function
$\mathtt{R}_\pi$	revenue of wholesale prosumer π
$\mathtt{W}$	welfare function

Other Symbols

Symbol	Description
$\mathbf{0}$	all-zeros matrix
$\mathbb{0}$	all-zeros vector
$\emptyset$	empty set
$\mathbb{1}$	all-ones vector
$\jmath$	imaginary unit
$\circ$	Hadamard product
$\times$	Cartesian product
$\mho$	penalty/barrier function

Indices and Exponents

Index	Description
$\square^*$	complex conjugate
$\square^\star$	optimizer
$\square^\dagger$	pseudoinverse
$\breve{\square}$	convexified
$\hat{\square}$	cellular
$\dot{\square}$	time derivative
$\vec{\square}$	complex variable
$\overline{\square}$	upper limit
$\underline{\square}$	lower limit
$\square_\mathrm{c}$	communication
$\square_\mathrm{d}$	d-axis
$\square_\mathrm{g}$	generation
$\square_i$	node (instance)
$\square_{ij}$	node-to-node (instance)
$\square_\mathrm{inj}$	injected
$\square_j$	node (instance)
$\square_k$	cell (instance)
$\square_{kl}$	cell-to-cell (instance)
$\square_\ell$	consumption
$\square_l$	cell (instance)
$\square_\mathrm{m}$	mechanical
$\square_m$	edge (instance)
$\square_\mathrm{n}$	nominal
$\square_\mathrm{p}$	physical system
$\square_\mathrm{q}$	q-axis

Index	Description
$\square_{\mathrm{R}}$	retail
$\square_{\mathrm{s}}$	stator
$\square_{\mathrm{sh}}$	shunt
$\square_{\mathcal{W}}$	wholesale
$\square_{\pi}$	power plant operator (instance)

Operators

Operator	Description
$\nabla\square$	gradient
$\neg\square$	not
$\arg\square$	argument
$\mathrm{bd}\square$	boundary
$\mathrm{cl}\square$	closure
$\mathrm{col}\square$	column vector
$\mathrm{diag}\square$	diagonal matrix
$\mathrm{im}\square$	image
$\mathrm{int}\square$	interior
$\ker\square$	kernel
$\mathrm{proj}\square$	canonical projection
$\Pi\square$	gradient projection
$\lvert\square\rvert$	absolute value
$\lVert\square\rVert_2$	Euclidean norm
$\lVert\square\rVert_\infty$	maximum norm

1 Introduction

The harnessing of electrical energy is among the most groundbreaking achievements in human history. Its permanent and widespread local availability by means of an interconnected power system is a cornerstone of our modern society and has become an indispensable condition for any civilizational, i.e. cultural, technological, and economic progress. As a result, this civilizational progress itself has consequences for the energy supply system: Stemming from a growing awareness of the negative impact of fossil energy production on the environment, and provoked by the climate debate of recent years, the idea of *sustainability* has gained an increasing presence. Accordingly, our natural environment has faced the repercussions of technological progress, thus in particular the emissions of greenhouse gases as well as the depletion of natural resources, should be given greater focus [RLP⁺15]. This understanding has led to a number of worldwide political declarations of intent, according to which the ecological footprint is desired to be drastically reduced. As part of the European Green Deal, in 2021 the European Climate Law [Cou21] was passed, according to which the EU member states are obliged to reduce their net greenhouse gas emissions to zero by 2050. Moreover, it enforces a reduction in greenhouse gas emissions of minus 55%[1] by the year 2030.

The power generation sector holds a prominent role in climate protection, as it not only accounts for the largest share of global greenhouse emissions to date [BBM⁺14, p. 516; MEH⁺17, p. 7], but also has the greatest quantitative potential for substitution compared with other sectors [DKH19, p. 3]. As a consequence, many countries have set ambitious goals for the next decades to shift away from conventional fossil-fueled power generation towards renewable and low-carbon energy sources. Germany's reduction targets are consolidated in the *Climate Protection Plan 2050.* For the power generation sector, this directive imposes a drastic reduction from 280 to 175 million tonnes of CO_2 equivalents between 2020 and 2030[2]. These actions are intended to increase the share of renewables in gross electricity consumption in Germany[3] from 32% in 2015 to 65% in 2030 [Deu14, KM17, Deu20].

From a systems viewpoint, these ambitious policy roadmaps are accompanied by significant techno-economic disruptions, namely, the increasing displacement of conventional, fossil-fueled power plants in favor of power generation from renewable energy sources (RESs), such as wind, photovoltaic, biomass or run-of-river plants, leading to a structural change in the feed-in characteristcs: Since the majority of RES power plants are based on energy sources whose availability is weather-dependent [SP18], no exact forecasts can be made about the expected feed-in. This growing uncertainty manifests itself in an increasing volatility of the *residual load*, meaning that the difference between the current demand and the current

1 related to the reference year 1990.

2 Sector-specific targets on an annual basis are imposed in a federal law, see [Deu19, Annex 2].

3 Likewise, there are national measures and legislative acts for stimulating green electricity in the U.S. [EMG⁺21], China [HZZ20], Canada [Gov16], India [Int20], and Scandinavia [Nor17]. An overview of the measures in the individual EU member states is given in [Eur20].

generation by RES power plants fluctuates to an increasingly large extent. However, it is important to note that this increasing volatility has not only a temporal, but also a geographical dimension. Unlike conventional power plants, which can be located deliberately at load centers [Sch14, p. 11], renewable generation units are usually located wherever geographic conditions allow an economically viable harvest of e.g. wind or solar supply. This creates the undesirable phenomenon of *locally non-integrable power* during periods of high solar exposure or high wind intensity [MGMH14]. Likewise, load centers with geographic conditions that are expected to be less profitable for power plant operators may suffer from ongoing deficits in local net generation. This consequently will result in significant and increasing amounts of compensation flows via the transmission lines [ZPM13].

In light of this upcoming scenario of increasing temporal and local volatility, there is serious reason among decision-makers as well as in the scientific community to expect that the existing power transmission and distribution system, in its current form, will not be able to keep pace with these developments. Furthermore, it is evident that the existing technical and regulatory system infrastructure is noticeably becoming stressed. Until now, this fact has already manifested itself e.g. in increasingly volatile spot market prices [RN20], increased incidence of line congestion leading to critical grid interventions [Sta19, p. 42], reverse power flows [NCB+18, HRR20], and increasingly high amounts of net imports and exports of electricity over long distances [KO20, p. 164].

The most evident technical means of counteracting these issues is through the installation of local energy storages (see [DBK+18, Gür18, BMC+20] for comprehensive reviews of prospective technologies and their economic viability), accompanied by a massive grid expansion. However, apart from the long time horizons from planning to launch[4], oversizing transmission lines according to a worst-case calculation is known to not be economically viable, especially if this worst-case scenario can only be roughly estimated. Apart from these technical challenges, there is also a strong need for redesign of the existing regulatory and market framework to accommodate the high intermittency of RESs while ensuring overall system stability [KO20, p. 164f.; Sch21, p. 6f.].

Moreover, increased *competition* among the flexible participants is a useful mechanism to stimulate the participants' own strive for innovation. At the same time, excesses in consumer prices have to be prevented in order to avoid undermining people's confidence in the viability of renewable energy generation and thus missing the targets of the whole energy transition. Consequently, a proper mechanism design should deliver *real-time signals* to all competitors that incentivize supply during times of particularly high demand and in geographic regions where it is of utmost necessity from a "global" network perspective. Although the current structure of the energy market, dominated by *normative interventions* such as feed-in tariffs or balancing group (BG) contracts, is not capable of providing such investment signals that internalize network effects, first studies also show that a mere adoption of locally differentiated prices may lead to false incentives as well, unless the slowly reacting settlement times of existing electricity markets are overcome (see e.g. [Con19]).

[4] Due to extensive tendering and approval processes, the average time required for a transmission line construction permit is 7 years, with an upper quartile of 14 years [BL11, p. 4]. Likewise, electrochemical storages with fast reaction times such as battery energy storage systems will presumably not be available on a large scale within the next 20 years [FSK+20, p. 2].

At this point, scientific contributions from the area of *automatic control theory* can contribute to "close the loop" between market and network (i.e., between the pricing mechanism and the physical interaction of participants across the power grid) and to rigorously analyze the interdependencies that arise in this process. The dissertation at hand is thus devoted to developing a holistic feedback control framework for large-scale interconnected power systems with competitive participants which ensures *stability* of the overall system. By applying a market-based architecture, the controller is intended to pursue *economically efficient* operating points without any need for additional supervision by a higher-level authority. At the same time, the control framework should be able to *support the grid* by taking the geographical component of momentary infeed into account. Thereby, continuous valuation of each participant's individual contribution to a stable network operation should provide instantaneous price incentives for local oversupply or undersupply. Despite this conceptual redesign towards real-time signaling, the proposed framework is intended to be a genuine *brownfield approach*, i.e., it should neither tighten the existing action space of competitive network participants nor require a redesign of *physical* infrastructure. Hence, the overall objective of this dissertation can be summarized as follows:

Objective. *Design of a unifying, market-based feedback control strategy to enforce a stable, economically efficient, and grid-supportive operation of future power systems.*

The outline of this dissertation is as follows (see Fig. 1.1). Chapter 2 reviews the current state of the art, formulates the existing research gaps, and outlines the specific contributions of the dissertation aligning with the above objective. Subsequently, these specific contributions are elaborated in Chapters 3–5. In Chapter 6, effectiveness and performance of the presented overall control framework is evaluated by means of case studies on a realistic IEEE 57-bus benchmark system. Finally, a conclusive discussion on the contributions is given in Chapter 7.

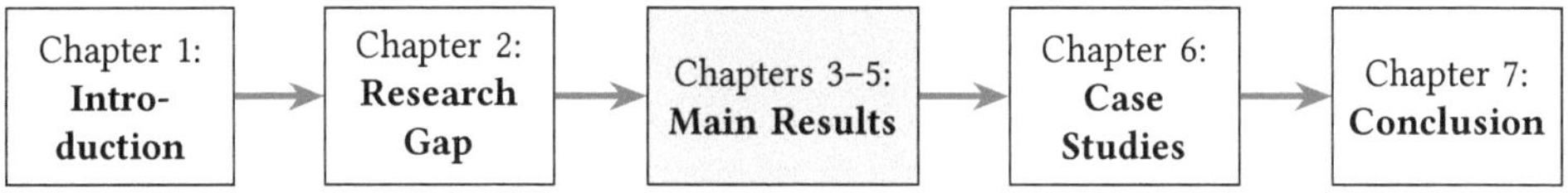

Figure 1.1: Outline of this dissertation.

2 Related Work and Research Contributions

This chapter forms the basis for the scientific contributions of the subsequent chapters by providing a comprehensive survey on recent advances in competitive power systems. Section 2.1 analyzes the status quo of today's electricity markets and outlines recent tendencies. Section 2.2 surveys the state of research. Section 2.3 summarizes the main findings of the previous sections by means of three specific research gaps. Finally, Section 2.4 presents the main contributions of this dissertation that result from these research gaps.

2.1 Today's Liberalized Electricity Markets

From the beginning of electricity production until the 1980s, the power grid was characterized by *vertically integrated* electricity companies, which concentrated all aspects of power supply, from generation and distribution to grid operation, under the same roof. However, as power systems became increasingly interconnected in the 1900s, this concept proved to be insufficient to take the increasing importance of power flows over long distances into consideration. For instance, a power plant operator (PPO) that simultaneously acts as a local system operator would always retain both the incentive and the opportunity to prevent competitors from entering the market. While it is economically desirable to create competition between PPOs, the system operators (i.e. owners of the transmission and distribution infrastructure), by contrast, constitute *natural monopolists* [GS17, p. 264f.] due to their high fixed costs and low variable costs[5]. From an economic perspective, the most effective way to handle natural monopolies is through *regulation*. For this purpose, the ownership of some part of the network is auctioned off by the state on the basis of long-term concessions, and a regulatory authority constantly monitors compliance with these contracts. This separation of different roles has led to the need of *unbundling*, which includes the following key aspects [BGM$^+$14, p. 7]:

a) *Legal Unbundling:* Corporate splitting of former integrated electricity companies into system operators, and generation and supply companies.

b) *Operational Unbundling:* Authority and operational decision-making of the generation and supply companies must not be influenceable by system operators.

c) *Informational Unbundling:* Information from each company must either be disclosed on a non-discriminatory basis to all other participants or treated confidentially. Information received from other participants must be treated confidentially as well.

[5] Practically speaking, it is not useful to construct multiple power line infrastructures from different companies in parallel.

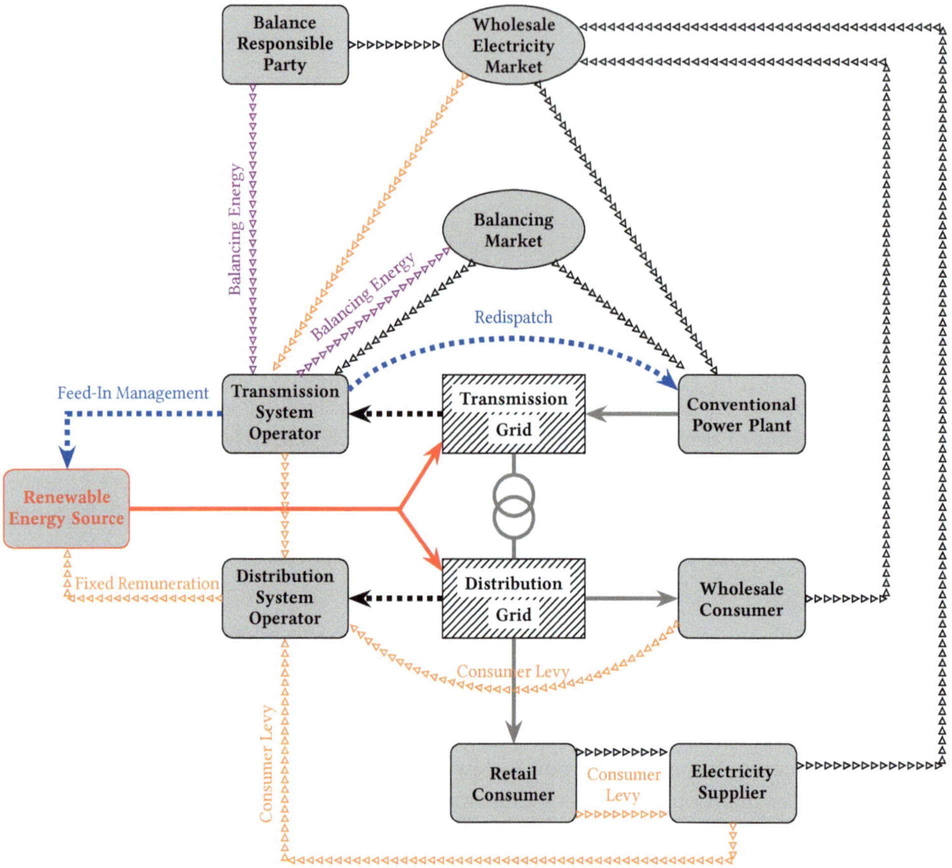

Figure 2.1: Status quo of key stakeholders in Germany's power system. The arrows mark the typical flow direction of physical power (solid lines), information (dashed lines), or capital (triangle lines). Color-coded components mark those effects that arise by increased infeed of RESs and are discussed in Subsection 2.1.2.

As a consequence, there exist individual companies with clearly defined roles and narrow information boundaries. In this process, power generation and supply continue to be left to free competition, while system operators are regulated by the state authority.

2.1.1 Overview and Participants

Fig. 2.1 shows a schematic overview of the relevant stakeholders in Germany's power system[6] as it exists today. The power network infrastructure can be divided into the *transmission network* (typically with nominal voltages above 100 kV) and the *distribution network* (nominal voltage below or equal to 100 kV). While the transmission network operates as a supraregional transport and interconnection layer, the distribution network is conceived for subordinate,

[6] Although nomenclature is based on the usual terminology in Germany, a fairly similar structure can also be found in other European countries as an outcome of EU-wide harmonization.

Table 2.1: Electricity markets in Germany.

Type	Product	Timeframe	Volume Tick
WEM	EEX Futures	$\geq$1 w	1 MW
WEM	EPEX SPOT Day-Ahead	1 h	0.1 MW
WEM	EPEX SPOT Intraday	15 min	0.1 MW
BM	Minute Reserve	15 min	1 MW
BM	Secondary Control Reserve	5 min	1 MW
BM	Primary Control Reserve	30 s	1 MW

regional distribution of electrical energy to the point of consumption. Consequently, transmission grids tend to be more meshed than distribution grids, which are typically operated as open ring or radial grids, and are characterized by an R/X ratio close to zero.

Large-scale *conventional power plants* (CPPs) are directly connected to the transmission grid via step-up transformers. The aim of each CPP is to deliver a certain amount of electrical power within a specific period of time. There are various opportunities for selling the produced power: On one hand, the own supply can be offered on the Wholesale Electricity Market (WEM). The WEM is an *energy-only market*, which means that only the energy actually to be supplied at a certain time is offered. On the other hand, there exists a *balancing market* (BM), where each PPO's production capacity can alternatively be offered to system operators as a means for stabilizing interventions, e.g. after disturbances or faults. The BM is a *capacity market*, which means that it trades the *option* of having a certain amount of electrical power delivered for a certain period of time. In fact, the offered power packages are standardized in terms of spatial and temporal quantization, and depending on the specific time scale, both WEM and BM divide into specific sub-markets. Table 2.1 provides an overview of the existing electricity markets in Germany. In a fully unbundled power system, CPPs are owned by independent companies which are purely profit-driven. Whether marketing its own electrical power is more profitable on the WEM or the BM depends on the current market forecast of the individual CPP.

The demand side can be divided into *wholesale consumers*, which cover their own demand directly via the WEM, and *retail consumers*, which pay a fixed electricity price quoted by an *electricity supplier* (ES). The measurement of the actually consumed power is done over a 5-minute interval for wholesale consumers and at an annual interval for retail consumers. Between measurement intervals, consumption of the latter group is estimated based on standard load profiles (SLPs). The ES has to purchase respective amounts of power on the WEM in the exact amount corresponding to the (estimated or actual) consumption.

Each supply or demand point of the power system is assigned to a BG. The *balance responsible party* (BRP) has to provide a zero balance between supply and demand for each quarter-hour interval. This can be achieved optionally by means of the own assets, or with the help of purchases or sales on the WEM.

The *transmission system operator* (TSO) is the owner of a certain part of the transmission network infrastructure[7] and is responsible for its maintenance and expansion. Furthermore, the TSO is responsible for stability of the own network. For this purpose, it uses its own measurements in the network as well as information from the subordinate *distribution system operators* (DSOs)[8]. By aggregating the transmitted schedules of the BRPs, the TSO performs an after-market power flow calculation for the upcoming day and, if necessary, has the authority to intervene in the schedules of the power plants in the event of forecast line overloads (*redispatch*) or to prevent non-integrable generation by RESs on times of high net generation surpluses (*feed-in management*). Furthermore, the TSO shall ensure that any current imbalance is remedied by requesting balancing power from the BM or by applying its own stabilizing measures, such as staging of transformers or installation of compensator elements, e.g. static VAR compensators or flexible AC transmission systems (FACTSs).

2.1.2 Recent Developments

The increased penetration of RESs into the transmission or distribution grid (see red components in Fig. 2.1) has a significant implication on the existing market. The main developments are reported below.

Fixed Remuneration of RESs Infeed

To encourage a more rapid investment in RESs, numerous financial incentive instruments have been created. For instance, RES operators are either paid a fixed (i.e. guaranteed) remuneration or an additional premium per amount of power fed in (see orange arrows in Fig. 2.1). The power fed into the grid is marketed by the TSOs. However, since revenue from sales on the WEM is typically far less than the guaranteed remuneration for RESs [CHMG14, p. 304], the resulting extra costs must be compensated by retail consumers and wholesale consumers[9] paying an additional consumer levy to the local DSO. Likewise, if feed-in management is required due to an oversupply by renewables, the respective RESs are compensated for the lost profit, which is also covered by the consumer levy. This levy is taking an ever-increasing share of the net electricity price. In Germany, it averaged 20.4% for retail consumers in 2021 and is considered a significant driver of current and future electricity prices [Bun21, p. 19.]. This is a paradoxical fact, since the actual variable costs of electricity generation by RESs are significantly lower than those from fossil energy sources.

[7] In Germany, there are 4 TSOs. All of them are part of the *European Supergrid* of the European Network of Transmission System Operators for Electricity (ENTSO-E), consisting of 42 TSOs from 35 countries in total.

[8] Within the ENTSO-E network there are more than 2300 DSOs, 880 of which are located in Germany [TM15, p. 4; PFA+19, p. 11].

[9] If wholesale consumers manage to be classified as "energy-intensive", they can exempt themselves from the consumer levy [Deu14].

Redispatch and Copper Plate Assumption

The remuneration policy for RESs discussed above has another side effect that arises particularly in the case of weakly developed grids or long transmission paths: If CPPs are displaced from the merit order during periods of high solar and/or wind generation, there is a local undersupply of load centers. While these may be compensated arithmetically by the respective RESs, significant spatial disparities between the locations of generation and consumption may occur, e.g. when offshore wind turbines ramp up, thereby superseding CPPs far from the coast. If these disparities exceed the physical transmission capacity limits of the network, redispatch has to be ordered by the responsible TSO (see blue arrow in Fig. 2.1). This redispatch mechanism is *cost-based*, i.e. both the down-regulated and up-regulated power generators are each compensated pairwise for their lost profits or additional costs to be incurred. In fact, central Europe has seen a surge in the occurence of redispatch events over the last decade [Fek21, p. 5f.; EE21], and this trend is assumed to continue [Sta19, p. 42].

Cost-based redispatch provides numerous inappropriate incentives for PPOs from a macroeconomic and environmental perspective: For an individual PPO, the compensation mechanism makes the specific location of generation and thus the role of a certain power plant to the overall network completely irrelevant[10]. Instead, the PPOs' investment decisions are only governed by minimizing their own fixed costs, which often results in a preference for rural areas far from load centers. At the same time, PPOs have no incentive to replace existing, low-efficiency power plants at critical load centers with higher-efficiency power plants, since cost reimbursement by the TSO is always guaranteed, even if the bid falls well outside the merit order. Furthermore, studies have shown that PPOs can anticipate the occurence of a redispatch event with very good accuracy solely based on seasonal patterns and local weather measurements, even without any knowledge of the prevailing network conditions [Sta19, Chapter 5]. This creates several speculative trading opportunities for PPOs that are unfavorable from a global systems view: For instance, if it can be anticipated that a specific power plant will be curtailed by the TSO, it is advantageous for this PPO to submit a lower bid in order to be scheduled, but then to be reimbursed for the curtailment without having to physically deliver the amount of power. Conversely, if a certain power plant is likely to be requested after-market by the TSO for redispatch, strategic PPOs will bid too high to raise their own expected compensation payments. Further speculative trading strategies following from good predictability are discussed in [Sta19, p. 53].

Balancing Energy

The increasing volatility of power supply increases the difficulty for BRPs to keep their BGs balanced. If a BRP delivers less power than scheduled, the TSO is obliged to procure the difference on the BM and later charges its own costs as a *balancing energy* payment (see violet arrows in Fig. 2.1). Likewise, if the BRP delivers more power than scheduled, it receives a balancing energy payment in the same amount by the TSO. This mechanism is thus another speculative trading opportunity: In times of high spot market prices it is advantageous for the

[10] This phenomenon is called *copper plate assumption*, which refers to the fact that from the PPO's point of view, the whole power network is reduced to a single copper plate with unlimited transmission capacity.

BRP to sell its own power on the WEM instead of keeping its own BG balanced. Analogously, in times of low spot market prices, it is preferable to purchase power on the WEM and earn the difference to the received balancing energy payment as arbitrage profit [MRF11]. This exploitation pattern counteracts the efforts of the TSO to keep the network balanced and is likely to endanger the stability of the network. In recent years, several critical incidents in the ENTSO-E network have been reported which trace back to strategic behavior of the BRPs (cf. [JW15, KH19, PLd20]).

2.1.3 Discussion

The above examples show that under the impact of increasing RESs infeed, the existing electricity market in its current form is tending towards undesirable interactions between stakeholders. Thereby, the (normative) role of the TSO on electricity markets is increasing, such that the original requirement for unbundling is only partially fulfilled. In particular, while legal unbundling is still given, there are an increasing number of situations where the TSO has to intervene in PPOs' decision making or where the receivable information empowers PPOs to exploit their knowledge to the disadvantage of other participants, thus violating both operational and informational unbundling. These trends result in a more frequent occurrence of critical network situations and thus in an increasing instability of the network. The economic inefficiencies also create substantial extra costs for consumers (see Fig. 2.1). In addition, cost-based redispatch and copper plate assumption act as key drivers for PPOs to act contrary to what would be necessary and desirable from a network perspective.

In order to resolve the shortfalls shown in Subsection 2.1.2, a certain tendency towards even stronger normative intervention by the regulatory authority can be observed [EE19]. However, such interventions do not adequately address the intrinsic problem and often reach their target far too late[11].

Moreover, numerous position papers suggest a shortening of the current time frames (see Table 2.1) towards a *real-time* interaction of the stakeholders. In a real-time scenario, the actions (i.e. outputs) of individual stakeholders are based on continuous measurements of the own state and environment. Thus the overall interaction of all participants on the physical network an via different markets can be interpreted as a *feedback control system* comprising a large number of internal control loops with different response speeds. However, there is also a wide consensus across the research community that retaining the *existing* WEM structure is not sufficient to prevent the tendencies described in Section 2.1.2. For instance, the occurence of redispatch is always an indicator of structural *market failure* [Ris21, p. 58], as it may trigger network participants at critical points to further aggravate the (anticipated) congestion.

Consequently, the following section provides a comprehensive overview over existing control-theoretic approaches which are able to make vital contributions towards a stable, economically efficient, and grid-supportive electricity market design.

[11] For example, after a particularly serious case of balancing energy speculation in Germany in June 2019, high penalties were imposed on the BRPs involved by the responsible Federal Network Agency [Bun20], albeit the majority of cases are likely to remain undetected [LC12, p. 259f.].

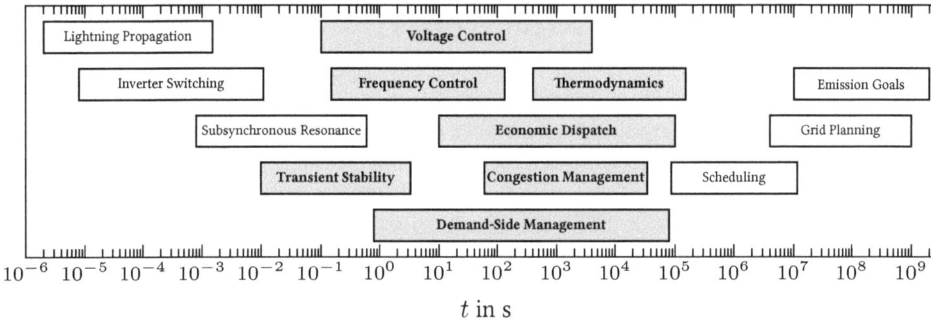

Figure 2.2: Overview of main phenomena in power systems [Sau11, MBB12, OW15, CLI+17]. The focus of this dissertation is on timescales between 10 ms and 1 d and is highlighted in gray.

2.2 State of Research

Fig. 2.2 provides an overview of the main phenomena in electrical power systems ranging from 10^{-6} s to 10^{9} s. Typically, the common term *power system stability* is associated with timescales smaller than 15 min. Stability is said to be given, *"if, after being subjected to a disturbance, all state variables [of the network] recover to (possibly new) steady-state values which satisfy operational constraints [...] without the occurence of involuntary load shedding"* [FCSP+20]. According to the latest IEEE-PES definition, power system stability encompasses the sub-tasks *frequency stability*, *voltage stability*, *rotor angle stability*, *converter-driven stability*, and *resonance stability* [HMR+20]. Among these sub-tasks, frequency and voltage stability play the most important role for system operators[12]. By contrast, measures for *economically efficient* and *grid-supportive* operation take place on timescales above 1 s (see Fig. 2.2). They range from supply-side optimization (*economic dispatch*) and demand-side management to *day-ahead scheduling* and long-term *grid planning*.

In the remainder of this section, the main research branches devoted to one or more of the three sub-objectives from Chapter 1 are presented. First of all, Subsection 2.2.1 provides an overview of the main control paradigms and system architectures that are being discussed in the power systems literature. Focusing on slower timescales, Subsection 2.2.2 outlines the main ideas and results for the economic operation on supply- and demand-side. Finally, Subsection 2.2.3 reviews the latest research on congestion management. Subsequently, an in-depth review on the state of research on faster timescales is given for frequency control (Subsection 2.2.4), voltage control (Subsection 2.2.5), and the combination of both (Subsection 2.2.6).

[12] In particular, rotor angle and converter-driven stability are typically maintained by the owner of the respective facility by means of inner control loops on a faster time scale [Tv16, WBL+14, RYI+16]. Accordingly, both categories are not considered for the remainder of this dissertation. Likewise, resonance stability encompasses subsynchronous phenomena being addressed by constructional measures (cf. [NFM+11, LS15, VG16]), which is beyond the scope as well.

2.2.1 Overview of Control Architectures

Historically, the power system followed a strict top-down approach not only in its physical structure but also concerning the responsibility of system operators. In particular, the TSO alone was responsible for the procurement of stability, while the DSO only undertook subordinate control tasks such as transformer staging and power line switching. However, triggered by the replacement of CPPs by RESs, the distribution network has evolved from a weakly equipped *supply grid* towards an interactive and self-sufficient *smart grid*, encompassing decentralized communication via broadband connections, bidirectional *prosumers* with local storages, sector coupling, *smart meter* instrumentation for retail customers, and advanced cyber-security protection [SRRD15, DEA$^+$17, KD18, ABAH$^+$20][13].

The current literature suggests different approaches for transitioning the former top-down control hierarchy between TSO and DSO towards a more agile and autonomous mode of operation. The concept most widely discussed is the *microgrid* (MG) topology[14]. Originally presented in [Las02], a single MG denotes a specific section of the network[15], which is connected to the superordinate grid by a single *point of common coupling*. In normal ("islanded") mode, each MG is isolated from the main network and thus both supply and stabilization are provided decentrally by means of local storage and generation capacity only. If islanded operation is no longer possible, e.g. in case of severe outages or load shifts, the MG switches to grid-connected mode, thus compensating the imbalance via the main grid. Through the principle of subsidiarity, each MG can be controlled separately without having to account for interdependencies with neighboring MGs. This factor has contributed to an increasing number of real-world studies where the MG concept is implemented[16]. A current overview of in-field experiences is given in [MMR$^+$21, p. 4]. Major challenge in MG operation, however, is the re-synchronization with the superordinate grid, which is highly nontrivial and yet in addition often occurs especially at that moment when the MG is in a severe state [DRVA10, ZS10]. Moreover, self-sufficient operation in islanded mode necessitates that generation and storage capacity be generously sized in order to always meet its own demand. It thus has to be critically acclaimed that the physical separation into isolated portions comes at the price of considerably higher investment costs to guarantee worst-case redundancy separately for each MG—in fact, computational overhead is simply shifted to resource overhead.

From the perspective of market participants, the increased availability of communication infrastructure has enabled the development of *virtual power plants* (VPPs) [Dv03, MR11, NY21], virtually aggregating geographically diverse small-scale *distributed energy resources* (DERs) to one single asset. This allows the VPP operator to internally reduce volatility and

---

13 All of these building blocks are either in process of field testing, or a nation-wide rollout in some countries is upcoming in the next few years. See the given references for practical evidence.

14 To date, no formal definition for microgrid exists within the research community. The one presented here is adopted from [OMSE$^+$14] and [SFO15, p. 52ff.]. However, there exist alternative definitions such as [TS12], which do not require a point of common coupling, but only require the MG to act as a single controllable entity with respect to the superordinate grid.

15 Typically on the scale of a medium voltage network. Similar structures at lower voltage levels, e.g. within households, are defined as *nanogrid* by a majority of authors.

16 In fact, even at the beginning of the electricity age in the late 19th century, all urban power grids were operated in islanded mode. Even today, this is still in practice in many remote areas e.g. in Canada or Russia. In this respect, it would be more appropriate to speak of a *revival* of the MG.

unpredictability through the effect of diversification, and thus be accessed similarly to large-scale CPPs for the purpose of stability procurement. In contrast to the MG concept, the overall network remains physically interconnected. Several authors distinguish between the two different roles of *Technical VPPs* (TVPPs) and *Commercial VPPs (CVPPs)*. In this notion, TVPPs pool the regional generation capacity of small-scale DERs within a certain area. They have detailed measurement information about the state of the local network and use their own resources to provide system services to TSOs and DSOs. Conversely, CVPPs are aggregators of DERs from possibly *distant* regions. They have no information about the network and place bids on the WEM based on their aggregated generation profile. The conceptual separation into CVPPs and TVPPs has the advantage that small- and micro-scale DERs gain access for both the WEM and for procurement of ancillary services on the BM. However, the previous structure of TSOs, DSOs, and BRPs which was criticized in Section 2.1 (cf. Fig. 2.1) typically remains unchanged. An alternative approach to organize a fully decentralized power system without the need for large redundancies is the *web-of-cells* (WoC) concept [MBR+17, LHK19][17]. Here, the overall network is partitioned into cells operated by autonomous *cell coordinators* (CCs), each exclusively responsible for stability within its own cell. The WoC concept is able to reduce complexity by means of information barriers between CCs, albeit without cutting the physical interconnections between cells. Furthermore, there exists no superordinate instance anymore, thus CCs communicate in a peer-to-peer manner. Although this approach is interesting, most studies have only focused on the conceptual formalism itself. Yet to date, the literature on specific control schemes tailored to the WoC scenario is scarce (see e.g. the contributions [RCT17, MRHD17]).

2.2.2 Real-Time Economic Dispatch and Demand-Side Management

A major implication of the copper plate assumption is that the existing WEM does not take into account the physical and operational conditions (e.g. power flow equations and line flow limits), but relies on the "*fiction of a congestion-free market area*" [HSMT19, p. 13]. Therefore, *real-time economic dispatch* approaches aim to replace the existing WEM with a (feedforward or feedback) controller, incorporating these physical and operational conditions in an *optimal power flow* (OPF) problem[18], which is solved at specific points in time.

In the remainder of this subsection, we distinguish between *centralized* approaches, which pursue a single overall objective, approaches from *noncooperative game theory*[19], where each network participant is modeled as a player, with no contracts possible between them, and approaches from *cooperative game theory*, which involve binding contracts between players. A comprehensive review of the recent state of research in economic dispatch and demand-side management is provided in the survey paper [WYL21]. For a detailed discussion of continuous-time methods, see [DBSPG19].

17 A number of sources refer this concept as *grid of MGs* or *networked MGs*, cf. [WL20].

18 OPF is a well-studied problem class in both theory and application. Early OPF algorithms [DT68, AS74, HG91] are based on Newton-type methods, while today's solvers usually rely on active set and interior point methods [NW06, Chapter 14; Tay15, Chapter 3; FR16, p. 1185] or on convex relaxation techniques of the OPF problem, such as linear relaxation (DC-OPF) [TCL12] or semidefinite relaxation [Low14].

19 An overview of basic concepts and main terminology of game theory used throughout this dissertation is summarized in Section A.2 in the appendix.

Centralized and Model-Predictive Approaches

Aggregating all objectives and constraints into a single, centralized OPF problem allows to circumvent the issues from Subsection 2.1.2, since no after-market interventions such as redispatch or consumer grid tariffs are necessary. *Model predictive control* (MPC) is an advanced control technique originally stemming from the process industry. It involves solving a repeated numerical solution of a snapshot of the optimization problem with a preferably fast sampling time, and then applying this numerical solution to the controller output within the current time interval upon repeating the calculation in the next time step. As the steady-state mappings derived from the dynamic constitutive equations of the physical system typically correspond to the AC power flow equations, the control architecture of MPC emerges as a repeated offline solution of static AC-OPF problems whose feasible set can be referred to as the nonconvex *power flow manifold* (see e.g. [BD15]). The ability to also incorporate predictive data like weather or load forecasts in the OPF problem makes MPC particularly suitable for optimal dispatch of power grids with high penetration of RESs [HMMD13, p. 5827f.]. Extensive research on the application of MPC for optimal dispatch of power networks is conducted for low voltage distribution networks [KPP+17, PRK+17, DSE+21], industrial consumers and households [SGAM18, SFC+18, FS21] and islanded power systems [PRG14, OCN+17, SSC+19]. While the authors in [PRG14] use computer simulations to show the effectiveness of the approach, power hardware-in-the-loop demonstrators are used in [PRK+17] and even field-site demonstrations are available in [FS21], which highlight the overall practicality of MPC for real-world optimal dispatch.

In spite of their widespread presence in the context of power networks, it must be critically acknowledged that the accuracy and thus feasibility of the obtained solutions is strongly dependent on the accuracy of the precomputed system model. Conversely, as the complexity of the underlying system model increases, so does the computational complexity [BBPS09]. This, in turn, prevents the choice of reasonably fast sampling times.

In response to these practical imperfections, which are very common in real-world scenarios (cf. [PZS+18]), *stochastic MPC* methods have gained a foothold in power system research over the last 20 years, employing the representation of model uncertainties as stochastic constraints. However, the robustness gained by including all possible model deviations and disturbances comes at the price of fairly "conservative" solutions that do not fully explore the entire feasible set and thus often yield profits which are significantly below the maximum achievable profits, whereby, the computational complexity for finding a solution increases strongly once again as compared to the deterministic framework. Beyond academic examples, the usability of stochastic feedforward models for real-time optimization of interconnected power systems with intermittent and volatile feed-in characteristics thus seems questionable.

Moreover, in both classical and stochastic MPC, a centralized, omniscient system operator is assumed to have both full knowledge about all system states and access to all actuators. This assumption is unfavorable for large-scale power systems both in terms of computational complexity and vulnerability, thus limiting the applicability of classical MPC to small-scale power systems with a moderate number of parameters.

Recent developments in *distributed model predictive control* (DMPC) methods [VHRW08, CSML13] allow to efficiently decompose the optimal dispatch problem into subproblems, such that local agents with geographically limited measurement and control access optimize only a part of the system. [SESJ13] proposes a DMPC algorithm for economic dispatch using Dantzig-Wolfe decomposition. [XXM19] presents a cost-minimal DMPC scheme for grid-connected low voltage networks, the performance of which is validated on a power hardware-in-the-loop platform. In [RCPP20], a dual decomposition method is applied for DMPC-based optimal scheduling of interconnected migrogrids. These approaches suffer from the fact that the distributed agents aim to jointly minimize a global objective function. However, in fully unbundled power systems, network participants usually are not committed to a common goal and do not take cooperative decisions. Instead, they operate selfishly by maximizing their own profits, i.e., try to minimize their own objective function. Distributed *continuous-time* controllers for radial power systems are presented in [ZP15]. In [ZLP15], the controllers are extended to meshed topologies. However, both approaches again rely upon the assumption of zero transmission losses.

In general, a proper accounting for transmission losses is a key challenge of all OPF-based methods, especially at distribution grid level, due to inherent nonconvexity of lossy AC-OPF problems and their resulting non-existence of appropriate Lyapunov candidate functions [SALB05; Sch15, Remark 6.2.2]. In [MNRQ17], losses are approximated by *robust loss coefficients* in order to provide a distributed solution to the economic dispatch problem. Improved loss coefficients formulas, which also consider reactive power flows, are presented in [HYC$^+$18].

Noncooperative Game-Theoretic Approaches

Noncooperative game theory is concerned with selfishly driven network participants and provides tools and methods on how to maximize their own profit while taking into account the other players' strategies. The main solution concept in noncooperative game theory is the Nash equilibrium (cf. Definition A.6), which is used to characterize a state in which no player can achieve a higher profit by unilaterally changing its strategy.

Former research focused on noncooperative approaches with highly simplified models of network participants, which allows Nash equilibrium seeking to be resolved using linear programming techniques, see e.g. [SLL02, WY13] for generator dispatch or [ZHB12] for demand-side management strategies. Afterwards, stronger emphasis was placed on tractability and practicability aspects by applying parallelization and approximation techniques based on e.g. proximal decomposition [AOS$^+$13a, AOS$^+$13b] or distributed agreement [CLLV14].

In [XZN13], the applicability of DMPC for a linearized model of interconnected power systems with different objective functions is shown. The authors in [FS12] propose a *noncooperative* DMPC method, which requires an information exchange about the agents' planned trajectories via a non-manipulable information path. In addition, [RYNC14] proposes a noncooperative DMPC procedure, where each agent is equipped with a distributed Kalman Filter to estimate the overall state of the system. Both approaches are restricted to linear systems only. [LKF15] presents a *contract-based* DMPC method for agents with differing objective fuctions, which is applicable to nonlinear non-affine systems. However, it needs guaranteed information about the future trajectories of coupling variables to be exchanged among agents. [VBGQ+20] presents a DMPC procedure for optimal dispatch of different network participants (in particular, conventional generators, RESs, and energy storage systems). However, the approach does not account for physical line limits and thus not prevent possible redispatch contingencies.

However, as the strategic interplay of network participants is typically represented by a full peer-to-peer competition among players, it must be assumed that each has complete information of the other players' strategies. Thus, a major drawback of classical noncooperative frameworks is their utterly high communication effort for broadcasting the relevant information in a non-discriminatory manner via non-manipulable information paths. Moreover, a calculation of the individual best response depending on each other player's strategy is computationally demanding and poorly scalable, which makes a real-time solution intractable, except for very simplified academical case studies.

In this context, *Stackelberg approaches* try to lighten this computational load by implementing a sequential sharing of information in a leader-follower manner. Stackelberg approaches are most common in the area of demand-side management. E.g. in [TZS+14, YH15, MWH+18, EES+19], a single-leader multiple-follower scenario is applied where a single ES is the price-setting leader and multiple consumers are the price-taking and utility-maximizing followers. In the area of VPP operation, [MR11] presents a Stackelberg model for the optimal interaction between a single VPP operator (leader) and multiple micro-cogeneration units (followers). Multiple-leader multiple-follower games are investigated in [ALCB15, LGCZ15, MMO17], allowing multiple price-setting ESs to compete on their part. However, due to its high computational effort, Stackelberg approaches require a time discretization with sufficiently large time intervals, thus there is a possible conflict between real-time computability and real-time adaptivity. Moreover, the underlying assumption that followers have no means at all to affect the leaders' strategies seems rather disputable.

A compromise between the fully decentralized information pattern of peer-to-peer competition and a centralized scheme is undertaken by *price-based* frameworks featuring different models of competition. Here, individual competition among players is replaced by a real-time bidding process against a single market instance. The two main directions studied in the context of power systems are *Cournot competition* and *Bertrand competition*, whose main distiguishing feature is which quantity constitutes the strategic decision variable. In Cournot competition, each participant bids *quantities of power* while respecting the given price per quantity. [NSH15] and [CBK17] present Cournot-based demand-side management schemes where a central control authority ensures supply-demand balancing. The authors of [DM19] propose a fully distributed pricing mechanism for a Cournot model of competition between price-setting generators and price-taking loads. In Bertrand competition, by contrast, the *bid price* is the strategic decision variable. Network participants send bids to an *Independent System Operator* (ISO), which evaluates the bids centrally and then dispatches them. Two recent papers [CC20] and [CSD+20] develop a Bertrand competition model between price-setting generators resulting in a continuous-time bidding process against a centralized system operator, which is shown to provide economic efficiency. Transmission line congestion can be considered only for radial networks. In [SCD+19], a Bertrand competition between generators and ISO was implemented by means of a time-triggered update scheme of bids and generator setpoints.

Another iterative algorithm is proposed in [MWY+19] for interaction between multiple price-setting consumers and a centralized generation company.

In spite of different terminology, e.g. *ISO* [CC20], *macro-station* [SHP11], *central distributor* [PLB+16], *non-profit central control authority* [CBK17], these approaches follow the same principle: The centralized instance acts as the only counterpart and antagonist for every single player, with all information on the other players' strategies being aggregated in the momentary price. In contrast to a generalized game-theoretic setup with peer-to-peer dependencies between all players, both models of competition provide better scalability, since the own individual strategy of each player depends only on the *aggregate* of the other strategies (Cournot) or on the strategy of the single ISO (Bertrand). This helps mitigate computational overhead, which is an advantage especially in large-scale power systems.

Cooperative Game-Theoretic Approaches

While noncooperative game theory supposes that no binding contracts are possible between players, *cooperative* game theory involves binding contracts between players and provides methods for allocating a jointly realized profit to the players involved. The two main branches in cooperative game theory are *coalition formation* and *bargaining*.

Within coalition formation, the goal is to find stable alliances of network participants that result in an individual improvement for all as compared to not collaborating. In the field of coalition formation, a conceptual profit allocation strategy invoking coooperation among independent power producers is presented in [JY03], and a first coalition formation algorithm for radial distribution grids is presented in [SHP11]. Moreover, there has been considerable research effort into using the three classical solution concepts *core*, *nucleolus* and *Shapley value*, see e.g. [LH16, CBP+19], [HMM19], and [LH16, CK19], respectively. These concepts represent different manifestations of a *stable equilibrium*, originated by different notions of individual gratification, which aim to prevent each participant from leaving the coalition. While calculation of core and nucleolus is computationally expensive and possibly non-unique, the Shapley value, in contrast, can be calculated considerably easier. In [NKR13], a conceptual approach for arbitration of conflicting interests between producers and system operators based on a coalitional game is proposed. Through an additional power routing function, the approach is capable of preventing network issues such as transmission congestion. [FMC17] develops a discrete-time model for dynamic coalition formation of smart grid prosumers including storage devices. The last two approaches discussed also allow to account for transmission losses. However, their high computational cost prevents real-time application to large-scale systems.

In contrast to coalition formation, bargaining approaches focus in particular on the sharing of the additional gains that exist as an outcome of cooperation. The classical minimum requirement for bargaining solutions is *Pareto efficiency* (cf. Definition A.5). Numerous papers deal with bargaining strategies among ESs, see e.g. [YTL12, FAP18, WH18]. In the area of power system dispatch, [WK09] presents a bargaining solution for small-scale DC power networks, where sources and loads may aim for different goals such as maximizing the own infeed of power or optimally following a certain voltage profile. However, since players do not act simultaneously but sequentially, the method is unsuitable for large-scale systems. Other solutions for economic generator dispatch based on simple cost functions are presented in [WLM15] and [WWM16]. One central issue in all bargaining approaches is the strong correlation between simplicity of the model and computational tractability of the bargaining problem. In particular, finding a solution for nonlinear or nonconvex problems is computationally expensive even for low-dimensional problems. Consequently, most papers dealing with bargaining approaches for economic dispatch rely on highly simplified models of both the market and its participants.

Generally speaking, approaches originating from cooperative game theory are subject to even more rigorous requirements e.g. with regard to revenue sharing mechanisms or multi-stage side payments. In practice, this implies the need for a centralized instance that has full knowledge about all states and control actions, and that is authorized to redistribute the profit over all players with respect to some predefined rules. However, this requirement is diametrically opposed to both operational and informational unbundling.

2.2.3 Price-Based Congestion Management

In the context of Cournot and Stackelberg approaches, it is shown e.g. in [YTN13, MWH⁺18] that the resulting *real-time pricing* controllers[20] incentivize stakeholders to shift their generation and consumption curves to off-peak hours and thus provoke a more balanced demand curve over time. However, a purely time-varying, but regionally uniform price is unable to reflect *regional* scarcity patterns since physical network constraints such as congestion cannot be internalized [TBW15, p. 212]. In particular, the methods discussed in the preceding subsection do not protect against after-market reallocation by the system operator due to transmission congestion. A remedy to this situation is the introduction of *locational*, i.e. regionally differentiated, prices in order to adequately signal local network conditions. Moreover, locational pricing is considered to reduce structural and regional mismatches in generation and consumption by providing long-term investment signals to deploy generation capacity in load centers [Sta19, p. 29ff.]. Initial conceptual studies were conducted in [SCTB88] and [Hog92]. Afterwards, nodal pricing frameworks have been implemented in numerous countries, e.g. the US, Scandinavia and Australia. Although being highly efficient in avoiding redispatch, nodal pricing tends to support the exploitation pattern of BRPs (cf. Section 2.1.2), since large generators may exploit their own impact on their nodal price e.g. by temporarily cutting supply to provoke high local spot market peaks [BHJ16]. Accordingly, there is an ongoing debate among both policymakers and researchers about the most effective grain size of price zones[21]. From a practical viewpoint, it has been found useful to choose zone boundaries along transmission lines which are prone to congestion [Sto97].

Besides numerous research contributions on a rather strategic level, which typically analyze the long-term perspective, a few papers deal with real-time pricing schemes to address congestion

[20] In energy-related literature, the term *real-time pricing* is typically used for closed-loop pricing strategies, while *time-of-use pricing* refers to open-loop pricing strategies.

[21] The grain size of price zones varies widely across countries and states: While the price zones in Norway (5), Sweden (4) and Denmark (2) are supra-regional, there are 11 price zones in the state of New York and even more than 4000 in the state of Texas.

management. [MY16] and [MSM17] present centralized strategies for cost-efficient generator rescheduling based on an offline calculation of sensitivity factors. In [SCA16, SBCA19], active power flow limits are included in a centralized real-time economic controller, while [ZP15, ZLP15] present a distributed controller representation and [SCD+19, CC20, CSD+20] incorporate the limits in a real-time bidding scheme. However, all of the above pricing approaches rely on simplified network models such as DC-OPF or decoupled AC-OPF, where the crucial effects of transmission losses and reactive power flows on congestion are both neglected, cf. [Wei17, p. 220ff.; JB20].

2.2.4 Frequency Control

In a power system fed by synchronous generators (SGs), deviations from the global active power balance cause a positive or negative mechanical torque on the rotating masses of the rotors[22] and thus a deviation of each nodal frequency from its nominal value. Vice versa, frequency control aims at restoring the nominal frequency by means of positive or negative active power injections. As of today, frequency control interventions are invoked by the TSO using a hierarchical sequence of actions based on the standard scenario of a large outage or load jump: Immediately after the failure, there is a change in the kinetic energy of the rotating masses and thus a steady decrease or increase in frequency. The rotational inertia prevents an immediate collapse of the grid, which is called *momentary reserve*. To stabilize the frequency, a decentralized proportional control (*primary control*) is activated within the first 30 s after the incident. Power plants participating in primary control are equally distributed over the interconnected grid. Using a specific *droop coefficient*, each power plant's own positive or negative active power injection can then be calculated directly from in-situ measurements of the nodal frequency deviation. Since the proportional controllers from primary control lead to a steady-state frequency deviation, *secondary control* is activated in the next step in order to regulate the frequency towards its nominal value. For this purpose, each TSO calculates its own *area control error* by integrating the frequency deviation and then centrally activates the responsible power plants so that the required active power is available in no later than 5 min. Lastly, *tertiary control* (sometimes also called *minute reserve*), which is available within 15 min, aims at restoring the pre-contingency balance among TSOs by activating additional positive or negative power injections or, if necessary, re-scheduling the setpoints of the secondary layer. All three control layers are marketed separately on the BM[23] (cf. Table 2.1), with a further separation between positive and negative power for the secondary and tertiary level.

With the ongoing transition from SG-dominated CPPs towards inverter-interfaced DERs, plenty of research has been undertaken to investigate how inverters can be seamlessly integrated into the existing hierarchy of frequency control. The most prominent approach is *inverter droop control*, where the droop characteristic of SGs with proportional relationship between active power infeed and frequency deviation is mimicked. Originally published in [CDA93], this approach has since been extensively studied, e.g. in [TJUM97, CCG02, BPP+08].

[22] The dynamic behavior during the self-synchronization process between the nodal frequencies of an interconnected power system can be analyzed in more detail by the theory of coupled Kuramoto oscillators [DCB13].

[23] Within the ENTSO-E supergrid, primary, secondary and tertiary frequency control are referred to as *frequency containment reserve* (FCR), *automatic frequency restoration reserve* (aFRR), and *manual frequency restoration reserve* (mFRR), respectively.

The main advantage of inverter droop control is its decentralized nature without the need for extensive communication and its full backward compatibility with a SG-dominated environment. However, as with the conventional droop control for SGs, a steady-state deviation from nominal frequency remains. Besides, the practical effort to correctly tune the droop parameters is supposed to be high [CS13]. Apart from droop-based approaches, several alternative control methods have been proposed, such as *virtual synchronous machine* [BH07, DSF15], *virtual oscillator control* [SDJD15], or *internal matching control* [JAD16, MDS$^+$18], all with the goal of adopting the rotational inertia concept familiar from SGs to inverters.

Owing to the faster operating time scales of inverter-interfaced DERs and new topological concepts such as MG, VPP, and WoC, numerous papers have investigated how to overcome the traditional hierarchy of primary, secondary, and tertiary frequency control by means of a *unifying* control strategy [GCLL13]. It was derived in [LCZL14] and [LZC16] that economic dispatch can be integrated into a unifying secondary control scheme. Moreover, in [DSPB16] it was shown that each minimizer of an economic dispatch problem can be reached by a decentralized droop controller with properly chosen droop coefficients. To combine the primary and secondary control level, the authors in [ADJS13, SPDB13] present *distributed averaging proportional-integral* (DAPI) controllers where zero frequency deviation is reached via a neighbor-to-neighbor communication path. A gather-and-broadcast approach is derived in [DG17], where the integration between economic dispatch and frequency control is performed in a semi-decentralized fashion with locally aggregated measurements of the nodal frequency deviations. In [FP10], *primal-dual saddle-point flow* is applied to frequency control. It allows to unify all three control levels together with economic dispatch by characterizing the desired equilibrium as a minimizer of a constrained optimization problem. Since the dual variables of the cost functions can be interpreted as price signals, no generation setpoints by a superordinate authority are needed. This understanding has led to the idea of *continuous-time dynamic pricing* controllers for frequency regulation [ZTL12, SDv15, TBD16, MZL17, SDv17a]. Furthermore, inequality constraints can be incorporated either by using extended Lagrangians [SDv17b], projection operators [ZP15, SDv17b, WLP$^+$19], or a saturation term [ZMLB18]. A rigorous performance analysis of primal-dual saddle-point flow applied to DC power flow can be found in [SPPMD16] and [CY17]. Detailed discussions of current research work on optimization-based frequency control are conducted in [DBSPG19].

Meanwhile, increased computing power has also permitted MPC-based and game-theoretic control methods to be considered for real-time or near-real-time purposes: A computationally tractable DMPC scheme for frequency control based on a linearized MG model is presented in [BT17]. The authors of [STY$^+$18] present a DMPC strategy for heterogeneous power systems fed by CPPs and wind power plants, which is able to achieve frequency restoration by incorporating ultra short-term wind forecasts. In addition, there exist noncooperative concepts to frequency regulation which address the individual goals of stability providers [CYWL15, NOS$^+$15, CSP$^+$19]. While [CYWL15] focuses on load frequency control using a linear-quadratic differential game, [NOS$^+$15] provides a real-time framework for frequency control by means of time-varying prices offered by a centralized ISO. Both approaches rely on a linearized system model. In [SCA16], an aggregate frequency error is used as additive feedback signal in a real-time WEM. In [CSP$^+$19], advanced meta-heuristic optimization techniques are applied to a nonlinear system model in order to improve the frequency response characteristic on device level.

2.2.5 Voltage Control

Voltage control aims to keep voltage magnitudes at each node within a tolerance band around the specific nominal voltage. Unlike frequency control, local actions of voltage control only have a local impact, thus control actions are typically[24] accomplished in a three-stage process with increasing geo-spatial coverage: At the lowest level (*primary control*), the local voltage is regulated to a given setpoint. For this purpose, excitation systems of the SGs (*automatic voltage regulation*) as well as active and passive network components, e.g. phase shifters and *flexible AC transmission systems* (FACTSs) are applied. At the *secondary control* level, local setpoints within an entire zone are adjusted so that a designated *pilot point* matches its setpoint and, in addition, there is no voltage band violation at any node. Lastly, *tertiary control* aims to centrally coordinate the individual zones and adjust the setpoint voltages at pilot points.

To enable seamless integration of inverter-inferfaced DERs into the existing control hierarchy, numerous decentralized droop-based approaches for voltage control have been proposed [CDA93, CC18, LSZ18]. Here, local deviations from the voltage setpoint are compensated by additional *reactive* power injections similarly to the active power injections of the frequency droop approach. Another similarity to frequency droop is the nonzero steady-state deviation due to the purely proportional control architecture. Namely, it can be shown that in order to keep the steady-state deviation within an acceptable range, R/X ratio of all lines must be small, ideally zero. Yet, this assumption is not admissible for distribution networks [MMMT12].

There are several alternative methods for voltage control based on the formulation of optimization problems. As shown in [FCL13], the standard voltage droop control is inversely optimal to a convex optimization problem. Based on this finding, [ZC16] and [KZGB16] propose an iterative procedure for adjusting the droop characteristic in order to keep all voltages of a radial network within an acceptable range. [SPDB17] presents a *quadratic droop* controller as the solution to an optimization problem which minimizes a trade-off between voltage setpoint deviation and reactive power dissipation. [HGK18] presents an approach how to optimally tune the droop parameters.

Furthermore, several publications have appeared in recent years documenting distributed *non-droop-based* methods with neighbor-to-neighbor communication among nodes. [CTFS18] proposes a combination of sliding mode and DAPI control, where the weighted average voltage of the MG is provably equal to the weighted average of all voltage references. [THL19] develops an optimization-based procedure for voltage control in radial distribution grids, where the objective function reflects the setpoint deviation of voltages and where upper and lower limits on reactive power injection are specified by box constraints. The proposed method is based upon the assumption of a uniform R/X ratio throughout the network. In addition, [LSZ18] allows to specify the overall amount of reactive power injection as additional soft constraint and a linearized relation between voltage and reactive power as additional equality constraint. To enable an online computation, [THL19] uses the distributed accelerated dual descent algorithm, while [LSZ18] is based on the alternating direction method of multipliers. Both approaches rely on a linearized system model.

[24] The degree of automation of each level as well as the responsibilities of TSO and DSO highly depend on the specific grid code. The interested reader is referred to [vV08, MC11] for further elaborations.

In the context of MGs, several authors propose *passivity-based* approaches to voltage stabilization. In [ABEPCR18, ABMEPG18], primary voltage controllers are developed on the basis of a linearized *port-Hamiltonian* model[25] of the MG, such that a specific energy function of the closed-loop system has its minimum at the desired setpoint voltages. [AvJ17] presents a secondary voltage controller based on a nonlinear port-Hamiltonian model which provides zero setpoint deviation in lossless networks.

More recently, a number of game-theoretic methods addressing voltage stability have emerged. Most papers focus on sequential approaches, where a trade-off between the grid-forming generators and, if applicable, consumers can be found either iteratively [GA20] or in a leader-follower manner [CZ18]. A simultaneous approach is proposed by [WHK18], where voltage stabilization is achieved by incentivizing PV owners with additional financial payments, if they follow a given desired voltage profile. Yet, the proposed method is applicable for radial networks only.

2.2.6 Combined Frequency and Voltage Control

Regulating frequency and voltage separately from each other always relies upon the assumption that line resistances (or line reactances[26]) are negligible. However, especially in medium voltage networks, which are neither dominantly inductive nor dominantly resistive, this assumption does not hold in practice. For this purpose, [GMG$^+$07, LK09] use *virtual impedance* techniques which allow a modification of the setpoint voltages by means of a second feedback loop in order to compensate the cross-term. In [RLBR12, BS13], a *generalized droop control* is introduced, which provides coupled droop characteristics with cross-terms for active and reactive power. With the resulting equations, it is required that all line and load parameters are known. Moreover, similar to decoupled droop control, generalized droop control cannot regulate the steady-state deviation of frequency and voltage to zero.

In [SPSD$^+$15], a DAPI controller for primary and secondary frequency and voltage control in inverter-based MGs is proposed. While the primary level relies upon a droop characteristic, the secondary controller uses a consensus protocol in order to restore the nominal frequency by means of an arbitrary connected communication path between the agents. Moreover, the presented controller in [SPSD$^+$15] provides a tuning functionality for the conflicting goals of reactive power sharing and voltage regulation. [DMSD16] augments the DAPI controller by a modified integral control law to minimize a quadratic optimization problem for the active power generations. In order to achieve better reactive power sharing, [WSI18] combines a consensus protocol for secondary frequency and voltage control with an additive correction of the voltage references. A slightly different approach is presented in [SGV14]. Here, distributed consensus is reached via all-to-all communication among all agents in the network. The approach is capable of providing both frequency and voltage regulation without knowing the

[25] The port-Hamiltonian modeling framework [vJ14] is a powerful modeling tool for complex multi-domain physical systems that has gained increased attention from various fields of application in the last years [MMS09, FZO$^+$13, AS17, ACRMA19, PCH$^+$20]. A brief introduction to the basic terminology used in the context of this dissertation is given in Appendix A.4.

[26] If the network is assumed to be purely resistive, then frequency and voltage are decoupled as well, where frequency can be controlled by reactive power and voltage by active power [LSK05].

physical network parameters. However, both the amount of tunable parameters as well as overall communication effort is substantial.

Among non-droop-based methods, a number of contributions employ passivity theory to obtain frequency and voltage controllers with zero setpoint deviation: In [RSFT15, TFRFT16, TFT17], stabilizing controller parameter are found via a centralized solution of linear matrix inequalities. An extension is presented in [NFT19], which allows to include the dynamics of the power lines into the underlying MG model. In [SMKH19], the criterion for stabilizing controller parameters is further simplified to sufficient conditions which can be calculated a priori. Besides, [DE18] presents a first differential game formalism for joint frequency and voltage control of inverter-based islanded MGs. However, as pointed out by the authors, a high computational effort as well as scalability issues limit its applicability to small-scale systems only.

2.3 Discussion

In light of the future developments and disruptions, the preceding section conducted an extensive review of current research trends towards a stable, economically efficient and grid-supportive operation as objected at the end of Chapter 1. Table 2.2 provides a summarizing overview of existing papers on distributed control methods that combine at least *two* of the subjects economic dispatch, frequency control, or voltage control.

First, we analyze the adequacy of the state of research in ensuring a *stable* operation of future power systems. Table 2.2 indicates that research in voltage control has tended to concentrate on internal and primary control methods specifically suited for inverter-based networks, and to overlook integrating secondary and tertiary voltage control strategies from an optimization viewpoint. Since the traditional control hierarchy described in Section 2.2.5 is typically kept untouched, it has to be assumed that a higher-level controller is able to provide optimal setpoints. Although a few papers, e.g. [LQD14, KZGB16, GA20, HBHD21a], have addressed the question of *autonomous* voltage control frameworks that unify the primary and secondary control layer, all of these approaches are subject to a centralized controller and thus exhibit a single point of failure.

In the area of frequency control, the rise of intermittent DERs and the new phenomenon of volatile residual loads have led to a new, more general understanding of the traditional hierarchy and its purpose: Originally defined as a scheduled cascade which has to be activated in a strict sequence in the event of a major power plant outage, the literature nowadays more generally defines primary control as frequency containment and secondary control as frequency restoration, which has led to many unifying control approaches in which both goals are addressed at the same time. Meanwhile, a strong connection between tertiary frequency control and economic dispatch has been exposed [LCZL14], which later resulted in the idea of real-time dynamic pricing.

However, it can be seen from Table 2.2 that all previous papers dealing with the combination of economic dispatch with frequency and voltage control are not capable in providing nonzero deviation of frequencies and voltages from their respective setpoints. This limitation is caused

Table 2.2: State of literature on multi-disciplinary real-time control methods for interconnected power systems.

NCs♣	ED‡	CM§	FC¶	VC♭	
	competitive agents / distributed controllers	meshed topology / active + reactive power	lossless networks / lossy networks	primary / secondary/tertiary	
$-\mathcal{I}-$	– –	– –	◐◐	◐–	[CDA93] [RLBR12] [BS13]
$-\mathcal{I}-$	– –	– –	●◐	◐–	[DMSD16] [DM16] [AvJ17]
$-\mathcal{I}-$	– –	– –	●●	◐–	[GCLL13] [SGV14] [DSPB16]
$-\mathcal{I}\mathcal{L}$	– –	– –	●●	●–	[SPSD+15] [TFRFT16] [TFT17] [NFT19] [SMKH19] [SNM+21]
$\mathcal{S}\mathcal{I}\mathcal{L}$	◐◐	●●	– –	– –	[MY16] [MSM17]
$-\mathcal{I}\mathcal{L}$	◐◐	– –	◐◐	●◐	[KJK+10] [HGK18]
$-\mathcal{I}-$	◐●	– –	– –	◐–	[FCL13] [WHK18]
$-\mathcal{I}\mathcal{L}$	◐●	– –	– –	–●	[ZC16]
$-\mathcal{I}\mathcal{L}$	◐●	– –	– –	◐–	[SPDB17] [CTFS18] [LSZ18]
$-\mathcal{I}\mathcal{L}$	◐●	– –	– –	●●	[LQD14] [KZGB16]
$\mathcal{S}\mathcal{I}\mathcal{L}$	◐●	– –	– –	●●	[HBHD21a]
$\mathcal{S}-\mathcal{L}$	◐●	– –	●◐	– –	[SDv17a]
$\mathcal{S}--$	◐●	– –	●◐	– –	[LCZL14] [LZC16] [TBD16]
$\mathcal{S}\mathcal{I}-$	◐●	– –	●◐	– –	[STY+18]
$-\mathcal{I}-$	◐●	– –	●●	– –	[MNRQ17]
$-\mathcal{I}\mathcal{L}$	◐●	– –	●●	●–	[WSI18]
$\mathcal{S}-\mathcal{L}$	◐●	– –	●●	– –	[MZL17] [ZMLB18] [WLP+19]
$\mathcal{S}-\mathcal{L}$	◐●	– –	◐◐	– –	[ZTL12]
$\mathcal{S}\mathcal{I}-$	◐●	– –	●●	– –	[DG17]
$\mathcal{S}--$	◐●	◐◐	●◐	– –	[SDv15] [SDv17b]
$\mathcal{S}-\mathcal{L}$	◐●	◐◐	●●	– –	[SBCA19]
$\mathcal{S}-\mathcal{L}$	◐●	●◐	◐◐	– –	[SCA16] [BT17]
$\mathcal{S}-\mathcal{L}$	◐●	●◐	●◐	– –	[ZP15]
$\mathcal{S}-\mathcal{L}$	◐●	●◐	●●	– –	[ZLP15]
$\mathcal{S}-\mathcal{L}$	●◐	– –	◐◐	– –	[NOS+15]
$\mathcal{S}-\mathcal{L}$	●◐	◐◐	●◐	– –	[SCD+19] [CSD+20]
$-\mathcal{I}\mathcal{L}$	●●	– –	– –	●●	[WCW+15] [GA20]
$\mathcal{S}\mathcal{I}-$	●●	– –	◐◐	– –	[WY13]
$\mathcal{S}-\mathcal{L}$	●●	– –	◐◐	– –	[CYWL15]

♣ Network Connectors: $\mathcal{S}$ = SMs, $\mathcal{I}$ = inverters, $\mathcal{L}$ = loads.
‡ Economic Dispatch: ●● = competitive goals with distributed controllers, ●◐ = competitive goals with centralized controller, ◐● = centralized goal with distributed controllers, ◐◐ = centralized goal with centralized controller, – – = not provided.
§ Congestion Management: ● = fully provided, ◐ = partly provided, – = not provided.
¶ Frequency Control: ● = zero frequency deviation, ◐ = nonzero frequency deviation, – = not provided.
♭ Voltage Control: ● = fully provided, ◐ = partly provided, – = not provided.

by the fact that the controllers are designed under the simplifying assumption that line conductances can be neglegted. Yet this assumption, as mentioned earlier, is inadmissible especially for distribution grids [MMMT12]. As a consequence, a practical implementation of all of these controllers would still require some kind of additional frequency or voltage restoration by a superordinate controller.

Moreover, it was pointed out in e.g. [Sch15] that the *heterogeneity* of future power systems requires that frequency and voltage control have to be provided by both conventional SMs and inverter-interfaced DERs. However, referring to the first column of Table 2.2, the existing controllers are applicable only to a limited group of network connectors, for instance conventional SMs or grid-forming inverters only. These findings together expose the following research gap.

Research Gap 1. *For interconnected power systems with nonzero transmission losses, there are no control strategies for optimal dispatch of heterogeneous network connectors which ensure both frequency and voltage stability.*

With regard to the second sub-objective of *economic efficiency*, the existing controllers reveal several weaknesses: As Table 2.2 suggests, most control approaches follow a common goal, i.e. all network participants are assumed to be non-competitive. On the contrary, price-based approaches such as [NOS+15, SCD+19, CSD+20], which allow for individual profit-maximizing agents, feature a centralized information and communication structure, with a single node that is responsible for the entire pricing process. The few existing approaches with both competitive agents and distributed communication all require to disclose their own strategy to all other competitors. Thus, no existing approach is able to fully amalgamate both market and control strategy while also achieving legal, operational, and informational unbundling. Even worse, due to their higher communication effort, these approaches are likely to further amplify the interferences and dependencies between competitive electricity providers (CPPs or inverter-interfaced RESs) and regulated infrastructure providers (system operators), that are already occuring at present (cf. Section 2.1.2).

Besides these structural inefficiencies, the employed game-theoretic solution concepts of Nash [SLL02, MWY+19], generalized Nash [AOS+13b, AOS+13a], or Stackelberg equilibrium [TZS+14, ALCB15, EES+19] only indicate the existence of *some* equilibrium, which no participant will unilaterally leave. In particular, these solution concepts are not able to properly assess whether the resulting equilibrium might be dominated by another equilibrium which is more desirable from the participants' points of view (i.e., generates higher profits for some of the participants).

As a further major drawback, most market-based control approaches also rely on very simplified models with zero line resistances. Due to the disregard of losses over transmission lines, the resulting market-clearing prices are infeasible, i.e. this again requires some sort of after-market mechanism with an exchange of side payments and/or additional grid tariffs, which would have had to be supervised by a higher-level authority. Therefore, to the best of the author's knowledge, all existing price-based control schemes are unable to balance the resulting net payments to zero at market clearing, whenever the underlying power network has nonzero transmission losses.

> **Research Gap 2.** *There is no systematic control framework which provides a trade-off between individual interests and market balancing while complying with the desirable information barriers between power plants and system operators.*

For the third sub-objective of *grid-supportive* operation of power systems, the existing price-based control methods suffer from a number of pitfalls that prohibit to incentivize generation close to load centers. In particular, the vast majority of price-based approaches developed so far result in temporally different, but geographically uniform prices. Therefore, it has no *spatial control effect* in any case. On the other hand, iterative bidding approaches such as [SDv19, CC20], which allow for geographically non-uniform prices, do not account for transmission losses throughout the network. In this way, the copper plate assumption, which was problematized in Section 2.1.2, persists, thus there is no incentive for local energy supply near load centers (cf. [HSMT19, p. 14]). Likewise, the conceptual approaches for nodal and zonal pricing surveyed in [Wei17] always rely on simplifying DC-OPF models not considering transmission losses, which are not able to accurately reflect the actual states and physical relationships in the network, especially if the network is heavily loaded [Bak21, p. 266].

Considerably, no price-based control approach is able to combat congestion in real time. The majority of congestion management strategies are on a larger time frame, e.g. with time slices at an hourly basis and a daily planning horizon. With regard to real-time approaches, Table 2.2 shows that several authors integrate congestion management strategies as additional constraints into the economic dispatch problem. However, in most cases this economic dispatch results in a centralized optimization problem which is solved by non-competitive agents. The few competitive approaches considering congestion management [SCD$^+$19, CSD$^+$20] are only applicable for power systems with a tree topology. Moreover, all of the existing real-time congestion management strategies neglect the effect of reactive power flows on network congestion. At the same time, none of the purely competitive approaches with distributed communication (last three rows in Table 2.2) are able to properly reflect network congestion.

All in all, existing dynamic pricing controllers fail to implement a real-time mechanism in order to incorporate network effects to the local prices. Within the literature, there is a large gap between long-term, investment signaling schemes, which are purely conceptual and do not provide a real-time mechanisms, and real-time pricing controllers that neglect the physical characteristics and limitations of the network and thus provide no means for incentive signaling. Therefore, a holistic short-term control framework is lacking, that is able to hold grid participants accountable for their own impact on the momentary state of the network.

> **Research Gap 3.** *There exists no constructive pricing mechanism to provide regional market incentives for power plants by evaluating the momentary positive or negative impact of local power generation.*

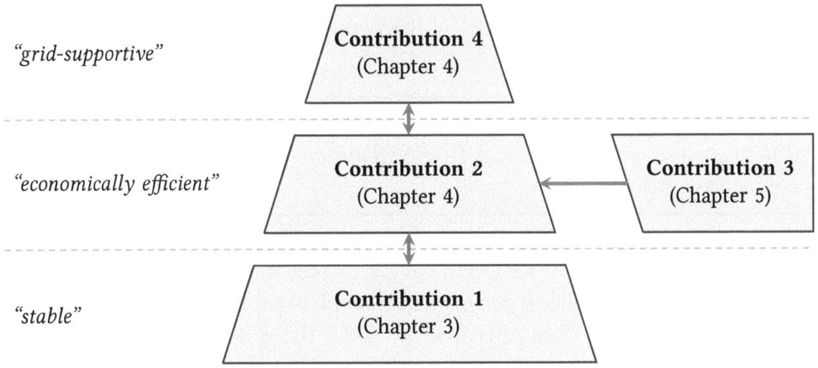

Figure 2.3: Outline of main contributions. "Higher-level" contributions on slower timescales are positioned at the top, while "lower-level" contributions acting on faster timescales are positioned at the bottom.

2.4 Statement of Contributions

To overcome the limitations and gaps of existing methods, this dissertation is dedicated to developing a distributed, continuous-time control framework for competitive power systems based on real-time dynamic pricing. By consistently accounting for transmission losses, the resulting control scheme shall not require any kind of settlement payments or after-market clearing procedure. Furthermore, due to a rigid compliance with information barriers between economic players and system operators, the presented brownfield mechanism shall be incentive-compatible in a fully unbundled environment. All control laws are expressed as explicit differential equations. The main contributions of this dissertation and their relationships to the revealed research gaps are illustrated in Fig. 2.3 and briefly outlined in the following.

Contribution 1. *This dissertation develops a distributed, market-based feedback control strategy for lossy and heterogeneous power grids, incorporating both, frequency and voltage control, with optimal dispatch.*

The control methods presented in Chapter 3 of this dissertation allow a real-time optimization of the connected DERs as well as conventional SGs, with the primary goal of maintaining the nominal frequency and keeping all voltage magnitudes within predefined limits whilst minimizing an individual user-defined cost function. The proposed controller is unifying in the sense that the former hierarchical division into primary, secondary, and tertiary frequency and voltage control tasks are combined within a single controller (see bottom panel in Fig. 2.3). Furthermore, frequency and voltage stabilization is maintained simultaneously. The resulting market-based control structure is fully compliant with legal, operational, and informational unbundling. In particular, motivated from an economic perspective, no communication between competitive players, no generation of exogenous setpoints by a higher-level authority, and no disclosure of individual cost functions is necessary. The controller design is based on a nonlinear, *network-preserving* model[27] of the power system, and allows for a distributed implementation.

> **Contribution 2.** *This dissertation presents a dynamic pricing framework for vertically unbundled power systems that enforces Pareto efficiency and market-clearing allocations among competitive participants.*

The real-time pricing approach presented in Chapter 4 (see middle panel in Fig. 2.3) enables each network participant to selfishly perform an individual profit maximization by means of internal control actions. In the light of [Ahl18, p. 324f.], the game-theoretic notion of *Pareto efficiency* (cf. Definition A.5) is employed to ensure economically efficient operation of all network participants. Accordingly, the resulting closed-loop equilibrium is such that there exists no other equilibrium where one or more network participants take a higher profit. A distributed market balancing scheme ensures that the total revenues of producers are always kept identical to the payments of consumers. This permits the additional costs arising from momentary transmission losses to be reflected directly in the real-time price, without creating net deficits in overall generation or an accumulation of capital. The resulting (optimal) market balancing problem gives rise to the problem class of *(dynamic) optimal control problems and differential games subject to port-Hamiltonian systems* (PHSs) (cf. Appendix A.4), for which no solution concept in the sense of an explicit stabilizing control strategy exists yet.

> **Contribution 3.** *This dissertation develops a continuous-time optimal control method for (nonlinear) port-Hamiltonian systems with N agents.*

In Chapter 5 of this dissertation, we propose a continuous-time adaptive control strategy for solving single- and multi-player noncooperative differential games with general (Lagrangian) performance indices and system dynamics modeled as general nonlinear input-state-output PHSs. The proposed control law (see right panel in Fig. 2.3) implements an online learning procedure which uses the Hamiltonian of the system as initial value function candidates. The feedback strategy of players is extended by adaptively weighting the individual value functions to ensure convergence to the Nash solution. Necessary and sufficient conditions for stability of the resulting controlled system are provided by employing Lyapunov stability theory. Since PHSs have received substantial attention as being a powerful tool for systematic modeling of dynamical multi-physics systems, the presented methodology qualifies for a broad range of applications apart from power systems.

> **Contribution 4.** *This dissertation presents feedback control strategies for incentivizing grid-supportive behavior of power plants by continuous generation of price signals.*

In order to contribute to a grid-supportive mechanism design, this dissertation provides a distributed framework for real-time zonal pricing. Being fully compatible with the WoC

[27] A power system model is said to be network-preserving if each bus is modeled separately (possibly as a dynamic subsystem) and connected to the other buses via power flow equations. Conversely, the model is said to be network-reduced, if buses within a certain geographical region are aggregated and treated as a single bus.

concept introduced in Section 2.2.3, local CCs are enabled to allocate cell-specific, temporally and spatially differentiated prices. Thereby, higher-level goals such as congestion management or local investment signaling can be implemented in both the short term and long term (see top panel in Fig. 2.3). In fact, the presented mechanism allows to pursue those equilibria that are desirable from a global perspective by means of a distributed consensus protocol.

The above contributions are systematically elaborated in the subsequent Chapters 3–5, which are organized as follows: Chapter 3 develops a network-preserving power system and market model conforming with the WoC concept, and presents a price-based control framework with distributed *intra-cell* communication to satisfy Contribution 1. Chapter 4 presents a distributed control framework at *inter-cell* level for an automatic regulation of spatial price differences, working towards Contribution 2 and Contribution 4. Finally, Chapter 5 provides the solution of dynamic optimal control problems for PHSs to complete the remaining tasks of Contribution 2 along with Contribution 3.

3 Optimization-Based Control of Cellular Power Networks

In this chapter, a WoC-based control framework for economic dispatch and power system stabilization of competitive network participants acting in an unbundled power system is developed. After introducing the general setup of the WoC-based infrastructure in Section 3.1, Section 3.2 presents a dynamic, network-preserving model of the physical power system. Section 3.3 investigates the competitive behavior of unbundled network participants and develops a conforming market mechanism. Section 3.4 derives price-based local controllers at intra-cell level and investigates the overall system dynamics resulting from the coupling with the physical system. Thereby Contribution 1 is provided. Section 3.5 examines the interaction between cells and gives an initial diretion towards inter-cell pricing strategies. Section 3.6 briefly discusses the main findings of this chapter[28].

3.1 General Setup

We consider an interconnected power system with meshed physical topology, which is modeled as a directed graph $\mathscr{G}_\mathrm{p} = (\mathcal{V}_\mathrm{p}, \mathcal{E}_\mathrm{p})$. The nodes $v \in \mathcal{V}_\mathrm{p}$ correspond to n_p buses and the edges $e \in \mathcal{E}_\mathrm{p}$ to m_p power lines connecting the buses, where the direction of the edges indicates the direction of positive power flow. The incidence matrix (cf. Definition A.10) of $\mathscr{G}_\mathrm{p}$ is denoted by $\boldsymbol{D}_\mathrm{p}$. The power system is divided into $n_\mathcal{Z} \in \mathbb{N}$ cells operated by a CC $k \in \mathcal{Z} = \{k_1, k_2, \ldots\}$. Within its own cell boundaries, each CC acts as an ISO, being exclusively responsible to ensure the main objectives of stability, economic efficiency and grid-supportive operation outlined in Chapter 1. The responsibilities and information patterns assigned to the CCs as well as their interactions with other network participants are detailed in Section 3.3.

Regardless of cell boundaries, the power system is connected with *wholesale prosumers* (WPs) and *retail prosumers* (RPs). Both WPs and RPs are generic groups for network participants that produce and/or consume electrical energy. A major difference between both groups, however, is the presence or absence of *flexibility*: For (flexible) WPs, the own power supply or demand is the result of economic considerations (i.e. maximization of own profit) and thus supposed to be affected by prices. For (inflexible) RPs, on the other hand, the supply or demand of RPs is assumed to be uncontrollable. These two classes thus cover all possible types of energy producers and consumers.

28 Preliminary versions of the results of this chapter have been published in the conference papers [KBKH19, KDB+20, KWKH20].

For the overall model, let $\mathcal{W} = \{\pi_1, \pi_2, \ldots\}$ denote the set of $n_{\mathcal{W}}$ WPs[29]. Each node is associated to *at least* one WP $\pi \in \mathcal{W}$. We denote the subset of nodes $i \in \mathcal{V}_\mathrm{p}$ associated to WP $\pi \in \mathcal{W}$ by $\mathcal{V}_{\mathcal{W},\pi} \subseteq \mathcal{V}_\mathrm{p}$. Moreover, each node $i \in \mathcal{V}_\mathrm{p}$ belongs to *exactly* one of the $n_{\mathcal{Z}}$ *cells* $\mathcal{V}_{\mathcal{Z},k} \subseteq \mathcal{V}_\mathrm{p}$, $k = 1, \ldots, n_{\mathcal{Z}}$, i.e. $\mathcal{V}_{\mathcal{Z},k} \cap \mathcal{V}_{\mathcal{Z},l} = \emptyset$ for all $k \neq l$ and $\bigcup_{k=1}^{n_{\mathcal{Z}}} \mathcal{V}_{\mathcal{Z},k} = \mathcal{V}_\mathrm{p}$. The edges in $\mathcal{E}_\mathrm{p}$ can further be divided into the set $\mathcal{E}_\mathrm{p}^\mathrm{b}$ of *inter-cell* or *boundary* edges representing power lines that connect buses located in different cells[30] and the set $\mathcal{E}_\mathrm{p}^\mathrm{nb}$ of *intra-cell* or *non-boundary* edges representing power lines that start and end within the same cell. The physical interconnection of different cells is described by the graph $\hat{\mathcal{G}}_\mathrm{p} = (\mathcal{Z}, \mathcal{E}_\mathrm{p}^\mathrm{b})$ with incidence matrix $\hat{\boldsymbol{D}}_\mathrm{p}$.

A fundamental principle of the WoC concept is that measurement and control information is only shared *locally*. This is expressed by a separate communication infrastructure for each cell $k \in \mathcal{Z}$ represented by directed graphs $\mathcal{G}_{\mathrm{c},k} = (\mathcal{V}_{\mathcal{Z},k}, \mathcal{E}_{\mathrm{c},k})$ with incidence matrices $\boldsymbol{D}_{\mathrm{c},k}$. Note that the direction of the edges of the communication infrastructures do not imply a one-way connection but rather the direction of positive "communication flows". The following assumption holds under mild conditions:

> **Assumption 3.1.** *The graph of the physical system $\mathcal{G}_\mathrm{p}$ and all communication graphs $\mathcal{G}_{\mathrm{c},k}$, $k \in \mathcal{Z}$ are weakly connected.*

Assumption 3.1 implies that $\hat{\mathcal{G}}_\mathrm{p}$ is weakly connected as well.

The totality of all communication between nodes is represented by the communication graph

$$\mathcal{G}_\mathrm{c} = \left(\mathcal{V}_\mathrm{p}, \bigcup_{k \in \mathcal{Z}} \mathcal{E}_{\mathrm{c},k} \right). \tag{3.1}$$

This leads to the following formal definition:

> **Definition 3.2 (Cell).** *A* cell *is a weakly connected component in $\mathcal{G}_\mathrm{c}$.*

Notably, the characteristics of the WoC scheme rely only on the *communication topology*, while the physical topology is not subject to any restrictions. In particular, power flows across the cell boundaries are allowed without any limits. In the following, we summarize the main properties and assumptions on the considered power system:

> **Definition 3.3 (Competitive Power System with Cell-Based Infrastructure).** *We consider an interconnected power system with the following characteristics.*
>
> - *We consider a lossy AC system with meshed topology,*

[29] Note that this framework is compatible to the VPP concept (cf. [NY21]), since the individual resources of prosumers may be located at geographically distant nodes in different cells.

[30] If there are multiple inter-cell power lines connecting the same cells, then $\hat{\mathcal{E}}_\mathrm{p}$ contains an edge for each of these power lines.

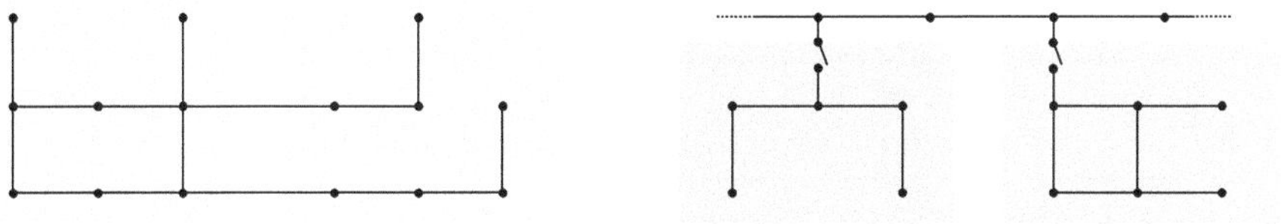

(a) WoC topology **(b)** MG topology

Figure 3.1: Differences between WoC and MG topology.

- *the topology (i.e. the interconnection by power and communication lines) is fixed during the examination period,*
- *there may exist multiple points of coupling between cells,*
- *WPs and/or RPs are distributed arbitrarily through the power system,*
- *the generation capacity available in the cell is not necessarily sufficient to cover its own demand,*
- *there exists no superordinate network layer.*

Remark 3.4. *Compared to our model from Definition 3.3, the definition of a* grid of MGs *[Sch15, Definition 3.2.1] differs by the following limiting characteristics:*

- *For each MG there exists a single* point of common coupling *to a superordinate grid layer,*
- *each MG can switch from islanded to grid-connected mode. In grid-connected mode, it behaves as a single controllable generator or consumer from the viewpoint of the superordinate grid,*
- *it possesses enough capacity to supply most of its loads autonomously.* ◊

Remark 3.4 emphasizes that the MG concept is subject to more structural restrictions as well as more conservative capacity requirements (which are hard to verify in advance) than the WoC concept pursued in this work. Fig. 3.1 schematically summarizes the topological differences between these two concepts. Using the above notation, the topology of an islanded MG is regarded as the special case $\hat{\mathcal{E}}_\text{p} = \emptyset$. In the following, however, we will generally assume that $\mathscr{G}_\text{c} \neq \mathscr{G}_\text{p}$ and $\hat{\mathcal{E}}_\text{p} \neq \emptyset$ hold in order to permit a global exchange of power. For reasons of comparison, though, the MG approach will occasionally serve as a reference scenario.

3.2 Modeling of the Physical Network

The set of nodes can be partitioned as $\mathcal{V}_\text{p} = \mathcal{V}_\mathcal{S} \cup \mathcal{V}_\mathcal{I} \cup \mathcal{V}_\mathcal{L}$, where each subset represents a certain type of network connectors, namely, *synchronous machine-type* ($\mathcal{S}$) nodes, *inverter-type* ($\mathcal{I}$), and *load-type* ($\mathcal{L}$) nodes. $\mathcal{S}$ nodes are connected to synchronous machines (SMs) of e.g.

Table 3.1: Network connectors.

	No. of Nodes	Independent Controllable	Independent Uncontrollable	Dependent
$\mathcal{S}$	$n_{\mathcal{S}}$	$p_{\mathrm{g},i}, U_{\mathrm{f},i}$	$p_{\ell,i}$	θ_i, q_i
$\mathcal{I}$	$n_{\mathcal{I}}$	$p_{\mathrm{g},i}, U_i$	$p_{\ell,i}$	θ_i, q_i
$\mathcal{L}$	$n_{\mathcal{L}}$		$p_{\ell,i}, q_{\ell,i}$	θ_i, U_i
Total No.	n_{p}	$2n_{\mathrm{p}}$		$2n_{\mathrm{p}}$

gas or hydro turbines. $\mathcal{I}$ nodes are connected to power electronics interfaced DERs. $\mathcal{L}$ nodes are connected to producers or consumers with a fixed active and reactive power supply or demand. $\mathcal{S}$ and $\mathcal{I}$ nodes $i \in \mathcal{V}_{\mathcal{S}} \cup \mathcal{V}_{\mathcal{I}}$ are each equipped with an active power generation input $p_{\mathrm{g},i}$ and an active power consumption input $p_{\ell,i}$. $\mathcal{L}$ nodes $i \in \mathcal{V}_{\mathcal{L}}$ are equipped with active and reactive power consumption inputs $p_{\ell,i}$ and $q_{\ell,i}$, respectively. Table 3.1 gives an overview of the three node types along with their specific independent (exogenous) and dependent (endogenous) variables. The set $\mathcal{N}_{\mathrm{p},i}$ denotes all *neighboring* nodes $j \in \mathcal{V}_{\mathrm{p}}$ of $i \in \mathcal{V}_{\mathrm{p}}$ with $(i, j) \in \mathcal{E}_{\mathrm{p}}$ or $(j, i) \in \mathcal{E}_{\mathrm{p}}$.

Remark 3.5. *$\mathcal{S}$ and $\mathcal{I}$ nodes correspond to the notion of* PV nodes *from classical power flow calculation. In the presented framework, the voltage magnitudes are either controlled directly (at $\mathcal{I}$ nodes) or indirectly via the excitation voltages $U_{\mathrm{f},i}$ (at $\mathcal{S}$ nodes). $\mathcal{L}$ nodes correspond to the notion of* PQ nodes, *where the active and reactive power generation or consumption is given and the voltage magnitudes and bus voltage angles θ_i follow from Kirchhoff's laws.*

In classical power flow calculation, apart from PV and PQ buses there typically exists one slack bus *$i \in \mathcal{V}_{\mathrm{p}}$, which represents a large power plant or a superordinate network with infinite capacity of real and reactive power. The slack bus is characterized by fixed U_i and θ_i as independent and p_i and q_i as dependent variables. As the resulting power flows are invariant with respect to a uniform shift of all bus voltage angles θ_i, the slack bus typically serves as bus voltage angle reference by setting $\theta_i = 0$. Yet in line with Definition 3.3, our model is intended to avoid any external sources of power. Thus, no slack bus is considered within our power system model. However, to provide for nonsingular solutions within calculations, all CCs and prosumers use bus voltage angle* differences *$\vartheta_{ij} = \theta_i - \theta_j$ instead of explicitly dealing with the absolute values θ_i and θ_j.* ◇

The following subsections systematically derive a common model for the physical interaction of the power system. Subsection 3.2.1 considers the power flows over lossy transmission lines. Subsections 3.2.2–3.2.4 present an appropriate model for SM-type, inverter-type, and load-type nodes, respectively. On this basis, Subsection 3.2.5 synthesizes the dynamic model of the overall physical plant system.

3.2.1 Power Lines

Regarding the physical infrastructure, Definition 3.3 does not restrict the voltage level of the underlying power grid. In the following, however, we will focus on *medium voltage networks* (with nominal voltages typically between 10 kV and 100 kV), as the majority of DERs are connected at this voltage level. In addition to the properties from Definition 3.3, medium voltage networks can be assumed to have the following characteristics under normal operating conditions.

Assumption 3.6. *a) The network is operating around the nominal frequency.*

b) The network is a three-phased AC system which is operated under symmetrical conditions.

Note that Assumption 3.6a) is only needed for the purpose of model validity and does not imply the (much stricter) requirement of frequency stability to be addressed in the following sections. Assumption 3.6b) implies that the network is both symmetrically configured and symmetrically operated, see [SFO15, Section 2.4.1] for basic terminology. With Assumption 3.6b), a single-phase equivalent circuit model [MBB12, p. 66] can be applied (see Fig. 3.2). Here,

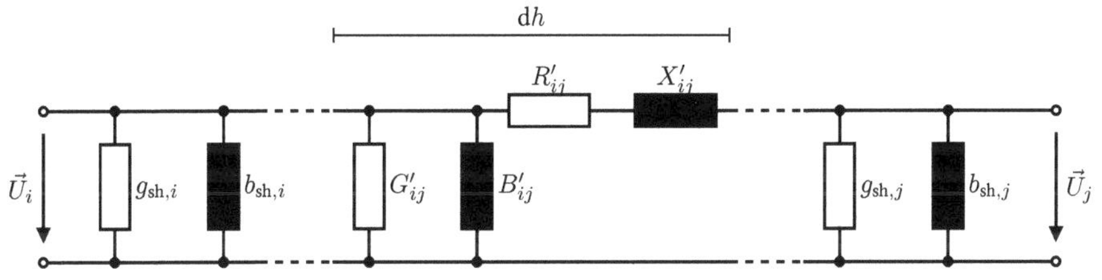

Figure 3.2: Infinitesimally short line segment of length dh.

each phase of a power line connecting two nodes $i, j \in \mathcal{V}_\mathrm{p}$ is decomposed into an infinite number of increments dh (see Fig. 3.2). Each of these increments can be represented by a specific series resistance R'_{ij} characterizing the thermal losses, a specific series reactance X'_{ij} characterizing the flux linkage within the power line and the flux linkage to the other phases, a specific parallel susceptance B'_{ij} describing the potential difference between the individual phases, and a specific parallel conductance G'_{ij}.

With Assumption 3.6b) and provided that no line length exceeds 300 km, the Π-equivalent circuit model can be applied [And, p. 101; MBB12, p. 67]. It provides a lumped-parameter representation (see Fig. 3.3) with

$$\vec{Z}_{\mathrm{L},ij} = \sqrt{\frac{\vec{Z}'_{ij}}{\vec{Y}'_{ij}}} \cdot \sinh\left(\sqrt{\vec{Z}'_{ij}\vec{Y}'_{ij}} \cdot h\right), \tag{3.2}$$

$$\vec{Y}_{\mathrm{L},ij} = \sqrt{\frac{\vec{Z}'_{ij}}{\vec{Y}'_{ij}}} \cdot \tanh\left(\sqrt{\vec{Z}'_{ij}\vec{Y}'_{ij}} \cdot h\right), \tag{3.3}$$

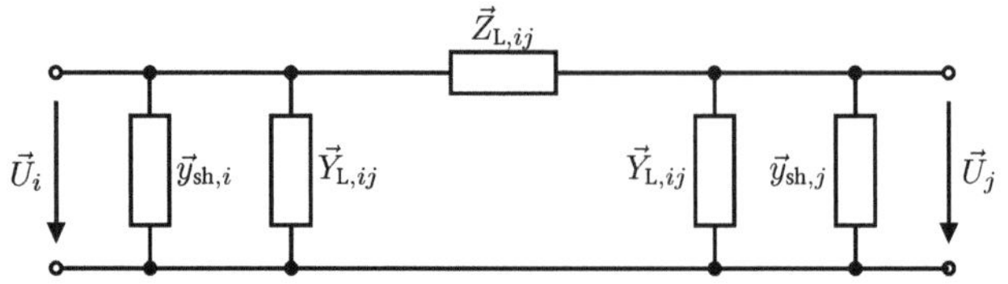

Figure 3.3: Π-equivalent circuit model for a power line between nodes i and j.

where $\vec{Z}'_{ij} := R'_{ij} + \jmath X'_{ij}$ and $\vec{Y}'_{ij} = G'_{ij} + \jmath B'_{ij}$. Together with the nodal shunt admittance $\vec{y}_{\mathrm{sh},i} = g_{\mathrm{sh},i} + \jmath b_{\mathrm{sh},i}$, we obtain $\vec{y}_i = \vec{y}_{\mathrm{sh},i} + \sum_{j \in \mathcal{N}_{\mathrm{p},i}} \vec{Y}_{\mathrm{L},ij}$. Consequently, the entire network of power lines can be described in vector-matrix notation by the *nodal network equation* [MBB12, p. 113]

$$\vec{\boldsymbol{I}}_{\mathrm{p}} = \vec{\boldsymbol{Y}}_{\mathrm{p}} \vec{\boldsymbol{U}}_{\mathrm{p}} \tag{3.4}$$

with $\vec{\boldsymbol{I}}_{\mathrm{p}} = \mathrm{col}_{i \in \mathcal{V}_{\mathrm{p}}}\{\vec{I}_i\}$ and $\vec{\boldsymbol{U}}_{\mathrm{p}} = \mathrm{col}_{i \in \mathcal{V}_{\mathrm{p}}}\{\vec{U}_i\}$. The nodal admittance matrix $\vec{\boldsymbol{Y}}_{\mathrm{p}}$ describes the relationship between the nodal voltages $\vec{\boldsymbol{U}}_{\mathrm{p}}$ and the currents $\vec{\boldsymbol{I}}_{\mathrm{p}}$ entering the nodes. If node i is connected to node j by a line, then

$$\left[\vec{\boldsymbol{Y}}_{\mathrm{p}}\right]_{ij} = \left[\vec{\boldsymbol{Y}}_{\mathrm{p}}\right]_{ji} = -\frac{1}{\vec{Z}_{\mathrm{L},ij}}, \tag{3.5}$$

$$\left[\vec{\boldsymbol{Y}}_{\mathrm{p}}\right]_{ii} = \vec{y}_i + \sum_{j \in \mathcal{N}_{\mathrm{p},i}} \frac{1}{\vec{Z}_{\mathrm{L},ij}}. \tag{3.6}$$

Using the nodal admittance matrix $\vec{\boldsymbol{Y}}_{\mathrm{p}}$, the apparent power balance can be determined to

$$\vec{s}_i = \sum_{i \in \mathcal{V}_{\mathrm{p}}} \sum_{j \in \mathcal{N}_{\mathrm{p},i}} \vec{U}_i \vec{U}_j^* [\vec{\boldsymbol{Y}}_{\mathrm{p}}]_{ij}^*, \tag{3.7}$$

where $\vec{s}_i$ denotes the net power flow from node $i \in \mathcal{V}_{\mathrm{p}}$ to neighboring nodes. Now the equation can be split into its real and imaginary part [SFO15, p. 90]:

$$p_i = \sum_{i \in \mathcal{V}_{\mathrm{p}}} \sum_{j \in \mathcal{N}_{\mathrm{p},i}} U_i U_j Y_{\mathrm{p},ij} \cos(\vartheta_{ij} - \beta_{ij}), \tag{3.8a}$$

$$q_i = \sum_{i \in \mathcal{V}_{\mathrm{p}}} \sum_{j \in \mathcal{N}_{\mathrm{p},i}} U_i U_j Y_{\mathrm{p},ij} \sin(\vartheta_{ij} - \beta_{ij}), \tag{3.8b}$$

where $Y_{\mathrm{p},ij} := |[\vec{\boldsymbol{Y}}_{\mathrm{p}}]_{ij}|$ and $\beta_{ij} := \arg([\vec{\boldsymbol{Y}}_{\mathrm{p}}]_{ij})$.

A widely used representation of the power flows (3.7) are the *hybrid network equations*, where the nodal admittance matrix is given in rectangular coordinates using $\vec{\boldsymbol{Y}}_{\mathrm{p}} =: \boldsymbol{G}_{\mathrm{p}} + \jmath \boldsymbol{B}_{\mathrm{p}}$, while the nodal voltages are given in polar coordinates using $\vec{\boldsymbol{U}} =: \boldsymbol{U} \circ \mathbf{exp}(\jmath \boldsymbol{\theta})$. This yields the network equations at node $i \in \mathcal{V}_\text{p}$ as

$$p_i = U_i^2 G_{ii} + \sum_{j\in\mathcal{N}_{\text{p},i}} U_i U_j \left(B_{ij}\sin(\vartheta_{ij}) + G_{ij}\cos(\vartheta_{ij})\right), \tag{3.9}$$

$$q_i = -U_i^2 B_{ii} + \sum_{j\in\mathcal{N}_{\text{p},i}} U_i U_j \left(G_{ij}\sin(\vartheta_{ij}) - B_{ij}\cos(\vartheta_{ij})\right). \tag{3.10}$$

The (sending-end) power flow from node $i \in \mathcal{V}_\text{p}$ to node $j \in \mathcal{V}_\text{p}$ is given by $\vec{S}_{ij} = P_{ij} + \jmath Q_{ij}$, where

$$P_{ij} = -U_i^2 G_{ij} + U_i U_j G_{ij}\cos(\vartheta_{ij}) + U_i U_j B_{ij}\sin(\vartheta_{ij}), \tag{3.11}$$

$$Q_{ij} = U_i^2 B_{ij} + U_i U_j B_{ij}\cos(\vartheta_{ij}) + U_i U_j G_{ij}\sin(\vartheta_{ij}). \tag{3.12}$$

Hence, the active transmission loss over line $(i,j) \in \mathcal{E}_\text{p}$ equals

$$\Phi_{ij} = P_{ij} + P_{ji} = -(U_i^2 + U_j^2)G_{ij} + 2U_i U_j G_{ij}\cos(\vartheta_{ij}). \tag{3.13}$$

3.2.2 Synchronous Machine-Type Nodes

SMs form the backbone of today's power systems. Besides their main application of converting mechanical energy into electrical energy, e.g. as turbogenerators in bulk CPPs, they can also serve as constant-speed drives, e.g. for centrifugal pumps or air compressors.

The constant magnetic field in the *rotor* (typically governed by a DC exciter winding) interacts with the rotating field of the *stator* by inducing a *rotor voltage* (or polar wheel voltage) in the stator windings, where in *motor mode* the rotor follows the stator rotating field, while in *generator mode*, the stator rotating field follows the rotor. The rotor field winding of the SM is aligned with the *direct* (d-) axis, while the axis perpendicular to the d-axis is called *quadrature* (q-) axis. Both axes form a reference system inside the machine that rotates with the rotor. Under Assumption 3.6a), the electrical side of the SM at node $i \in \mathcal{V}_\mathcal{S}$ can be adequately described by the *sixth-order model* (3.14)–(3.16) from [MBB12, p. 455]

$$U_{\text{gd},i} = U''_{\text{d},i} - R_{\text{s},i} I_{\text{d},i} - X''_{\text{q},i} I_{\text{q},i}, \tag{3.14a}$$

$$U_{\text{gq},i} = U''_{\text{q},i} + X''_{\text{d},i} I_{\text{d},i} - R_{\text{s},i} I_{\text{q},i}, \tag{3.14b}$$

$$T'_{\text{d}0,i}\dot{U}'_{\text{q},i} = U_{\text{f},i} - U'_{\text{q},i} + I_{\text{d},i}\cdot(X_{\text{d},i} - X'_{\text{d},i}), \tag{3.14c}$$

$$T'_{\text{q}0,i}\dot{U}'_{\text{d},i} = -U'_{\text{d},i} + I_{\text{q},i}\cdot(X_{\text{q},i} - X'_{\text{q},i}), \tag{3.14d}$$

$$T''_{\text{d}0,i}\dot{U}''_{\text{q},i} = U'_{\text{q},i} - U''_{\text{q},i} + I_{\text{d},i}\cdot(X'_{\text{d},i} - X''_{\text{d},i}), \tag{3.14e}$$

$$T''_{\text{q}0,i}\dot{U}''_{\text{d},i} = U'_{\text{d},i} - U''_{\text{d},i} + I_{\text{q},i}\cdot(X'_{\text{q},i} - X''_{\text{q},i}), \tag{3.14f}$$

where $U''_{(\cdot)}$ and $U'_{(\cdot)}$ denote the subtransient and transient electromotive forces (internal voltages) induced at the stator coils and $U_{\text{g}(\cdot)}$ denotes the *terminal voltage* of the SM. Due to time-varying screening effects in the rotor, usually a distinction is made between *subtransient state*, *transient state*, and *steady state*, which is represented in (3.14) by the (sub-)transient and steady-state reactances $X^{(\cdot)}_{(\cdot),i}$ and time constants $T''_{\text{d}0,i} \ll T'_{\text{d}0,i}$ and $T''_{\text{q}0,i} \ll T'_{\text{q}0,i}$, each modeled separately for d- and q-axis. For a detailed derivation of the model (3.14), the reader is referred to the comprehensive textbooks [Kun94, BV00, MBB12].

The mechanics of the rotor is described by a second-order system following from Newton's equation of motion

$$\dot{\gamma}_i = \Omega_{\mathrm{m},i}, \tag{3.15}$$

$$J_i \dot{\Omega}_{\mathrm{m},i} + D_{\mathrm{d},i} \Omega_{\mathrm{m},i} = M_{\mathrm{T},i} - M_{\mathrm{e},i}, \tag{3.16}$$

where γ_i denotes the *rotor angle* between the q-axis and a static (non-rotating) reference frame, J_i is the moment of inertia of the turbogenerator set, $D_{\mathrm{d},i}$ is the damping-torque coefficient, $M_{\mathrm{T},i}$ is the turbine torque, and $M_{\mathrm{e},i}$ is the (counteracting) electromechanical torque.

We make the following simplifying assumptions for the SM model.

Assumption 3.7. *a) Subtransient dynamics of the SMs are neglected.*

b) The screening effect of the rotor body eddy currents in the q-axis and the stator resistance is neglected.

c) Saliency of the rotor is neglected.

Assumptions 3.7a) and 3.7b) are justified by the timescale limitation adopted in Section 2.2 (cf. Fig. 2.2). Assumption 3.7c) reflects the most common configuration of turbogeneraors in CPPs.

With Assumption 3.7a), all effects of the damper windings are neglected, yielding $T''_{\mathrm{d0},i} = T''_{\mathrm{q0},i} = 0$. Typically, the asynchronous torque of the damper windings is approximated by an additional damping term $+d_{\mathrm{d},i}\omega_{\mathrm{m},i}$ in (3.16), which is proportional to the deviation $\omega_{\mathrm{m},i} = \Omega_{\mathrm{m},i} - \Omega^0_{\mathrm{m},i}$ of the mechanical angular velocity of the rotor from its nominal value $\Omega^0_{\mathrm{m},i}$. In particular, inserting (3.14e) with $T''_{\mathrm{d0},i} = 0$ into (3.14a), and (3.14f) with $T''_{\mathrm{q0},i} = 0$ into (3.14a) yields

$$U_{\mathrm{gd},i} = U'_{\mathrm{d},i} - I_{\mathrm{q},i} X'_{\mathrm{q},i} - R_{\mathrm{s},i} I_{\mathrm{d},i}, \tag{3.17a}$$

$$U_{\mathrm{gq},i} = U'_{\mathrm{q},i} + I_{\mathrm{d},i} X'_{\mathrm{d},i} - R_{\mathrm{s},i} I_{\mathrm{q},i}, \tag{3.17b}$$

$$T'_{\mathrm{d0},i} \dot{U}'_{\mathrm{q},i} = U_{\mathrm{f},i} - U'_{\mathrm{q},i} + I_{\mathrm{d},i} \cdot (X_{\mathrm{d},i} - X'_{\mathrm{d},i}), \tag{3.17c}$$

$$T'_{\mathrm{q0},i} \dot{U}'_{\mathrm{d},i} = -U'_{\mathrm{d},i} + I_{\mathrm{q},i} \cdot (X_{\mathrm{q},i} - X'_{\mathrm{q},i}), \tag{3.17d}$$

$$\dot{\gamma}_i = \Omega_{\mathrm{m},i}, \tag{3.17e}$$

$$J_i \dot{\Omega}_{\mathrm{m},i} + D_{\mathrm{d},i} \Omega_{\mathrm{m},i} + d_{\mathrm{d},i} \omega_{\mathrm{m},i} = M_{\mathrm{T},i} - M_{\mathrm{e},i}. \tag{3.17f}$$

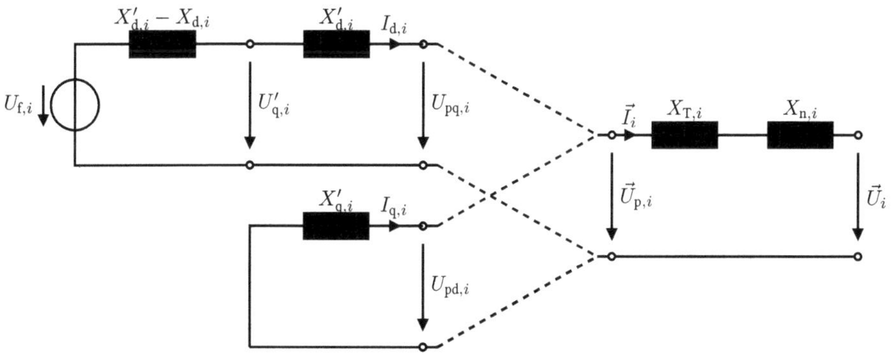

Figure 3.4: Equivalent circuit diagram of a SM in transient state connected to the grid.

With Assumption 3.7b), it holds that $X_{q,i} = X'_{q,i}$, $U'_{d,i} = 0$, and $R_{s,i} = 0$. This leads to

$$U_{gd,i} = -I_{q,i}X'_{q,i}, \tag{3.18a}$$
$$U_{gq,i} = U'_{q,i} + I_{d,i}X'_{d,i}, \tag{3.18b}$$
$$T'_{d0,i}\dot{U}'_{q,i} = U_{f,i} - U'_{q,i} + I_{d,i} \cdot (X_{d,i} - X'_{d,i}), \tag{3.18c}$$
$$\dot{\gamma}_i = \Omega_{m,i}, \tag{3.18d}$$
$$J_i\dot{\Omega}_{m,i} + D_{d,i}\Omega_{m,i} + d_{d,i}\omega_{m,i} = M_{T,i} - M_{e,i}. \tag{3.18e}$$

The dynamics of the electrical subsystem is illustrated in the left half of Fig. 3.4.

Electromechanical Coupling

On the mechanical side, it usually holds that $D_{d,i} \ll d_{d,i}$ [MBB12, p. 455], thus $D_{d,i}$ can be neglected in (3.18e). With the above assumption and $\dot{\Omega}^0_{m,i} = 0$, we get $J_i\dot{\omega}_{m,i} + d_{d,i}\omega_{m,i} = M_{T,i} - M_{e,i}$. Due to the fact that power equals torque times angular velocity, this yields

$$J_i\dot{\omega}_{m,i} + d_{d,i}\omega_{m,i} = \frac{1}{\Omega_{m,i}}\left(p_{T,i} - p_{e,i}\right). \tag{3.19}$$

Multiplication by $\Omega^0_{m,i}$ and Assumption 3.6a) lead to

$$\Omega^0_{m,i}J_i\dot{\omega}_{m,i} + \Omega^0_{m,i} \cdot d_{d,i}\omega_{m,i} = p_{T,i} - p_{e,i}. \tag{3.20}$$

Now we switch from mechanical to electrical domain by choosing

$$\Omega_i = \frac{2\Omega_{m,i}}{n_{pp}}, \qquad \Omega_{n,i} = \frac{2\Omega^0_{m,i}}{n_{pp}}, \tag{3.21}$$

where $n_{pp} \in \mathbb{N}$ denotes the number of pole pairs. The deviation of the electrical angular velocity from its nominal value is thus given by $\omega_i = \Omega_i - \Omega_{n,i} = \frac{2\omega_{m,i}}{n_{pp}}$. Accordingly, we get $\dot{\omega}_i = \dot{\Omega}_i = \frac{2\dot{\Omega}_{\mathrm{m},i}}{n_{\mathrm{pp}}} = \frac{2\dot{\omega}_{\mathrm{m},i}}{n_{\mathrm{pp}}}$. Expressing the mechanical velocity in (3.20) by means of the electrical velocity ω_i yields thus

$$\underbrace{J_i \cdot \frac{\Omega_{\mathrm{n},i} \cdot n_{\mathrm{pp}}}{2} \cdot \frac{n_{\mathrm{pp}}}{2}}_{=:\Gamma_i} \cdot \dot{\omega}_i + \underbrace{d_{\mathrm{d},i} \cdot \frac{\Omega_{\mathrm{n},i} \cdot n_{\mathrm{pp}}}{2} \cdot \frac{n_{\mathrm{pp}}}{2}}_{=:A_i} \cdot \omega_i = p_{\mathrm{T},i} - p_{\mathrm{e},i}, \tag{3.22}$$

which is the classical second-order *swing equation.*

Connection with the Power Network

The SM is connected to the power network as depicted in the right half of Fig. 3.4, where $\vec{U}_i$ denotes the phasor of the "external" nodal voltage U_i at the coupling point with the power network. The active and reactive power injection at the coupling point are given by[31]

$$p_{\mathrm{inj},i} = \frac{U_i' U_i}{X_{\mathrm{dn},i}} \sin(\vartheta_{\mathrm{n},i}) + \frac{U_i^2}{2} \frac{X_{\mathrm{dn},i} - X_{\mathrm{qn},i}}{X_{\mathrm{dn},i} X_{\mathrm{qn},i}} \sin(2\vartheta_{\mathrm{n},i}), \tag{3.23}$$

$$q_{\mathrm{inj},i} = \frac{U_i' U_i}{X_{\mathrm{dn},i}} \cos(\vartheta_{\mathrm{n},i}) - \frac{U_i^2}{X_{\mathrm{dn},i} X_{\mathrm{qn},i}} \left(X_{\mathrm{dn},i} \sin^2(\vartheta_{\mathrm{n},i}) + X_{\mathrm{qn},i} \cos^2(\vartheta_{\mathrm{n},i}) \right). \tag{3.24}$$

In case of non-salient pole SMs, we have $X_{\mathrm{d},i} = X_{\mathrm{q},i}$ and thus $X_{\mathrm{dn},i} = X_{\mathrm{qn},i}$. Therefore, the above equations simplify to

$$p_{\mathrm{inj},i} = \frac{U_i' U_i}{X_{\mathrm{dn},i}} \sin(\vartheta_{\mathrm{n},i}), \tag{3.25}$$

$$q_{\mathrm{inj},i} = \frac{U_i' U_i}{X_{\mathrm{dn},i}} \cos(\vartheta_{\mathrm{n},i}) - \frac{U_i^2}{X_{\mathrm{dn},i}}. \tag{3.26}$$

The coupling between mechanical and electric domain is given by

$$p_{\mathrm{e},i} = p_{\mathrm{inj},i}, \qquad \Omega_i = \frac{2\Omega_{\mathrm{m},i}}{n_{\mathrm{pp}}}. \tag{3.27}$$

With $\Omega_{\mathrm{m},i} = \Omega_{\mathrm{m},i}^0 + \omega_{\mathrm{m},i}$ and $\dot{\gamma}_i = \Omega_{\mathrm{m},i}$, we get $\dot{\delta}_i = \omega_{\mathrm{m},i}$. With Assumption 3.6a), it holds that $n_{\mathrm{pp}} \cdot \dot{\delta}_i = \dot{\theta}_i$, thus (3.18d) is equivalent to $\dot{\theta}_i = \omega_i$. This allows to write the overall equations compactly as

$$\dot{\theta}_i = \omega_i, \tag{3.28a}$$

$$\Gamma_i \dot{\omega}_i = -A_i \omega_i + p_{\mathrm{T},i} - p_{\mathrm{inj},i}, \tag{3.28b}$$

$$T_{\mathrm{d0},i}' \dot{U}_i' = U_{\mathrm{f},i} - U_i' + (X_{\mathrm{d},i} - X_{\mathrm{d},i}')(U_i')^{-1} q_{\mathrm{inj},i}. \tag{3.28c}$$

31 A detailed derivation of this physical relationship is provided in Appendix B.1.

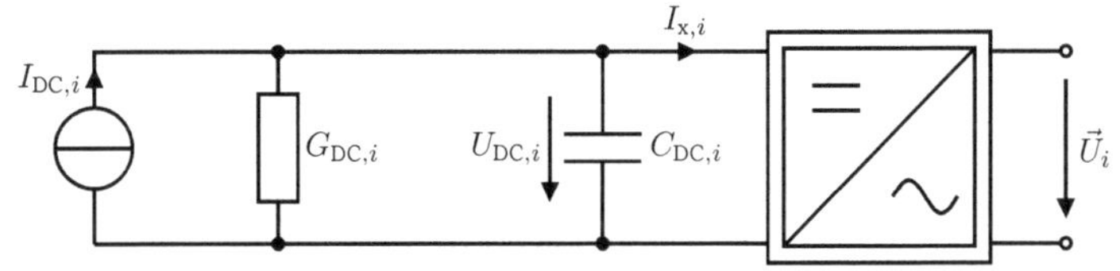

Figure 3.5: Single-phase circuit diagram of a grid-forming inverter with capacitive inertia.

3.2.3 Inverter-Type Nodes

The vast majority of DERs fed by RESs are connected to the grid via inverter interfaces. The model for $\mathcal{I}$ nodes derived in the following is able to both represent the main physical characteristics of the inverter dynamics while also ensuring a seamless integrability into the existing SM infrastructure.

At each $\mathcal{I}$ node, we consider a *grid-forming*[32] inverter [RLBR12] with capacitive inertia and controllable DC current $I_{DC,i}$. We assume that each inverter $i \in \mathcal{V}_\mathcal{I}$ fulfills the following assumptions regarding its inner control loop:

Assumption 3.8. *a) At AC side, the inverter is able to produce a perfect sinusoidal AC voltage* $\vec{U}_i$.

b) There are no switching losses in the inverter.

c) The inner control loop of the inverter is equipped with a $U_{DC} \sim \omega$ *matching controller [JAD16].*

The validity of these assumptions is ensured as follows: To comply with Assumption 3.8a), the AC current can be processed through an appropriate LCL low-pass filter to suppress the harmonics caused by switching dynamics of the involved insulated gate bipolar transistors. Assumption 3.8b) is justified by the fact that the losses caused by the DC-side conductance $G_{DC,i}$ are significantly larger. The $U_{DC} \sim \omega$ matching controller that was established in Assumption 3.8c) allows to exploit the structural similarities between kinetic energy of the rotor of a SM and electric energy stored in the DC-side capacitor of the inverter.

A key difference to alternative concepts of mimicking SM dynamics, such as the virtual synchronous machine concept, is that the $U_{DC} \sim \omega$ matching controller directly uses the physically available electric capacitance $C_{DC,i}$ instead of relying on a separate storage and/or a microprocessor-based external emulation of SM dynamics which may be more susceptible to delays. By contrast, using the $U_{DC} \sim \omega$ matching controller as inner control loop for frequency and voltage control purposes has been explicitly promoted in the recent paper [DBSPG19].

32 Inverters are said to be *grid-forming* if they operate as controllable voltage source. By contrast, the inverter is said to be in *grid-following* mode, if it acts as an (active and reactive) power source while U_i and θ_i are the dependent variables [SFO15, p. 69f.].

With Assumption 3.8a), the inverter can be represented by its single-phase equivalent circuit which is depicted in Fig. 3.5. The constitutive equation for the DC-side voltage $U_{\mathrm{DC},i}$ is

$$C_{\mathrm{DC},i}\dot{U}_{\mathrm{DC},i} = -\frac{1}{R_{\mathrm{DC},i}}U_{\mathrm{DC},i} - I_{\mathrm{x},i} + I_{\mathrm{DC},i}. \tag{3.29}$$

In grid-forming mode, the DC-side voltage $U_{\mathrm{DC},i}$ serves as independent variable which can be chosen freely. With Assumption 3.8c), the phase angle θ_i of $\vec{U}_i$ is chosen according to the integral controller scheme [JAD16, MDS+18] by

$$\dot{\theta}_i = k_i U_{\mathrm{DC},i} \tag{3.30}$$

with $k_i = \Omega_{\mathrm{n},i}/U^\star_{\mathrm{DC},i}$, where $U^\star_{\mathrm{DC},i}$ denotes the nominal DC link voltage. With (3.30), the DC side follows the dynamics

$$\frac{C_{\mathrm{DC},i}}{k_i}\cdot\dot{\Omega}_i + \frac{1}{k_i R_{\mathrm{DC},i}}\cdot\Omega_i = -I_{\mathrm{x},i} + I_{\mathrm{DC},i}. \tag{3.31}$$

With Assumption 3.8b), it holds that $I_{\mathrm{x},i}U_{\mathrm{x},i} = I_{\mathrm{x},i}U_{\mathrm{DC},i} = p_{\mathrm{DC},i} = p_{\mathrm{inj},i}$. Furthermore, we choose

$$I_{\mathrm{DC},i} = \frac{\Omega_{\mathrm{n},i}}{R_{\mathrm{DC},i}\cdot k_i} + \frac{p_{\mathrm{g},i}}{\Omega_i}, \tag{3.32}$$

where $p_{\mathrm{g},i}$ is a desired active power input or output. Since $\dot{\Omega}_i = \dot{\omega}_i$, (3.31) reads as

$$\frac{C_{\mathrm{DC},i}}{k_i^2}\cdot\dot{\omega}_i + \frac{1}{k_i^2\cdot R_{\mathrm{DC},i}}\cdot\omega_i = \frac{1}{\Omega_i}\left(p_{\mathrm{g},i} - p_{\mathrm{inj},i}\right). \tag{3.33}$$

Multiplication with $\Omega_{\mathrm{n},i}$ along with Assumption 3.6a) leads to

$$\underbrace{\frac{C_{\mathrm{DC},i}(U^\star_{\mathrm{DC},i})^2}{\Omega_{\mathrm{n},i}}}_{=:\Gamma_{\mathcal{I},i}}\cdot\dot{\omega}_i + \underbrace{\frac{(U^\star_{\mathrm{DC},i})^2}{\Omega_{\mathrm{n},i}R_{\mathrm{DC},i}}}_{=:A_{\mathcal{I},i}}\cdot\omega_i = p_{\mathrm{g},i} - p_{\mathrm{inj},i}, \tag{3.34}$$

where we can define $\Gamma_{\mathcal{I},i} := \frac{C_{\mathrm{DC},i}(U^\star_{\mathrm{DC},i})^2}{\Omega_{\mathrm{n},i}}$ as the *virtual inertia* and $A_{\mathcal{I},i} := \frac{(U^\star_{\mathrm{DC},i})^2}{\Omega_{\mathrm{n},i}R_{\mathrm{DC},i}}$ as the *virtual damping coefficient*. In summary, this yields the inverter AC side dynamics

$$\dot{\theta}_i = \omega_i, \tag{3.35a}$$

$$\Gamma_{\mathcal{I},i}\dot{\omega}_i = -A_{\mathcal{I},i}\omega_i + p_{\mathrm{g},i} - p_{\mathrm{inj},i}. \tag{3.35b}$$

In particular, we clearly see the structural equivalence between (3.35) and the swing dynamics (3.28). However, both the resulting (virtual) inertia $\Gamma_{\mathcal{I},i}$ and (virtual) damping $A_{\mathcal{I},i}$ are typically much smaller in magnitude than Γ_i and A_i, respectively (see [JAD16, Remark 2] and [MDS+18, Remark 2]). As a result, a higher damping would necessitate a larger parallel conductance $G_{\mathrm{DC},i}$ and thus provoke a higher power loss on the DC side. Likewise, a larger virtual inertia $\Gamma_{\mathcal{I},i}$ requires a larger capacitance $C_{\mathrm{DC},i}$ and/or a larger nominal DC voltage $U^\star_{\mathrm{DC},i}$.

3.2.4 Load-Type Nodes

Load-Type nodes represent the power supply or demand of retail producers or consumers. These network participants are connected via step-down transformers or grid-feeding inverters, which are each modeled by a constant impedance $\vec{y}_{\ell,i} = g_{\ell,i} + \jmath b_{\ell,i}$ plus an uncontrollable active and reactive power demand[33] along with a frequency-dependent part with an unknown damping coefficient $A_{\mathcal{L},i} \geq 0$. Accordingly, we get

$$\dot{\theta}_i = \omega_i, \tag{3.36}$$

$$p_{\text{inj},i} = -g_{\ell,i} U_i^2 - A_{\mathcal{L},i}\omega_i - p_{\ell,i}, \tag{3.37}$$

$$q_{\text{inj},i} = -b_{\ell,i} U_i^2 - q_{\ell,i}. \tag{3.38}$$

Remark 3.9. *Distributor nodes with $p_{\ell,i} = q_{\ell,i} = A_{\mathcal{L},i} = 0$ can be eliminated from the model by Kron reduction [CT14; SFO15, p. 45ff.].* ◇

Despite the sign convention for $\mathcal{L}$ nodes, $p_{\ell,i}$ and $q_{\ell,i}$ may be negative as well.

3.2.5 Overall Model

In the following, the dynamic equations of $\mathcal{S}, \mathcal{I}$, and $\mathcal{L}$ nodes are combined to a novel, network-preserving overall model. For all $\mathcal{S}$ and $\mathcal{L}$ nodes, we choose $p_i = p_{\text{inj},i}$ and $q_i = q_{\text{inj},i}$. $\mathcal{I}$ nodes are connected via $p_i = p_{\text{inj},i}$ and $U_{\text{AC},i} = U_i$. Moreover, we assume that all shunt admittances $\vec{y}_{\text{sh},i}$ of $\mathcal{L}$ nodes are incorporated in the diagonal elements of the nodal admittance matrix $\vec{\boldsymbol{Y}}_{\text{p}}$.

Now we define the *extended* physical graph $\mathscr{G} = (\mathcal{V}, \mathcal{E})$ with $\mathcal{V} = \mathcal{V}_\text{p} \cup \mathcal{V}'_\mathcal{S}$ and $\mathcal{E} = \mathcal{E}_\text{p} \cup \mathcal{E}'_\text{p}$, where $\mathcal{V}'_\mathcal{S}$ represents the SMs' internal voltages U'_i, and $\mathcal{E}'_\text{p}$ represents the SMs' connections between internal voltage U'_i and external voltage U_i. For convenience of notation, $X_{\text{dn},i}$ is included in the line impedance. Then, we can define the extended nodal admittance matrix $\vec{\boldsymbol{Y}} = \boldsymbol{G} + \jmath\boldsymbol{B}$ incorporating both the admittances of power lines and the reactances of SMs.

[33] According to the node type, we use the passive sign convention.

This finally leads to the compact notation

$$\begin{aligned}
\dot{\vartheta}_{ij} &= \omega_i - \omega_j, && ij \in \mathcal{E}, && \text{(3.39a)}\\
\dot{L}_i &= -A_i\omega_i + p_{\mathrm{g},i} - p_{\ell,i} - U_i^2 G_{ii} \\
&\quad - \sum_{j\in\mathcal{N}_i} U_i U_j \left(B_{ij}\sin(\vartheta_{ij}) + G_{ij}\cos(\vartheta_{ij})\right), && i \in \mathcal{V}_{\mathrm{g}}, && \text{(3.39b)}\\
T'_{\mathrm{d0},i}\dot{U}_i &= U_{\mathrm{f},i} - U_i - (X_{\mathrm{d},i} - X'_{\mathrm{d},i}) \\
&\quad \cdot \Bigg(- U_i B_{ii} + \sum_{j\in\mathcal{N}_i} U_j \left(G_{ij}\sin(\vartheta_{ij}) - B_{ij}\cos(\vartheta_{ij})\right)\Bigg), && i \in \mathcal{V}'_{\mathcal{S}}, && \text{(3.39c)}\\
0 &= -A_i\omega_i - p_{\ell,i} - U_i^2 G_{ii} \\
&\quad - \sum_{j\in\mathcal{N}_i} U_i U_j \left(B_{ij}\sin(\vartheta_{ij}) + G_{ij}\cos(\vartheta_{ij})\right), && i \in \mathcal{V}_\ell, && \text{(3.39d)}\\
0 &= -q_{\ell,i} + U_i^2 B_{ii} - \sum_{j\in\mathcal{N}_i} U_i U_j \left(G_{ij}\sin(\vartheta_{ij}) - B_{ij}\cos(\vartheta_{ij})\right), && i \in \mathcal{V}_\ell, && \text{(3.39e)}
\end{aligned}$$

where $\mathcal{V}_{\mathrm{g}} = \mathcal{V}_\mathcal{S} \cup \mathcal{V}_\mathcal{I}$, $\mathcal{V}_\ell = \mathcal{V}_\mathcal{L} \cup \mathcal{V}_\mathcal{S}$, $L_i = \Gamma_i\omega_i$, $G_{ij} = \Re\{[\vec{\boldsymbol{Y}}]_{i,j}\}$, $B_{ij} = \Im\{[\vec{\boldsymbol{Y}}]_{i,j}\}$, and $\mathcal{N}_i$ denoting the set of neighboring nodes $j \in \mathcal{V}$ of the extended graph $\mathscr{G}$ without regard to edge directions.

By defining the *plant state vector* $\boldsymbol{x}_\mathrm{p} = \mathrm{col}\{\boldsymbol{\vartheta}, \boldsymbol{L}_\mathcal{S}, \boldsymbol{L}_\mathcal{I}, \boldsymbol{U}_\mathcal{S}, \boldsymbol{\omega}_\ell, \boldsymbol{U}_\ell\}$ of (3.39), where for all $m \in \mathcal{E}$, $j \in \mathcal{V}'_\mathcal{S}$, $k \in \mathcal{V}_\mathcal{I}$, $l \in \mathcal{V}_\ell$:

$$\begin{aligned}
\boldsymbol{\vartheta} &:= \mathrm{col}_m\{\vartheta_m\}, & \boldsymbol{L}_\mathcal{S} &:= \mathrm{col}_j\{L_j\}, && \text{(3.40)}\\
\boldsymbol{L}_\mathcal{I} &:= \mathrm{col}_k\{L_k\}, & \boldsymbol{U}_\mathcal{S} &:= \mathrm{col}_j\{U_j\}, && \text{(3.41)}\\
\boldsymbol{\omega}_\ell &:= \mathrm{col}_l\{\omega_l\}, & \boldsymbol{U}_\ell &= \mathrm{col}_l\{U_l\}, && \text{(3.42)}
\end{aligned}$$

we can set up the *plant Hamiltonian* as

$$\begin{aligned}
H_\mathrm{p}(\boldsymbol{x}_\mathrm{p}) &= \frac{1}{2}\sum_{i\in\mathcal{V}'_\mathcal{S}} \left(\Gamma_i^{-1}L_i^2 + \frac{U_i^2}{X_{d,i} - X'_{d,i}}\right) \\
&\quad + \frac{1}{2}\sum_{i\in\mathcal{V}_\mathcal{I}} \Gamma_i^{-1}L_i^2 \\
&\quad - \frac{1}{2}\sum_{i\in\mathcal{V}} B_{ii}U_i^2 - \sum_{(i,j)\in\mathcal{E}} B_{ij}U_iU_j\cos(\vartheta_{ij}) \\
&\quad + \frac{1}{2}\sum_{i\in\mathcal{V}_\ell} \omega_i^2 && \text{(3.43)}
\end{aligned}$$

to describe the total energy stored in the system. The first row of (3.43) represents the shifted kinetic energy of the rotors and the magnetic energy of the SM circuits, the second row represents the "virtual" kinetic energy at inverter-type nodes, the third row represents the magnetic energy of transmission lines and the fourth row represents the local deviations of load nodes from nominal frequency.

Using the plant Hamiltonian (3.43) along with its *co-state* $\boldsymbol{z}_{\mathrm{p}} = \nabla H(\boldsymbol{x}_{\mathrm{p}})$ and $\mathcal{V}_{\mathrm{g}} = \mathcal{V}_{\mathcal{S}} \cup \mathcal{V}_{\mathcal{I}}$, with

$$\frac{\partial H_{\mathrm{p}}}{\partial \vartheta_{ij}} = B_{ij} U_i U_j \sin(\vartheta_{ij}), \qquad (i,j) \in \mathcal{E}, \tag{3.44}$$

$$\frac{\partial H_{\mathrm{p}}}{\partial L_i} = L_i / \Gamma_i = \omega_i, \qquad i \in \mathcal{V}_{\mathrm{g}}, \tag{3.45}$$

$$\frac{\partial H_{\mathrm{p}}}{\partial U_i} = \frac{U_i}{X_{d,i} - X'_{d,i}} - B_{ii} U_i - \sum_{j \in \mathcal{N}_i} B_{ij} U_j \cos(\vartheta_{ij}), \qquad i \in \mathcal{V}'_{\mathcal{S}}, \tag{3.46}$$

$$\frac{\partial H_{\mathrm{p}}}{\partial U_i} = - \sum_{j \in \mathcal{N}_i} B_{ij} U_j \cos(\vartheta_{ij}), \qquad i \in \mathcal{V}_{\ell}, \tag{3.47}$$

$$\frac{\partial H_{\mathrm{p}}}{\partial \omega_i} = \omega_i, \qquad i \in \mathcal{V}_{\ell}, \tag{3.48}$$

we can set up a port-Hamiltonian representation for the novel physical plant model (3.39) as

$$\underbrace{\begin{bmatrix} \dot{\boldsymbol{\vartheta}} \\ \dot{\boldsymbol{L}} \\ \dot{\boldsymbol{U}}_{\mathcal{S}} \\ \mathbb{0} \\ \mathbb{0} \end{bmatrix}}_{\dot{\boldsymbol{x}}_{\mathrm{p}}} = \left[\underbrace{\begin{bmatrix} \mathbf{0} & \boldsymbol{D}_{\mathrm{pg}}^{\top} & \mathbf{0} & \boldsymbol{D}_{p\ell}^{\top} & \mathbf{0} \\ -\boldsymbol{D}_{\mathrm{pg}} & \mathbf{0} & \mathbf{0} & \mathbf{0} & \mathbf{0} \\ \mathbf{0} & \mathbf{0} & \mathbf{0} & \mathbf{0} & \mathbf{0} \\ -\boldsymbol{D}_{p\ell} & \mathbf{0} & \mathbf{0} & \mathbf{0} & \mathbf{0} \\ \mathbf{0} & \mathbf{0} & \mathbf{0} & \mathbf{0} & \mathbf{0} \end{bmatrix}}_{\boldsymbol{J}_{\mathrm{p}}} - \underbrace{\begin{bmatrix} \mathbf{0} & \mathbf{0} & \mathbf{0} & \mathbf{0} & \mathbf{0} \\ \mathbf{0} & \boldsymbol{A}_{\mathrm{g}} & \mathbf{0} & \mathbf{0} & \mathbf{0} \\ \mathbf{0} & \mathbf{0} & \boldsymbol{R}_{\mathcal{S}} & \mathbf{0} & \mathbf{0} \\ \mathbf{0} & \mathbf{0} & \mathbf{0} & \boldsymbol{A}_{\ell} & \mathbf{0} \\ \mathbf{0} & \mathbf{0} & \mathbf{0} & \mathbf{0} & \hat{\boldsymbol{U}}_{\ell} \end{bmatrix}}_{\boldsymbol{R}_{\mathrm{p}}} \right] \boldsymbol{z}_p$$

$$- \underbrace{\begin{bmatrix} \mathbb{0} \\ \boldsymbol{\rho}_{\mathrm{g}} \\ \boldsymbol{\varrho}_{\mathcal{S}} \\ \boldsymbol{\rho}_{\ell} \\ \boldsymbol{\varrho}_{\ell} \end{bmatrix}}_{\boldsymbol{r}_{\mathrm{p}}} + \begin{bmatrix} \mathbf{0} & \mathbf{0} & \mathbf{0} & \mathbf{0} \\ \boldsymbol{I} & \mathbf{0} & \mathbf{0} & -\hat{\boldsymbol{I}}_{\mathrm{g}} \\ \mathbf{0} & \hat{\boldsymbol{T}}_{U} & \mathbf{0} & \mathbf{0} \\ \mathbf{0} & \mathbf{0} & \mathbf{0} & -\hat{\boldsymbol{I}}_{\ell} \\ \mathbf{0} & \mathbf{0} & -\boldsymbol{I} & \mathbf{0} \end{bmatrix} \begin{bmatrix} \boldsymbol{p}_{\mathrm{g}} \\ \boldsymbol{U}_{\mathrm{f}} \\ \boldsymbol{q}_{\ell} \\ \boldsymbol{p}_{\ell} \end{bmatrix}, \tag{3.49}$$

where

$$\boldsymbol{L} = \operatorname{col}\{\boldsymbol{L}_{\mathcal{S}}, \boldsymbol{L}_{\mathcal{I}}\}, \tag{3.50}$$

$$\boldsymbol{A}_{\mathrm{g}} = \operatorname{diag}_i\{A_i\}, \quad i \in \mathcal{V}_{\mathrm{g}}, \tag{3.51}$$

$$\boldsymbol{A}_{\ell} = \operatorname{diag}_i\{A_i\}, \quad i \in \mathcal{V}_{\ell}, \tag{3.52}$$

$$\boldsymbol{R}_{\mathcal{S}} = \operatorname{diag}_i\left\{(X_{di} - X'_{di})/\tau_{U,i}\right\}, \quad i \in \mathcal{V}'_{\mathcal{S}}, \tag{3.53}$$

$$\hat{\boldsymbol{U}}_{\ell} = \operatorname{diag}_i\{U_i\}, \quad i \in \mathcal{V}_{\ell}, \tag{3.54}$$

$$\boldsymbol{\rho}_{\mathrm{g}} = \operatorname{col}_i\Big\{G_{ii}U_i^2 + \sum_{j\in\mathcal{N}_i} G_{ij}U_iU_j\cos(\vartheta_{ij})\Big\}, \quad i \in \mathcal{V}_{\mathrm{g}}, \tag{3.55}$$

$$\boldsymbol{\rho}_{\ell} = \operatorname{col}_i\Big\{G_{ii}U_i^2 + \sum_{j\in\mathcal{N}_i} G_{ij}U_iU_j\cos(\vartheta_{ij})\Big\}, \quad i \in \mathcal{V}_{\ell}, \tag{3.56}$$

$$\boldsymbol{\varrho}_{\mathcal{S}} = \operatorname{col}_i\Big\{R_{g,i}\sum_{j\in\mathcal{N}_i} G_{ij}U_iU_j\sin(\vartheta_{ij})\Big\}, \quad i \in \mathcal{V}'_{\mathcal{S}}, \tag{3.57}$$

$$\boldsymbol{\varrho}_{\ell} = \operatorname{col}_i\Big\{\sum_{j\in\mathcal{N}_i} G_{ij}U_iU_j\sin(\vartheta_{ij})\Big\}, \quad i \in \mathcal{V}_{\ell}, \tag{3.58}$$

$$\hat{\boldsymbol{T}}_U = \operatorname{diag}_i\{1/T_{\mathrm{d0},i}\}, \quad i \in \mathcal{V}'_{\mathcal{S}}, \tag{3.59}$$

$$\hat{\boldsymbol{I}}_{\mathrm{g}} = \begin{bmatrix}\boldsymbol{I}_{n_{\mathrm{g}}\times n_{\mathrm{g}}} & \boldsymbol{0}_{n_{\mathrm{g}}\times n_{\ell}}\end{bmatrix}, \tag{3.60}$$

$$\hat{\boldsymbol{I}}_{\ell} = \begin{bmatrix}\boldsymbol{0}_{n_{\ell}\times n_{\mathrm{g}}} & \boldsymbol{I}_{n_{\ell}\times n_{\ell}}\end{bmatrix}. \tag{3.61}$$

The input vector in (3.49) is composed of the control input $\boldsymbol{u}_{\mathrm{p}} = \operatorname{col}\{\boldsymbol{p}_{\mathrm{g}}, \boldsymbol{U}_{\mathrm{f}}\}$ and the disturbance input $\boldsymbol{d} = \operatorname{col}\{\boldsymbol{q}_{\ell}, \boldsymbol{p}_{\ell}\}$.

Remark 3.10. *Note that* (3.49) *is a nonlinear system, whereby the main nonlinearities appear in the co-state equations* (3.44), (3.46), *and* (3.47), *caused by nonlinear electromagnetic relations. Moreover, there is a modulation of the dissipation matrix* $\boldsymbol{R}_{\mathrm{p}}$ *with the state vector* $\boldsymbol{U}_{\ell}$. *The interconnection matrix* $\boldsymbol{J}_{\mathrm{p}}$, *however, contains only entries which are constant in time.* ◇

Remark 3.11. *Even in the case that certain parts of Assumption 3.7 do no longer apply, the port-Hamiltonian framework can principially be applied to formulate a corresponding (higher-order) system model for S nodes. In [SDv16], a port-Hamiltonian model for multi-machine power networks composed of* sixth-order *SM models* (3.14) *is established. A systematic derivation of port-Hamiltonian representations for* all *classical SG models based on first-principles modeling is conducted in [SDv19].* ◇

3.3 Modeling of Participants' Interaction

This section presents a model for the network participants' interaction and competition on the energy market. The conceptual classification into WPs and RPs is based on the premise that WPs can shift their own production or consumption strategically to some extent[34] based on a profit-maximizing strategy. Thus, WPs are competing players on a cross-zone electricity market which pursue to maximize their own profits $\mathtt{P}_{\mathrm{g},i}$ through (positive or negative) power injections $p_{\mathrm{g},i}$, while incurring power production costs $\mathtt{C}_{\mathrm{g},i}$.

Assumption 3.12. *The cost functions* $\mathtt{C}_{\mathrm{g},i}(p_{\mathrm{g},i})$ *are strictly convex.*

Remark 3.13. *Assumption 3.12 is typically mild for both consumers and producers [BV00, ZP15]. This is due to the fact that WPs tend to internally work with a strictly convex surrogate function in order to apply their pursuit of profit maximization and to avoid being stuck in sub-optimal operating points (see the discussion in Appendix B.3)* ◇

By contrast, RPs typically perform their decision-making concerning power production or consumption independently of other RPs or WPs and have no means to displace their instantaneous (positive or negative) demand. Thus, RPs are assumed to be inelastic.

Remark 3.14. *The role of WPs is congruent with the definition of CVPPs (cf. Subsection 2.2.1), although the formalism presented in the following also allows negative power injections. Moreover, the duties and responsibilities of CCs as well as their geographic coverage strongly resemble those of a TVPP. However, the latter terminology is avoided in the context of this dissertation, since the attribute "virtual" negates the active role of CCs in the mutual coordination of supply and demand under consideration of physical network conditions. Furthermore, our concept of WPs does not require unidirectional power flows as might be suggested by the term "power plant".* ◇

We summarize the number of WP nodes as $n_{\mathrm{g}} = n_{\mathcal{S}} + n_{\mathcal{I}}$. WPs inject electrical power $\boldsymbol{p}_{\mathrm{g}} \in \mathbb{R}^{n_{\mathrm{g}}}$ into the power system and are able to influence the nodal voltages $\boldsymbol{U} \in \mathbb{R}^{n_{\mathrm{g}}}_{>0}$. Both $\mathcal{S}$ and $\mathcal{I}$ nodes have to meet upper and lower limits on power generation $\overline{p}_{\mathrm{g},i}$ and $\underline{p}_{\mathrm{g},i}$. Note that all buses $i \in \mathcal{V}$ may be equipped with RPs and thus with an uncontrollable active and reactive power demand $p_{\ell,i}$ and $q_{\ell,i}$.

For compactness of notation, we introduce cell-aggregated variables for the power generation $\hat{\boldsymbol{p}}_{\mathrm{g}} = \mathrm{col}_{k\in\mathcal{Z}}\{\hat{p}_{\mathrm{g},k}\} \in \mathbb{R}^{n_{\mathcal{Z}}}$ where $\hat{p}_{\mathrm{g},k} = \sum_{i\in\mathcal{V}_{\mathcal{Z},k}} p_{\mathrm{g},i}$, the power consumption $\hat{\boldsymbol{p}}_{\ell} = \mathrm{col}_{k\in\mathcal{Z}}\{\hat{p}_{\ell,k}\} \in \mathbb{R}^{n_{\mathcal{Z}}}$ where $\hat{p}_{\ell,k} = \sum_{i\in\mathcal{V}_{\mathcal{Z},k}} p_{\ell,i}$, and the active power loss $\hat{\boldsymbol{\Phi}} = \mathrm{col}_{k\in\mathcal{Z}}\{\hat{\Phi}_k\} \in \mathbb{R}^{n_{\mathcal{Z}}}$, where $\hat{\Phi}_k = \sum_{i\in\mathcal{V}_{\mathcal{Z},k}} \Phi_i$.

The following subsections pave the way towards a market-based control scheme with zonal (i.e. spatially differentiated) prices $\lambda_k \in \mathbb{R}^{n_{\mathcal{Z}}}$. Subsection 3.3.1 derives an incentive-compatible market mechanism which respects the requirement of unbundling. Subsections 3.3.2–3.3.4 model the behavior of WPs, RPs, and CCs, respectively, under the chosen market mechanism. Finally, Subsection 3.3.5 summarizes the consequences of market interactions on the upcoming controller design.

[34] Hence, spatially differentiated prices also provide medium- and long-term incentives for shifting capacity to other cells that, on average, provide higher prices for generators or lower cost fro consumers.

3.3.1 Design of Market Mechanism

In the classical model of energy economics, a *market equilibrium* of the competitive power system from Definition 3.3 is attained if the market price λ resulting from competition yields a balance between total demand $\mathbb{1}^\top \boldsymbol{p}_\ell$ and total generation $\mathbb{1}^\top \boldsymbol{p}_\mathrm{g}$ (see the common Definition A.23 in Appendix A.6). However, this definition makes no statement whether the achieved market equilibrium is "valuable" from the invididual participants' point of view or from a global perspective. The *individual* utility of an equilibrium is often certified by means of the Pareto efficiency (or Pareto optimality), cf. Definition A.5. According to the first theorem of welfare economics (cf. Theorem A.25), each market equilibrium is also Pareto efficient if the participants compete over a *perfect market*. This prototypical notion of an idealized market contains numerous simplifying assumptions and is summarized in Definition A.24.

For the normative evaluation of the *global* efficiency of an attained equilibrium solution, there exist different metrics such as the *utilitarian welfare function*[35] or the *Rawlsian welfare function* (cf. Appendix A.6). When adopting the utilitarian welfare function as a measure for global efficiency, the market equilibrium achieved in a perfect market is not only individually, but also globally efficient (cf. Remark A.27). It can thus be stated that such a perfect market should always be targeted by means of any appropriate market design. However, certain properties of real-world power systems are diametrically opposed to those required in Definition A.24:

(1) The economic model of market interaction is based upon the assumption that both suppliers and demanders are price-elastic. However, our above model resulting from Definition 3.3 also incorporates inelastic participants. In particular, we have adopted the natural assumption that *the supply or demand of inelastic RPs triggers the supply of elastic WPs.*

(2) In real-world electricity markets, supply and demand is settled over an interconnected, dynamic power system (3.49), which may be subject to a number of additional operational constraints, e.g. due to transmission capacity limits or security constraints to prevent voltage collapse. By contrast, the plain economic model does not take into account any underlying physical structure (see the copper plate assumption discussed in Section 2.1.2).

(3) The plain economic model does not reflect any transmission losses that always occur in real-world power systems.

These points emphasize that, when dealing with actual electricity markets under realistic operating conditions, a mere reference to the existing theorems of welfare economics is insufficient. In the subsequent sections each of the above challenges is therefore intensively studied when deducing a viable (i.e. network-preserving) market-based control scheme.

In the remainder of this subsection, we first address the issue of inelastic market participants (challenge (1) above). As outlined in Subsection 2.2.2, the two main philosophies for modeling distributed market-based interaction of noncooperative agents are Bertrand and Cournot competition (see Subsection 2.2.2 for a discussion of current research approaches related to

[35] The maximum of the utilitarian welfare function is called *social optimum* and is characterized by equal marginal profits of each participant.

electricity markets and [dD21, Chapter 1] for a more general economic framing). We thus investigate whether these two models of competition are able to generate (individually and/or globally) efficient market equilibria in case of inflexible RPs. For didactical reasons (yet without loss of generality), we exemplify these considerations on an atomistic model by requiring $p_{g,i} > 0$ and $p_{\ell,i} > 0$, and by temporarily neglecting both the system dynamics (challenge (2)) and the line losses (challenge (3)). Both challenges are then considered again from Subsection 3.3.2 onwards.

Bertrand Competition with Inelastic Demand

In Bertrand competition each WP $i \in \mathcal{W}$ bids a price $\lambda_i \in \mathbb{R}$ per amount of power to maximize its own profit

$$\mathtt{P}_{g,i}(\lambda_i, \lambda_{\neg i}) = \lambda_i p_{g,i}(\lambda_i, \lambda_{\neg i}) - \mathtt{C}_{g,i}(p_{g,i}(\lambda_i, \lambda_{\neg i})), \tag{3.62}$$

where $\mathtt{C}_{g,i} : \mathbb{R} \to \mathbb{R}_{\geq 0}$ represents the costs of WP i to generate $p_{g,i}$. The power generation invoked by the system operator is given by

$$p_{g,i}(\lambda_i, \lambda_{\neg i}) = \begin{cases} 0, & \exists j \in \mathcal{W} \backslash \{i\} : \lambda_j < \lambda_i, \\ \frac{\mathbb{1}^\top \boldsymbol{p}_\ell}{N}, & \text{otherwise}, \end{cases}, \tag{3.63}$$

where N is the number of WPs that bid the lowest price $\lambda = \min_i\{\lambda_i\}$. By (3.63), WP i is not able to sell any power if there exist other WPs that bid a lower price. Likewise, if WP i bids the lowest price $\lambda_i = \lambda$, then the total demand $\mathbb{1}^\top \boldsymbol{p}_\ell$ is equally shared among all WPs that bid the lowest price λ.

Bertrand competition has many undesirable properties. In [Das95], it is shown that even for strictly convex cost functions $\mathtt{C}_{g,i}(p_{g,i})$ (cf. Assumption 3.12), Bertrand competition may exhibit non-unique market equilibria. In [Wei06], it is elaborated that the range of acceptable prices is determined by $\lambda_i = [\lambda_i^\circ, \lambda_i^\bullet]$, where the *outside option price*

$$\lambda_i^\circ = \frac{\mathtt{C}_{g,i}\left(\frac{\mathbb{1}^\top \boldsymbol{p}_\ell}{N}\right)}{\frac{\mathbb{1}^\top \boldsymbol{p}_\ell}{N}} \tag{3.64}$$

is the price below which player i earns a negative profit when sharing the market, and the *competitive price limit* $\lambda_i^\bullet$ defined by

$$\mathtt{P}_{g,i}\left(\frac{\mathbb{1}^\top \boldsymbol{p}_\ell}{N}, \lambda_i^\bullet\right) = \mathtt{P}_{g,i}(\mathbb{1}^\top \boldsymbol{p}_\ell, \lambda_i^\bullet). \tag{3.65}$$

Bids $\lambda_i' > \lambda_i^\bullet$ would always attract the other players to bid $\lambda_{\neg i} = \lambda_i' - \varepsilon$, $\varepsilon > 0$, such that they share the total market exclusively among each other. Hence, the intersection of all those intervals constitutes the set of all possible Nash equilibria (cf. [Wei06])

$$\mathbb{L}_{\text{nash}} = \{\lambda : \lambda_{\max}^\circ \leq \lambda \leq \lambda_{\min}^\bullet\}, \tag{3.66}$$

where $\lambda_{\max}^\circ = \max_{i \in \mathbb{P}}\{\lambda_i^\circ\}$ and $\lambda_{\min}^\bullet = \min_{i \in \mathbb{P}}\{\lambda_i^\bullet\}$. This is demonstrated by the following two-player example.

Example 1

Consider two WPs competing in a Bertrand competition, whose individual cost functions are given by $\mathsf{C}_{g,1}(p_{g,1}) = 4p_{g,1}^2 - p_{g,1}$ *and* $\mathsf{C}_{g,2}(p_{g,2}) = 4p_{g,2}^2$*. This results in the profit functions*

$$\mathsf{P}_{g,1}(\lambda_1, \lambda_2) = \lambda_1 p_{g,1}(\lambda_1, \lambda_2) - 4p_{g,1}^2(\lambda_1, \lambda_2) + p_{g,1}(\lambda_1, \lambda_2), \tag{3.67}$$

$$\mathsf{P}_{g,2}(\lambda_2, \lambda_1) = \lambda_2 p_{g,2}(\lambda_2, \lambda_1) - 4p_{g,2}^2(\lambda_2, \lambda_1). \tag{3.68}$$

Let the total demand be $\mathbb{1}^\top \boldsymbol{p}_\ell = 1$*, then* (3.63) *takes the form*

$$p_{g,1}(\lambda_1, \lambda_2) = \begin{cases} 1, & \lambda_1 < \lambda_2, \\ 0.5, & \lambda_1 = \lambda_2, \\ 0, & \lambda_1 > \lambda_2, \end{cases}, \quad p_{g,2}(\lambda_2, \lambda_1) = \begin{cases} 1, & \lambda_2 < \lambda_1, \\ 0.5, & \lambda_2 = \lambda_1, \\ 0, & \lambda_2 > \lambda_1, \end{cases} \tag{3.69}$$

and inserting (3.69) *into* (3.68) *yields*

$$\mathsf{P}_{g,1}(\lambda_1, \lambda_2) = \begin{cases} \lambda_1 - 3, & \lambda_1 < \lambda_2, \\ 0.5\lambda_1 - 0.5, & \lambda_1 = \lambda_2, \\ 0, & \lambda_1 > \lambda_2, \end{cases}, \quad \mathsf{P}_{g,2}(\lambda_2, \lambda_1) = \begin{cases} \lambda_2 - 4, & \lambda_2 < \lambda_1, \\ 0.5\lambda_2 - 1, & \lambda_2 = \lambda_1, \\ 0, & \lambda_2 > \lambda_1. \end{cases} \tag{3.70}$$

The profit functions (3.70) *are depicted in Fig. 3.6.*

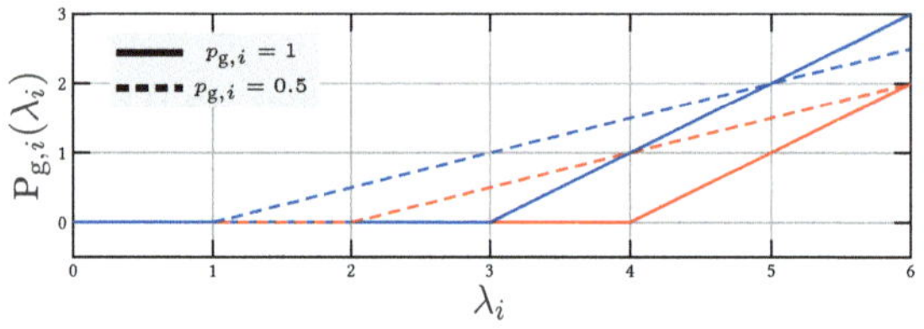

Figure 3.6: Exemplary profit functions for WPs 1 (blue) and 2 (red).

The blue lines correspond to WP 1 and the red lines correspond to WP 2. The solid lines show the WPs' profit if they were to capture the full market ($p_{g,i} = 1$*), while the dashed lines show their profit if they were to share the market (*$p_{g,i} = 0.5$*). Possible market equilibria can now be determined by construction: From Fig. 3.6 it can be seen that WP 2 will never choose* $\lambda_2 < 2$ *as this would not result in a positive profit. Moreover, since for all* $\lambda_1 \in [2, 5]$ *it is more profitable for WP 2 to share the market than to capture the full market or to capture nothing, the best option for WP 2 is to choose* $\lambda_2 = \lambda_1$*. The same reasoning holds if WP 1 were to choose its price* λ_1 *freely given* $\lambda_2 \in [2, 5]$*. Consequently, all* $\lambda \in [2, 5]$ *are market equilibria. By contrast, if WP 2 were to choose* $\lambda_2 > 5$*, WP 1 could make a higher profit by choosing* $\lambda_1 < \lambda_2$ *and capturing the full market, which is why* $\lambda > 5$ *cannot be a market equilibrium. The outside option prices (zeros of the dashed lines in Fig. 3.6) are* $\lambda_1^\circ = 1$ *and* $\lambda_2^\circ = 2$*, and the competitive price limits (intersection of dashed and solid lines) are* $\lambda_1^\bullet = 5$ *and* $\lambda_2^\bullet = 6$*. The intersection of the intervals* $[\lambda_1^\circ, \lambda_1^\bullet]$ *and* $[\lambda_2^\circ, \lambda_2^\bullet]$ *is thus equal to* $[2, 5]$*.*

Remark 3.15. *If all players are aware of each other (i.e., know the number of all competitors as well as well as their individual cost functions), it is the best choice for them to choose* $\lambda_i = \lambda^{\bullet}_{\min}$. *If the players are not aware of each other, then* any *equilibrium* $\lambda_i = [\lambda^{\circ}_{\max}, \lambda^{\bullet}_{\min}]$ *is stable. This can be justified as follows: Let* $\lambda^{\sharp} \in [\lambda^{\circ}_{\max}, \lambda^{\bullet}_{\min}]$ *be some equilibrium. If player* i *performs a so-called* ε-undercutting *[Cab17, p. 105] by choosing* $\lambda^{\heartsuit}_i = \lambda^{\sharp}_i - \varepsilon$ *to test if it can share the total market, this will result in lower profits* $\mathrm{P}_{\mathrm{g},i}(\lambda^{\heartsuit}_1, \lambda^{\sharp}_2) < \mathrm{P}_{\mathrm{g},i}(\lambda^{\star}_1, \lambda^{\sharp}_2)$[36]*, thus no player has an incentive to refrain from* $\lambda_i = \lambda^{\sharp}_i$ *and hence* $\lambda^{\sharp}$ *is a Nash equilibrium.* ♢

The fact that there exists no unique market equilibrium is undesirable, as it causes the actions of market participants to be unpredictable and path-dependent. Consequently, it becomes difficult to incorporate tangible higher-level objectives into the market mechanism. Another issue with Bertrand competition is that the resulting equilibria may not be (individually) economically optimal. We show this by means of a second example.

Example 2

Consider two WPs with cost functions

$$\mathrm{C}_{\mathrm{g},1}(p_{\mathrm{g},1}) = p^2_{\mathrm{g},1}, \qquad \mathrm{C}_{\mathrm{g},2}(p_{\mathrm{g},2}) = \frac{1}{3}p^2_{\mathrm{g},2}, \tag{3.71}$$

which compete over a total demand $\mathbb{1}^\mathsf{T}\boldsymbol{p}_\ell = 10$. *Following the same procedure as in Example 1 above, it can be shown that* $\lambda_1 = \lambda_2 = 5$ *is the unique market equilibrium. Both WPs generate* $p^{\star}_{\mathrm{g},1} = p^{\star}_{\mathrm{g},2} = 5$ *and their corresponding profits are* $\mathrm{P}_{\mathrm{g},1}(\lambda^{\star}_1, \lambda^{\star}_2) = 0$ *and* $\mathrm{P}_{\mathrm{g},2}(\lambda^{\star}_2, \lambda^{\star}_1) = 16.33$. *However, there exists an infinite amount of alternative equilibria which are Pareto dominating (both in terms of higher profits and lower costs for WPs). For example, if the generation were distributed* $p^{\sharp}_{\mathrm{g},2} = 2.5$ *and* $p^{\sharp}_{\mathrm{g},2} = 7.5$*, the individual profits were* $\mathrm{P}^{\sharp}_{\mathrm{g},1}(\lambda^{\star}_1, \lambda^{\star}_2) = 6.25$ *and* $\mathrm{P}^{\sharp}_{\mathrm{g},2}(\lambda^{\star}_1, \lambda^{\star}_2) = 18.75$.

Through the equal allocation of power generation[37] as in (3.63), Bertrand competition may cause large market barriers: If a new WP enters the market with $n_{\mathcal{W}}$ WPs already competing in a Bertrand competition with equilibrium price $\lambda^{\star}$, this new WP now has to bid $\lambda^{\star}$ and to generate $\mathbb{1}^\mathsf{T}\boldsymbol{p}_\ell/(N+1)$ (which might be very large[38]), in order to still obtain positive profits. This requires WPs which want to enter the market to make large investments beforehand and therefore prevents many small-scale WPs such as RESs to participate in the energy market, thereby creating wrong incentives that are considered anti-competitive.

---

[36] since, figuratively speaking, the solid profit curve is below the dashed profit curve (cf. Fig. 3.6).

[37] Some recent approaches, e.g. [CC20], overcome the problems of a pure Bertrand competition by introducing a second step after the price bidding where the power generation $\boldsymbol{p}_\mathrm{g}$ is divided *unequally* according to the individual preferences of WPs. In particular, WPs disclose how much power they would prefer to generate given the equilibrium price λ. Afterwards, the total demand $\mathbb{1}^\mathsf{T}\boldsymbol{p}_\ell$ is then distributed accordingly. This modification however has the main disadvantage that WPs have to disclose their preferences towards a centralized system operator, which is a contradiction to the overarching concept of a competitive power system with cell-based infrastructure which was adopted in Definition 3.3.

[38] If there is no WP that can meet the total demand at its marginal costs, it can be shown that there exists no equilibrium (*Edgeworth paradox*, cf. [KO93, p. 328]).

Cournot Competition with Inelastic Demand

In Cournot competition, each WP $i \in \mathcal{W}$ sets its power generation $p_{\mathrm{g},i}$ to maximize its own profit

$$\mathrm{P}_{\mathrm{g},i}(p_{\mathrm{g},i}, p_{\mathrm{g},\neg i}) = \lambda(\mathbb{1}^\top \boldsymbol{p}_\mathrm{g}) p_{\mathrm{g},i} - \mathrm{C}_{\mathrm{g},i}(p_{\mathrm{g},i}). \tag{3.72}$$

In contrast to Bertrand competition, where the power generation of each WP $i \in \mathcal{W}$ is a function of the price bids $\boldsymbol{\lambda}$, in Cournot competition the unique market price λ is set by the system operator as a function of the total power generation, while each WP chooses its own power generation $p_{\mathrm{g},i}$ by itself.

For the classic Cournot competition, it is required that all consumers are *price-elastic* in order to arrive at a meaningful market equilibrium and deduce the price function $\lambda(\mathbb{1}^\top \boldsymbol{p}_\mathrm{g})$ (see e.g. [DM19]). However, as discussed at the beginning of Section 3.3, we choose to model consumers as price-inelastic. In a classic Cournot competition, this would trigger all WPs to tactically choose their $p_{\mathrm{g},i}$ such that $\mathbb{1}^\top \boldsymbol{p}_\mathrm{g} < \mathbb{1}^\top \boldsymbol{p}_\ell$ in order to maximize profits. As a result, the prices would eventually grow infinitely large (yielding infinite profits for all WPs) and thus no market equilibrium could be reached at all. However, in order to still achieve a market equilibrium in a Cournot competition with price-inelastic consumers, we make the following assumption:

Assumption 3.16. *WPs do not know the total demand $\mathbb{1}^\top \boldsymbol{p}_\ell$ nor how the market price λ depends on the power injections $\boldsymbol{p}_\mathrm{g}$.*

Assumption 3.16 is met if strict observance of information barriers between WPs and CCs can be ensured (cf. [LPWL07, p. 24ff.]). As a result of Assumption 3.16, WPs act as pure *price-takers* as they are unaware of their own influence on the price λ. Note that it is not uncommon for WPs to be forced to act as price-takers in order to keep competition equitable, especially on critical services like power generation (cf. e.g. [Con14, p. 82]).

Owing to a price-taking behavior, each WP $i \in \mathcal{W}$ now solves

$$\max_{p_{\mathrm{g},i}} \quad \mathrm{P}_{\mathrm{g},i}(p_{\mathrm{g},i}) = \lambda p_{\mathrm{g},i} - \mathrm{C}_{\mathrm{g},i}(p_{\mathrm{g},i}) \tag{3.73}$$

in order to maximize its profit given the price λ, which is the solution of

$$\mathbb{1}^\top \boldsymbol{p}_\mathrm{g}(\lambda) = \mathbb{1}^\top \boldsymbol{p}_\ell. \tag{3.74}$$

As soon as the exact function $\lambda(\mathbb{1}^\top \boldsymbol{p}_\mathrm{g})$ is *not* known, Cournot competition has some very appealing features: First of all, the resulting market equilibrium is unique. Consider WPs with profit functions as in (3.73). Since the sensitivity of $p_{\mathrm{g},i}$ on λ is not known, WPs will always produce at marginal cost such that $\nabla \mathrm{C}_{\mathrm{g},i}(p_{\mathrm{g},i}) = \lambda$. With Assumption 3.12, $\nabla \mathrm{C}_{\mathrm{g},i}(p_{\mathrm{g},i})$ is strictly monotonically increasing and thus $p^\star_{\mathrm{g},i}(\lambda) = (\nabla \mathrm{C}_{\mathrm{g},i})^{-1}(\lambda)$. Under the condition of supply-demand matching, we therefore have $\sum_{i\in\mathbb{P}} (\nabla \mathrm{C}_{\mathrm{g},i})^{-1}(\lambda) = \mathbb{1}^\top \boldsymbol{p}_\ell$, which is again strictly monotonically increasing, thereby leading to a unique solution λ. More general statements on uniqueness of Cournot equilibria are provided e.g. by [MQ18, Theorem 4.1].

Example 2 (*cont'd*)
Recall Example 2, where two WPs with cost functions (3.71) *compete over a total demand* $\mathbb{1}^\top \boldsymbol{p}_\ell = 10$. *With Assumption 3.16, the market price* λ *is calculated as follows: For a fixed* λ, *the profit functions of WPs 1 and 2 with cost functions* (3.71) *are*

$$\mathrm{P}_{\mathrm{g},1}(p_{\mathrm{g},1}) = \lambda p_{\mathrm{g},1} - p_{\mathrm{g},1}^2, \qquad \mathrm{P}_{\mathrm{g},2}(p_{\mathrm{g},2}) = \lambda p_{\mathrm{g},2} - \frac{1}{3} p_{\mathrm{g},2}^2. \tag{3.75}$$

With (3.75), *the optimal power generations in terms of* (3.73) *are*

$$p_{\mathrm{g},1}^\star = \frac{1}{2}\lambda^\star, \qquad p_{\mathrm{g},2}^\star = \frac{3}{2}\lambda^\star. \tag{3.76}$$

Inserting (3.76) *into* (3.74) *with* $\mathbb{1}^\top \boldsymbol{p}_\ell = 10$ *yields*

$$\frac{1}{2}\lambda^\star + \frac{3}{2}\lambda^\star = 10 \qquad \Longrightarrow \qquad \lambda^\star = 5. \tag{3.77}$$

Although $\lambda^\star$ *is the same as with Bertrand competition (cf. page 51), the resulting power generations are now given by* $p_{\mathrm{g},1}^\star = 2.5$ *and* $p_{\mathrm{g},2}^\star = 7.5$, *which are Pareto dominating among all market-clearing solutions fulfilling* (3.74).

Despite the fact that both classical Bertrand and Cournot competition are unsuitable as soon as some fraction of the prosumers is inelastic, the above example suggests that using information barriers (in terms of unknown price sensitivities) to a Cournot model of competition result in the following benefits:

1. The resulting equilibrium is stable, since WPs have no incentive to supply more or less than the profit-maximizing input which is located at marginal cost (cf. [NL85]).
2. The resulting equilibrium is unique and causes minimal overall costs (cf. [MQ18, Theorem 4.1; Var10, p. 636ff.]).

Note that both findings will be formalized to the case of general lossy networks with additional operational contraints later in Sections 3.4 and 3.5, respectively.

As will be elaborated in the remainder of this subsection, those information barriers are not only adequate to allow Cournot competition with inelastic demand, but also to approximately satisfy the conditions of a perfect market (cf. Definition A.24).

Measures to Achieve a Perfect Market under Cournot Competition with Inelastic Demand

Definiton A.24 is based upon idealized assumptions that can never be fully satisfied in reality. For instance, each real-world WP always has a certain (i.e. finitely small) influence on the overall market, and both information and power flows can never propagate at infinitely high speed through the network. However, through a reasonable design of market mechanisms and interactions, the characteristics of perfect competition can be pursued to a large extent:

Conditions 1) and 2) in Definition A.24 are trivially fulfilled. To satisfy condition 3), it has to be ensured that price formation over the network occurs sufficiently fast so that each market participant is equipped with the same information. To meet condition 4), we have to assume that the portions of electrical power which can be offered by suppliers are not quantized. From condition 5) it can be concluded that the resulting market mechanism must not incure any costs in total that are charged to the participants, i.e., the net profit of all CCs has to be zero. To comply with condition 6), WPs should at least act *as fast as possible*, such that no internal technical limits, e.g. start-up times, turbine dynamics, or ramp-up constraints are violated. Finally, condition 7) has the following implication:

Assumption 3.17. *No network participant is large enough to exercise market power.*

Remark 3.18. *Assumption 3.17 is justified if the overall number of WPs is high, since a high percentage of inelastic demand is always accompanied by an increased tendency to form monopolies and thus consequently higher prices [RT06]. Another means to satisfy Assumption 3.17 is the requirement that no WP should have a substantial market share in a specific cell*[39] *[Sta19, SHH19]. Alternatively, if there is a small number of market participants in a specific cell or throughout the network, these participants must not be aware that they have a relevant influence on the prices. Another implication of condition 7) is that any collusion among WPs has to be prevented [Küh01].* ◇

Remark 3.19. *With the increasing penetration by intermittent RESs, the price dynamics becomes unpredictable [RN20]. Thus, it is increasingly cumbersome for WPs to correctly predict the future price dynamics as well as the price sensitivities needed in* (3.72)*. This warrants, to a greater extent, the assumption that the momentary supply of WPs is at marginal cost, even if no WP has zero market power under real-world conditions.* ◇

Remark 3.20. *While Assumption 3.17 allows no oligopoly among WPs, there is no such requirement for the overall amount of CCs as they are assumed to be non-strategic network participants acting as the network's* neutral market facilitators *[dQQA19, p. 16]. See also Section 7.2 and references therein for conclusive statements on optimal cell partitioning.* ◇

Fig. 3.7 gives a summarizing overview of the obtained key stakeholders and their physical, informational, and monetary interaction. Compared to the status quo shown in Fig. 2.1, there is a drastic reduction in communication complexity along with rigorous information barriers between CCs as *sovereigns over the network* (complete information about their own network, but no means to physically interact with it) and WPs as *sovereigns over generation capacity* (able to physically interact with the network, but no knowledge about the network state). Moreover, it is most notable that any communication from CCs to WPs is exclusively via

[39] The establishment of distributed CVPPs is thus not only welcome in terms of portfolio diversification for the individual WPs, but also from an overall welfare point of view to prevent local monopoly formation.

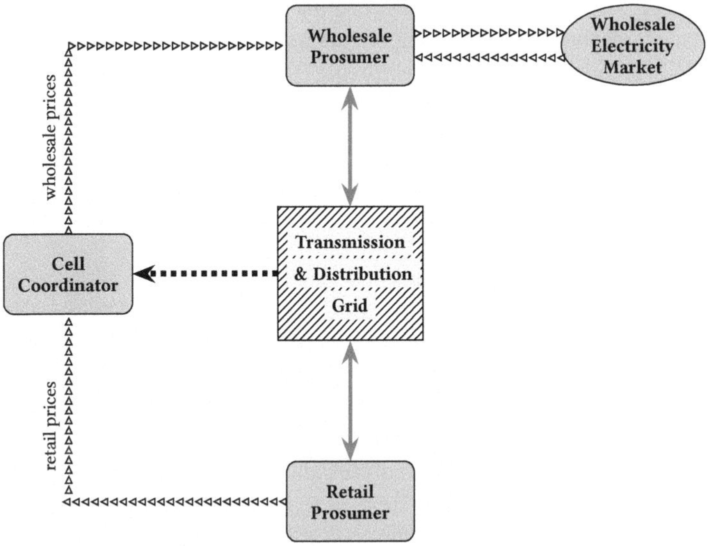

Figure 3.7: Schematic diagram of key stakeholders arising from the proposed control scheme. The yellow box encompasses those components that are elaborated in detail within the scope of this dissertation. The arrows mark the typical flow direction of physical power (solid lines), information (dashed lines), or capital (triangle lines).

(wholesale) prices, Consequently, a full separation between the market-driven world (upper pane in Fig. 3.7) and the non-market-driven world (middle and lower pane in Fig. 3.7) is reestablished.

In the next Subsections 3.3.2–3.3.4, detailed and appropriate cost and communication structures are derived for WPs, RPs, and CCs, respectively, such that both requirements for perfect competition (cf. Definition A.24) as well as challenges (1)–(3) from Subsection 3.3.1 are properly addressed.

3.3.2 Wholesale Prosumers

WPs are *flexibility providers* [AP21], where the flexibility currently provided to the network corresponds to the momentary upper and lower power limits $\overline{p}_{g,i}$ and $\underline{p}_{g,i}$. Throughout the following considerations, we suppose that all WPs behave *rationally* on an individual basis (cf. Assumption A.4) by means of a pure profit maximization strategy. To comply with the unbundling paradigm, we require the following:

> **Claim 3.21.** *There should be no need for WPs to disclose their own cost or profit function to any opponent or CC.*

From Assumption 3.17 and Claim 3.21 it immediately follows that each WP $\pi \in \mathcal{W}$ neither has knowledge about the individual components nor about its own influence and thus seeks to maximize the momentary profits, since there is no personal benefit in pursing a dumping or markup strategy. Another beneficial implication of Claim 3.21 is that there is an intrinsic motivation for WPs to minimize their own cost structure. Namely, a flattening of $\mathtt{C}_{\mathrm{g},i}$ leads to WP π being allocated *more shares of total demand* due to the merit order effect [ZPE17, p. 279f.]. Consequently, every cost reduction of WP $\pi \in \mathcal{W}$ also increases the own revenue $\mathtt{R}_\pi$ (while reducing the revenues of competitors).

To prevent any additional side payments, we further assume that all external costs are internalized appropriately into the cost functions $\mathtt{C}_{\mathrm{g},i}$, e.g. that the individual greenhouse gas emissions are settled by a suitable emissions trading scheme (ETS) such as *EU-ETS* from the European Commission [ZW10]. These internalized costs can then be represented by an additional linear-affine component in the cost function. Typically it has a positive gradient for fossil-fueled CPPs that have to purchase allowances and a zero or negative slope for power generators from RESs that can sell their surplus allowances on the market[40]. Thus, typically, the actual costs for CPPs are higher than those of RESs. Note that these additional internalized costs do not affect the strict convexity assumption (cf. Assumption 3.12).

CCs purchase power from each WP of the own cell by paying the *zonal price* Λ_k per unit of electrical power. To incorporate the marginal contribution of WPs on frequency stability, we propose the following remuneration scheme for WP $\pi \in \mathcal{W}$:

$$\mathtt{R}_\pi = \sum_{i \in \mathcal{V}_{\mathcal{W},\pi}} \lambda_i \cdot p_{\mathrm{g},i} - \omega_i \cdot p_{\mathrm{g},i}, \tag{3.78}$$

where $\lambda_i = \Lambda_k$ for $i \in \mathcal{V}_{\mathcal{Z},k}$. The revenue $\mathtt{R}_\pi$ consists of the frequency-independent payment for active power production at price λ_i and a frequency-dependent part. The latter serves as an instantaneous *penalization* for active power generation during an overfrequency period and as an additional *reward* for active power generation during an underfrequency period. Subtracting the node-individual costs $\mathtt{C}_{\mathrm{g},i}$ incurred for power generation from (3.78), we get the overall profit

$$\mathtt{P}_\pi = \sum_{i \in \mathcal{V}_{\mathcal{W},\pi}} \mathtt{P}_{\mathrm{g},i} = \sum_{i \in \mathcal{V}_{\mathcal{W},\pi}} -\mathtt{C}_{\mathrm{g},i}(p_{\mathrm{g},i}) + \lambda_i \cdot p_{\mathrm{g},i} - \omega_i \cdot p_{\mathrm{g},i} \tag{3.79}$$

for WP $\pi \in \mathcal{W}$. In addition to maximizing $\mathtt{P}_\pi$, WPs must respect their internal operating limits

$$\underline{p}_{\mathrm{g},i} \leq p_{\mathrm{g},i} \leq \overline{p}_{\mathrm{g},i}, \tag{3.80}$$

$$\underline{U}_i \leq U_i \leq \overline{U}_i \tag{3.81}$$

at each node $i \in \mathcal{V}_{\mathrm{g}}$ (cf. challenge (2) in Subsection 3.3.1). Constraints (3.80) represent the remaining power generation capacity of WPs that is not contracted on the WEM (cf. Fig. 3.7),

40 EU-ETS is based on the "cap and trade" principle, i.e., the quantity of emission allowances is capped and will be reduced over time. If a certain company cuts its emissions, it can sell its remaining emission certificates on the secondary market.

whereas constraints (3.81) are typically imposed by the grid code and monitored by the local CC.

Remark 3.22. *For the sake of simplicity we assume that the individual cost functions* $\mathtt{C}_{\mathrm{g},i}$ *and thus the profits* $\mathtt{P}_\pi$ *are not dependent on the own voltage magnitudes* U_i*. This is a common assumption for both CPPs and inverter-interfaced DERs, since changes applied to their output voltage usually do not incur additional cost [BV00, p. 405].* ♢

Remark 3.23. *Despite the fact that the above expressions feature a* producer-centric *notation, they also allow to incorporate elastic consumers at node* i *which aim to maximize their profits*

$$\mathtt{P}_{\ell,i} = \mathtt{U}_i(p_{\ell,i}) - \lambda_i \cdot p_{\ell,i} + \omega_i \cdot p_{\ell,i} \tag{3.82}$$

with $\mathtt{U}_i(p_{\ell,i})$ *being a strictly concave utility function. These consumers can be modeled in a straightforward manner as producers with negative generation* $p_{\mathrm{g},i} = -p_{\ell,i} \leq 0$*, since the profit maximization problem building upon* (3.82) *is structurally identical to* (3.79). ♢

With Assumption 3.17, Claim 3.21, and Assumption 3.12, each WP $\pi \in \mathcal{W}$ thus aims at solving the constrained optimization problem

$$\max_{\boldsymbol{p}_{\mathrm{g},\pi}, \boldsymbol{U}_{\mathrm{f},\pi}, \boldsymbol{U}_{\mathcal{I},\pi}} \quad \mathtt{P}_\pi \tag{3.83a}$$

$$\text{subject to} \quad \underline{p}_{\mathrm{g},i} \leq p_{\mathrm{g},i} \leq \overline{p}_{\mathrm{g},i}, \quad i \in \mathcal{V}_{\mathcal{W},\pi}, \tag{3.83b}$$

$$\underline{U}_i \leq U_i \leq \overline{U}_i, \qquad i \in \mathcal{V}_{\mathcal{W},\pi}. \tag{3.83c}$$

Remark 3.24. *The reaction time of a particular WP to output a certain power* $p_{\mathrm{g},i}$ *or voltage* U_i *varies in a wide range, depending on the specific electromechanical or electrochemical energy conversion technology used. For example, modern grid-forming inverters can provide the required quantities within less than one second [PEH18, p. 3662], while the turbogenerator of a classical large-scale thermal power plant may take several minutes until reaching maximum power [KRK+ 18]. We will account for such dynamic constraints directly in the WPs' controller equations to be derived in Section 3.4, so they need not be considered in the formulation of the optimization problem.* ♢

Lemma 3.25. *For each equilibrium of* (3.49)*, optimization problem* (3.83) *is equivalent to*

$$\max_{\boldsymbol{p}_{\mathrm{g},\pi}, \boldsymbol{U}_{\mathrm{f},\pi}, \boldsymbol{U}_{\mathcal{I},\pi}} \quad \mathtt{P}_\pi \tag{3.84a}$$

$$\text{subject to} \quad \underline{p}_{\mathrm{g},i} \leq p_{\mathrm{g},i} \leq \overline{p}_{\mathrm{g},i}, \quad i \in \mathcal{V}_{\mathcal{W},\pi}, \tag{3.84b}$$

$$\underline{U}_i \leq U_i \leq \overline{U}_i, \qquad i \in \mathcal{V}_{\mathcal{W},\pi} \cap \mathcal{V}_{\mathcal{I}}, \tag{3.84c}$$

$$\underline{\Psi}_i \leq U_{\mathrm{f},i} \leq \overline{\Psi}_i, \qquad i \in \mathcal{V}_{\mathcal{W},\pi} \cap \mathcal{V}_{\mathcal{S}}, \tag{3.84d}$$

where

$$\underline{\Psi}_i = U_i' - (X_{\mathrm{d},i} - X_{\mathrm{d},i}') \cdot \left(\frac{\underline{U}_i}{X_{\mathrm{n},i}} \cos(\vartheta_{\mathrm{n},i}) - \frac{\underline{U}_i^2}{U_i' X_{\mathrm{n},i}} \right), \tag{3.85a}$$

$$\overline{\Psi}_i = U_i' - (X_{\mathrm{d},i} - X_{\mathrm{d},i}') \cdot \left(\frac{\overline{U}_i}{X_{\mathrm{n},i}} \cos(\vartheta_{\mathrm{n},i}) - \frac{\overline{U}_i^2}{U_i' X_{\mathrm{n},i}} \right). \tag{3.85b}$$

Proof. Let $\square^\star$ denote an equilibrium of (3.49) and let $i \in \mathcal{V}_\mathcal{S}$. From (3.28c) it follows that

$$U_{\mathrm{f},i}^\star = (U_i')^\star - (X_{\mathrm{d},i} - X_{\mathrm{d},i}') \cdot q_{\mathrm{inj},i}^\star / (U_i')^\star =: \Psi_i(U_i', U_i). \tag{3.86}$$

Inserting (3.26) in (3.86) and comparison with (3.85) yields $\underline{\Psi}_i = \Psi_i(U_i', \underline{U}_i)$ and $\overline{\Psi}_i = \Psi_i(U_i', \overline{U}_i)$. With

$$\frac{\partial \Psi_i}{\partial U_i} = -\frac{X_{\mathrm{d},i} - X_{\mathrm{d},i}'}{X_{\mathrm{n},i}} \left(\cos(\vartheta_{\mathrm{n},i}) - \frac{2U_i}{U_i'} \right) > 0 \tag{3.87}$$

under normal operating conditions, it holds that Ψ_i is strictly monotonically increasing with respect to U_i, thus $U_i^\star \leq \overline{U}_i \iff \Psi_i \leq \overline{\Psi}_i$ and $\underline{U}_i \leq U_i^\star \iff \underline{\Psi}_i \leq \Psi_i$. Accordingly, each optimizer of (3.84) is an optimizer of (3.83), and vice versa. □

3.3.3 Retail Prosumers

RPs at node $i \in \mathcal{V}$ are characterized by an inelastic active and reactive power demand $p_{\ell,i}$ and $q_{\ell,i}$. The cumulative cost for consumption in cell $k \in \mathcal{Z}$ are set to

$$\mathrm{C}_k^\ell = \Lambda_{\mathrm{R},k} \cdot \sum_{i \in \mathcal{V}_{\mathcal{Z},k}} p_{\ell,i}, \tag{3.88}$$

where $\Lambda_{\mathrm{R},k}$ denotes the *retail price* for electricity in cell k (cf. Fig. 3.7).

Note that small-scale power plants relying on uncontrollable power sources (e.g. single photovoltaic power stations or wind power plants that are not integrated into a VPP) can be modeled as RPs with $p_{\ell,i} \leq 0$. In addition, $p_{\ell,i}$ and $q_{\ell,i}$ are used to cover all positive or negative physical supplies arising from other contracts such as forward contracts or obligations due to the day-ahead market. However, from these markets only the current value of the resulting market clearing quantities is known to the local CC, while none of the other network participants are aware of it. Furthermore, it is completely unknown for all network participants which active and reactive power flows are to be expected in the future.

3.3.4 Cell Coordinators

The CCs' exclusive responsibility for stable, economically efficient and grid-supportive operation which was outlined in Section 3.1 has two connotations: On one hand, *exclusive* means that no other network participant (neither CC nor WP) shares the same information or has the same responsibilities. On the other hand, it means that maintaining the above three objectives is the *sole purpose* of each CC. In particular, it does not have any economic goals. As pure service providers, CCs solely own a portion of the transmission and distribution infrastructure (cf. challenge (3) in Subsection 3.3.1) and, apart from that, neither storages nor any kind of generation capacity. This is a direct consequence from the unbundling requirement elaborated in Section 2.1. The geographic region that is covered by CC $k \in \mathcal{Z}$ is characterized by the weakly connected component $\mathscr{G}_{\mathrm{c},k}$ in $\mathscr{G}_{\mathrm{c}}$.

Assumption 3.26. *CCs have full knowledge about their cell's own line parameters (G_{ij}, B_{ij}, g_i, b_i) and maintain continuous, delay-free measurements of the voltage magnitudes U_i, the power injections $p_{\mathrm{g},i}$, the power consumptions $p_{\ell,i}$, $q_{\ell,i}$, and the power flows P_{ij}, Q_{ij} within their own cell.*

Assumption 3.26 can be justified as follows: The line parameters can be derived from manufacturers' specifications using (3.5) and (3.6). The power flows and voltage amplitudes can be measured directly in-situ by means of *phasor measurement units* (PMUs) [MBB12, p. 61f.]. In particular, note that the voltage angles θ_i or ϑ_{ij} do not need to be measured. Moreover, CCs are only aware of the current "grid-side" power flows as well as calculations of the momentary resistive transmission losses

$$\hat{\Phi}_k = \sum_{i \in \mathcal{V}_{\mathcal{Z},k}} \left(G_{ii} U_i^2 + \sum_{j \in \mathcal{N}_i} G_{ij} U_i U_j \cos(\vartheta_{ij}) \right) \tag{3.89}$$

within cell k. However, CCs do not know the WPs' or RPs' internal parameters nor any of the individual network participants' profits or optimization strategies. Furthermore, if some of these quantities are not directly measurable, *state observers* can be used to reconstruct suitable *estimators* for each of the respective quantities. Such state observes are already being used successfully in today's power systems. Recently, researchers have also increased their scrutiny of new methodological research on state observers tailored specifically for power systems. Comprehensive overviews of current research branches in this field are provided in [ARO+18, WGCS19, KKŠ20].

For all $i, j \in \mathcal{V}_{\mathcal{Z},k}$, the responsible CC $k \in \mathcal{Z}$ seeks to choose the cell-specific price $\Lambda_k = \lambda_i = \lambda_j$ in such a way that its own (virtual) profit P_k, determined by the net active power balance within the own cell, is zero:

$$\mathsf{P}_k = \Lambda_k \cdot \left(\hat{\Phi}_k + \sum_{i \in \mathcal{V}_k^{\mathcal{Z}}} p_{\ell,i} \right) - \Lambda_k \cdot \sum_{\substack{i \in \mathcal{V}_{\mathcal{Z},k} \\ i \notin \mathcal{V}_{\mathcal{L}}}} p_{g,i} \overset{!}{=} 0. \tag{3.90}$$

3.3.5 Summary

Using the prototypical network participants WPs, CCs, and RPs, Subsections 3.3.2–3.3.4 have modeled the different operating maxims of stakeholders in a fully unbundled power network.

Thereby, WPs $\pi \in \mathcal{W}$ constitute the players of an *electricity market game* with the goal of maximizing their own profit while complying with operating constraints e.g. for their own power injections and voltage magnitudes. While choosing their own actions as *best response* to given environmental conditions, WPs rely on information about their own cost structure and machine parameters as well as the current cell-specific price Λ_k. In the presented framework, there is no hierarchy in the execution of individual actions. As a consequence of the requirement of information barriers and unbundling, the individual players are not aware of the formation of prices. Thus, the profit functions of WPs are completely decoupled.

At the same time, CCs as pure service providers must ensure adherence with balancing constraint (3.90) through appropriate choice (i.e. real-time adjustment) of cell-specific prices Λ_k while considering real-time measurements of the RPs' power consumptions as well as cell-specific transmission losses.

The dynamic interplay of these different network participants with time-varying environmental conditions (induced e.g. by load jumps or changes in the network structure) leads to a superposition of constrained optimization problems (3.83) and (3.90). The individual constraints include *input constraints* (e.g. limitations of power $p_{\mathrm{g},i}$ (3.83b) or inverter voltages U_i in (3.83c)), *output constraints* such as limitations of voltage magnitudes at generator terminals in (3.83c), and *dynamic system constraints* which are enforced by the physical relationships of the interconnected plant system (3.49).

As already shown in Lemma 3.25, output constraints (here: constraints (3.83c) on the terminal voltages U_i) can be translated into input constraints (here: constraints on the excitation voltages $U_{\mathrm{f},i}$ (3.84d)) provided that a suitable *steady-state map* (here $\Psi_i(U_i)$ as in (3.85)) is known. While for input and output constraints *asymptotic* satisfaction is permissible, i.e., a temporary constraint violation is allowed in the course of solving the optimization problem, system constraints, on the other hand, are complex environmental conditions that typically have to be satisfied even during transients in order to achieve a result that is physically admissible at any point in time, i.e., in accordance with the laws of physics. However, due to the numerous incorporated information barriers, there is no central authority that has complete knowledge of the whole plant state $\boldsymbol{x}_\mathrm{p}$. A suitable controller design must thus take these information barriers into account, but at the same time must exhibit favorable properties from an overall system perspective such as robustness to disturbances and sufficiently low computational effort of local calculations. With this in mind, the next section provides the main result in the sense of Contribution 1. In particular, the derivation of distributed controller equations enables network participants to achieve the overall goal of a stable overall power system by means of individual and local control actions as required by Contribution 1.

3.4 Controller Development

For optimization-based control on constrained physical systems, arising from the interconnection of the physical plant system from Section 3.2 with the network participants from Section 3.3, two different branches are known in literature. The first branch emanates from a classical understanding of solving optimization problems in a *feedforward* manner, whereby there exists a preferably exact steady-state model for the plant system, allowing its direct representation as an additional set of constraints. MPC (see Section 2.2.2) is a classical representative of a feedforward optimization approach. As discussed in Section 2.2.2, there is always a trade-off between constraint satisfaction (i.e. real-world feasibility of the resulting OPF solution) and computational complexity. Thereby, feedforward optimization methods for power system applications typically scale poorly not only with model order and number of iteration steps, but also with the number of nodes per WP or CC. However, imperfectly estimated parameters, non-modeled side effects, or sudden structural changes not captured in the model always cause physically infeasible solutions at the plant input without any further possibility of compensation.

As a second branch, *feedback*-based optimization converts the optimization algorithm into a continuous-time control law based on online measurements of the current plant state $\boldsymbol{x}_{\mathrm{p}}$, while driving the physical system to its (constrained) optimum value[41]. By directly relying on measurement information, no perfect model knowledge is required. Instead, the physical constraints (along with possible disturbances or structural changes) manifest themselves directly in the measurement signal as they arise via the feedback branch, where *"the physical system [itself] acts as a constraint enforcer"* [HBHD21b, p. 2]. Since the system equations no longer need to be calculated separately by each local agent, this significantly reduces the problem size of the individual optimization problems, and thus also the computational effort for each network participant[42]. Furthermore, it is no longer necessary to conservatively model *all* possible realizations of a disturbance. Instead, only the *actual* materialization of the underlying disturbance process is employed for the participants' decision making via real-time measurements of the system state.

Due to these evident advantages, feedback optimization strategies will form the methodological basis of all controller design in the remainder of this dissertation. Subsection 3.4.1 presents the main types of feedback optimization strategies known in the literature and discusses their advantages and disadvantages. Subsection 3.4.2 then deduces distributed feedback control laws for WPs and CCs. Subsection 3.4.3 analyzes stability of the closed-loop equilibrium. Finally, Subsection 3.4.4 illustrates the main conclusions of this section by means of an academic example.

3.4.1 Introduction to Feedback Optimization

Feedback optimization methods interpret the governing optimization algorithm as a dynamic control system. The basic idea of re-thinking the classical discrete-time steepest descent

41 Although terminologically related, *extremum seeking* refers to optimization-based control methods for the *model-free* exploration of the feasible domain by means of approproate dither signals, see e.g. [TMM+10] for an insightful literature review. However, since the necessary excitation with these dither signals is contrary to the requirement of economic efficiency and, moreover, an appropriate choice is highly difficult for higher-dimensional systems [HBHD21a, p. 6], extremum seeking-based control schemes will not be considered any further in the context of this dissertation.

42 For quantitative robustness statements, see [CSPB19]. Practical experiments are conducted in [PBD20] and [BD19]. For an experimental validation of (feedforward) AC-OPF vs. feedback AC-OPF, the reader is referred to [OHC+20]. A general discussion of feedback vs. feedforward optimization of power systems is given e.g. in Section III of [DBSPG19].

approach for minimizing a certain (unconstrained) objective function $J(\boldsymbol{u}, \boldsymbol{y})$ as a time-continuous *gradient flow* was already proposed in the seminal publications [Kos56, AHU58] dating back to the 1950s, and has just been extensively revisited and improved[43]. Typically, it relies upon the assumption of a fast-decaying plant system whose input-output behavior can be represented by a *steady-state input-output map* $\boldsymbol{y} = \boldsymbol{h}(\boldsymbol{u}) + \boldsymbol{d}$. For an arbitrary objective function $J(\boldsymbol{u}, \boldsymbol{y})$ which depends on both the control and system output, we then get

$$\min_{\boldsymbol{u}} \quad J(\boldsymbol{u}, \boldsymbol{y}) \tag{3.91a}$$

$$\text{subject to} \quad \boldsymbol{y} = \boldsymbol{h}(\boldsymbol{u}) + \boldsymbol{d}. \tag{3.91b}$$

Remarkably, from the controller's point of view, constraint (3.91b) does not need to be considered explicitly. As each output $\boldsymbol{y}$ applied to the controller already satisfies the system equation, it automatically holds that any input-output pair $(\boldsymbol{u}, \boldsymbol{y})$ is a solution of (3.91b). Thus, by defining $\widetilde{J}(\boldsymbol{u}) = J(\boldsymbol{u}, \boldsymbol{h}(\boldsymbol{u}) + \boldsymbol{d})$, the own control strategy can be calculated by a gradient descent on $\widetilde{J}$. With the chain rule, we get

$$\dot{\boldsymbol{u}} = -\frac{\partial \widetilde{J}}{\partial \boldsymbol{u}} = -\left(\frac{\partial J}{\partial \boldsymbol{y}}\right)^{\mathsf{T}} \frac{\partial \boldsymbol{h}}{\partial \boldsymbol{u}} - \frac{\partial J}{\partial \boldsymbol{u}}. \tag{3.92}$$

It is evident that only the sensitivities $\nabla \boldsymbol{h}(\boldsymbol{u})$ need to be known[44].

Beyond the conceptually simplistic idea of continuous-time gradient flow, there are different approaches to dealing with possible additional input or output constraints of the form

$$\boldsymbol{u} \in \mathcal{U}', \tag{3.93a}$$

$$\boldsymbol{y} \in \mathcal{Y}', \tag{3.93b}$$

where $\mathcal{U}'$ and $\mathcal{Y}'$ denote the sets of feasible inputs and outputs, respectively. First of all, by defining $\mathcal{U} := \mathcal{U}' \cap \boldsymbol{h}^{-1}(\mathcal{Y}')$, (3.91)–(3.93) can be written as

$$\min_{\boldsymbol{u}} \quad \widetilde{J}(\boldsymbol{u}) \tag{3.94a}$$

$$\text{subject to} \quad \boldsymbol{u} \in \mathcal{U}. \tag{3.94b}$$

The approaches discussed in the literature differ in their strictness of *constraint enforcement*, in particular, whether or not each closed-loop equilibrium is always an "exact" optimizer of (3.94), and whether or not temporary violations of the constraint (3.94b) are allowed. Table 3.2 gives an overview of the main research branches, which are briefly discussed in the following.

---

43 Interestingly, the main impetus for the renaissance of this general control scheme came from the power system community, see e.g. [DBSPG19] for a historical perspective. The main reasons for their popularity in power system applications are their robustness properties as well as their ease of implementation with relatively low computational effort for the individual controllers (cf. [CSPB19]).

44 Trivially, if J only depends on the input, the gradient descent simplifies to $\dot{\boldsymbol{u}} = -\nabla J(\boldsymbol{u})$. For optimization problems which solely rely on $\boldsymbol{y}$, see [JLv09, BDE12].

Table 3.2: Constraint enforcement schemes for continuous-time feedback optimization.

	transient constraint violation	**no transient constraint violation**
no asymptotic optimality	• augmented gradient flow with penalty function [SDS88]	• augmented gradient flow with barrier function [HBHD21a, p. 10f.]
asymptotic optimality	• primal-dual saddle-point flow [Kos56, AHU58] • augmented saddle-point flow [Pol09a] • projected saddle-point flow [FP10] • mixed saddle-point flow [HBHD21a, p. 14] • Newton saddle-point flow [HBHD21a, p. 12]	• projected gradient flow [HBHD16] • projected Newton flow [HBD21]

Augmented Gradient Flow

By augmenting $\widetilde{J}(\boldsymbol{u})$ with a scalar *penalty* or *barrier* summand $\mho(\boldsymbol{u})$ indicating constraint violation, the constrained optimization problem (3.94) turns into the unconstrained optimization problem

$$\min_{\boldsymbol{u}} \quad \widetilde{J}(\boldsymbol{u}) + \mho(\boldsymbol{u}). \tag{3.95}$$

If the feasible domain $\mathcal{U}$ in (3.94b) can be characterized via inequality constraints of the form $\boldsymbol{g}(\boldsymbol{u}) \leq 0$, then for instance, $\mho(\boldsymbol{u}) = \tau \|\mathbf{min}\{\mathbb{0}, \boldsymbol{g}(\boldsymbol{u})\}\|^2$, $\tau > 0$ provides a suitable penalty term and $\mho(\boldsymbol{u}) = -\tau \mathbf{ln}(\boldsymbol{g}(\boldsymbol{u}))$, $\tau > 0$ can be used as barrier term. As $\tau \to \infty$, possible constraint violations become more pronounced in the augmented objective function, and thus constraint satisfaction is imposed more forcefully. However, while penalty and barrier functions always impose some sort of outer and inner approximation of the feasible set, respectively, the augmentation of $\widetilde{J}(\boldsymbol{u})$ typically leads to steady-state deviations between the closed-loop equilibrium and the optimizer $\boldsymbol{u}^\star$ of (3.94), especially if this optimizer is located at or near the boundary of $\mathcal{U}$. But at the same time, τ must still be sufficiently low to ensure a reasonably timescale separation between a "fast" plant system and "slow" optimization dynamics.

Projected Gradient Flow

A refinement of the gradient flow was introduced by *projected gradient flow* [HBHD16]. The remarkable feature of this method is that $\boldsymbol{u}(t) \in \mathcal{U}$ can be guaranteed at any time t, thus all constraints are always fulfilled even during transients. This is accomplished by requiring that all directional changes of $\boldsymbol{u}$ should point into the feasible set. For this purpose, the right-hand side of (3.92) is replaced by the *gradient projection* onto $\mathcal{U}$, leading to

$$\dot{\boldsymbol{u}} = -\Pi_{\mathcal{U}}(-\nabla \widetilde{J}(\boldsymbol{u})) = -\arg\min_{\boldsymbol{\xi} \in \mathcal{T}_{\mathcal{U}}} \| -\nabla \widetilde{J}(\boldsymbol{u}) - \boldsymbol{\xi} \|, \tag{3.96}$$

where $\mathcal{T}_{\mathcal{U}}(\boldsymbol{u}) := \text{cl}\{\boldsymbol{d} \,:\, \boldsymbol{d} = \alpha(\boldsymbol{u}' - \boldsymbol{u}), \boldsymbol{u}' \in \mathcal{U}, \alpha \geq 0\}$ is the *tangent cone* of $\mathcal{U}$ at $\boldsymbol{u}$. By definition, if $\boldsymbol{u} \in \text{int}\,\mathcal{U}$, then $\mathcal{T}_{\mathcal{U}}(\boldsymbol{u}) = \mathbb{R}^p$, while if $\boldsymbol{u} \in \text{bd}\,\mathcal{U}$, then $\mathcal{T}_{\mathcal{U}}(\boldsymbol{u})$ represents the set of all vectors starting from $\boldsymbol{u}$ and pointing into the feasible set $\mathcal{U}$. The ability of *strict transient constraint enforcement* is a salient feature of projected gradient flow. In [HBHD21b] quantitative statements about asymptotic stability of the closed-loop system with projected gradient are provided, which, however, require a priori knowledge of appropriate Lipschitz constants of the system with respect to both the input $\boldsymbol{u}$ and the state $\boldsymbol{x}$.

However, despite these advantages, projected gradient flow exhibits some major drawbacks, which hamper a practical application, especially in the light of the electricity market game mentioned in Subsection 3.3.5 with its intermittent nature of power consumption and/or production and a possibly large amount of distributed agents: Besides the fact that explicit expressions for the gradient projection (3.96) are often hard to find and generally not amenable to a distributed implementation, the main practical drawback of projected gradient flow is its tendency to "false equilibria" (i.e. equilibria of the closed-loop system that are no optimizers of (3.94)), whenever the current solution $\boldsymbol{x}(t)$ is infeasible with respect to (3.94b) at some point in time. We illustrate this effect by means of the following non-example:

Example 3 (Projected Gradient Flow)
Consider the optimization problem

$$\min_{u} \quad \frac{1}{2}u^2 \tag{3.97a}$$

$$\text{subject to} \quad u + 10 \leq 0. \tag{3.97b}$$

Note that (3.97) *is a convex optimization problem with strictly convex objective function. Hence,* $u^\star = -10$ *is the unique optimizer of* (3.97). *With* $\mathcal{U} = \{u \in \mathbb{R} \,:\, u \leq -10\}$, *we get*

$$\mathcal{T}_{\mathcal{U}}(u) = \begin{cases} \mathbb{R}_{<0}, & u \geq -10, \\ \mathbb{R}, & u < -10. \end{cases} \tag{3.98}$$

Now consider the trajectory of (3.96) *starting at* $u_0 = 10$. *With* $\mathcal{T}_{\mathcal{U}}(10) = \mathbb{R}$, *we get*

$$\dot{u} = \arg\min_{\xi \in \mathbb{R}} \| -u - \xi \| = -u. \tag{3.99}$$

Accordingly, a solution is given by $u(t) = 10\exp(-t)$ *and thus* $\lim_{t\to\infty} u(t) = 0$, *which means that* $u(t)$ *never converges to its optimum value* $u^\star = -10$. *Trivially, the same result holds for any other* u *starting in the positive orthant.*

With respect to the CCs' balance constraints (3.90) for instance, this shortcoming of projected gradient flow is highly problematic, since temporary constraint violations e.g. due to sudden load jumps are always possible, potentially causing the resulting closed-loop equilbrium to no longer satisfy (3.90).

Saddle-Point Flow

Besides penalty methods and projected gradient flow, another means for incorporating constraints is given by *primal-dual saddle-point flow*, originally proposed in [Kos56] and [AHU58]. It can be applied to optimization problems in *standard form* (cf. Definition A.18). The naming of primal-dual saddle-point flow stems from the fact that each optimizer of such optimization problems is a saddle point of the respective *(optimization) Lagrangian.* In particular, if (3.94b) can be represented by a set of $r \in \mathbb{N} \cup \{0\}$ equality and $s \in \mathbb{N} \cup \{0\}$ inequality constraints of the form $\boldsymbol{h}_1(\boldsymbol{u}) = \mathbb{0}$ and $\boldsymbol{h}_2(\boldsymbol{u}) < \mathbb{0}$, respectively, then the Lagrangian of (3.94) takes the form $\mathscr{L}(\boldsymbol{u}, \boldsymbol{\lambda}, \boldsymbol{\mu}) = \widetilde{J}(\boldsymbol{u}) + \boldsymbol{\lambda}^\mathsf{T} \boldsymbol{h}_1(\boldsymbol{u}) + \boldsymbol{\mu}^\mathsf{T} \boldsymbol{h}_2(\boldsymbol{u})$, where $\boldsymbol{\lambda}$ and $\boldsymbol{\mu}$ denote the vector of dual variables for $\boldsymbol{h}_1(\boldsymbol{u})$ and $\boldsymbol{h}_2(\boldsymbol{u})$, respectively. Provided that some constraint qualification is fulfilled, the Karush-Kuhn-Tucker (KKT) conditions (cf. Definition A.20 in Appendix A.5) constitute first-order necessary conditions for the primal-dual optimizer $(\boldsymbol{u}^\star, \boldsymbol{\lambda}^\star, \boldsymbol{\mu}^\star)$ with gradient descent for the primal variables $\boldsymbol{u}$ and gradient ascent for the dual variables $\boldsymbol{\lambda}$, yielding

$$\dot{\boldsymbol{u}} = -\frac{\partial \mathscr{L}(\boldsymbol{u}, \boldsymbol{\lambda}, \boldsymbol{\mu})}{\partial \boldsymbol{u}}, \tag{3.100a}$$

$$\dot{\boldsymbol{\lambda}} = +\frac{\partial \mathscr{L}(\boldsymbol{u}, \boldsymbol{\lambda}, \boldsymbol{\mu})}{\partial \boldsymbol{\lambda}}. \tag{3.100b}$$

The inequality constraints $\boldsymbol{h}_2(\boldsymbol{u}) \leq 0$ translate into complementary slackness condition (A.20c) and the nonnegativity of $\boldsymbol{\mu}$ (cf. (A.20d)). Therefore, in [FP10], an additional projected gradient flow for the dual variables $\boldsymbol{\mu}(t)$ onto the tangent cone of the nonnegative orthant

$$\dot{\boldsymbol{\mu}} = +\Pi_{\mathbb{R}^s_{\geq 0}}(\nabla \mathscr{L}(\boldsymbol{\mu})) = +\Pi_{\mathbb{R}^s_{\geq 0}}(\boldsymbol{h}_2(\boldsymbol{u})) \tag{3.101}$$

is used to enforce $\boldsymbol{\mu}(t) \geq \mathbb{0}$ for all t. Since the projection of $h_{1,i}$, $i \in \{1, \ldots, s\}$ onto the tangent cone $\mathcal{T}_{\mathbb{R}_{\geq 0}}$ allows for the explicit expression

$$\Pi_{\mathbb{R}_{\geq 0}}(h_{1,i}(\boldsymbol{u})) = \langle h_{1,i} \rangle^+_{\mu_i} := \begin{cases} h_{1,i}(\boldsymbol{u}), & h_{1,i} > 0 \vee \mu_i > 0, \\ 0, & \text{otherwise}, \end{cases} \tag{3.102}$$

the dual ascent (3.101) of $\boldsymbol{\mu}$ can be written compactly as

$$\dot{\boldsymbol{\mu}} = \langle \boldsymbol{h}_1(\boldsymbol{u}) \rangle^+_{\boldsymbol{\mu}}, \tag{3.103}$$

where $\langle \cdot \rangle^+_{\boldsymbol{\mu}} := \mathrm{col}_{i \in \{1,\ldots,s\}}\{\langle \cdot \rangle^+_{\mu_i}\}$.

In contrast to the "pure" projected gradient flow (3.96) without dualization, this *projected saddle-point flow* (3.100), (3.103) yields a constraint satisfaction only at steady-state[45]. Particularly,

[45] Despite this temporary constraint violation of primal-dual saddle-point flow, it should be emphasized that the resulting controller always yields "physically feasible" results $\boldsymbol{u}(t)$ at all times t, since the actual constitutive equations of the plant system are exactly respected at all times due to the feedback interconnection and thus do not need to be part of the constraints, as discussed above.

even the gradient projection (3.102) does not prevent temporary constraint violations of $\boldsymbol{h}_1(\boldsymbol{u}) \geq 0$.

Despite this tendency towards temporary constraint deviation, however, especially in recent years, several control approaches in the area of networked and power systems rely on primal-dual saddle-point flow and its derivations (see the list of most prevalent methods in Table 3.2). Likewise, some of the real-time economic dispatch and optimization-based frequency control approaches discussed in Subsections 2.2.2 and 2.2.4 resort to one of the above methods or a combination thereof for the real-time feedback optimization of their arising optimization problems. [MZL14, LZC16, TBD16, SDv17a, CY17] use the primal-dual saddle-point flow for optimization-based frequency control of $\mathcal{S}$ and/or $\mathcal{L}$ nodes. [SDv17b] incorporates additional inequality constraints on active power injections and power line capacities (for networks with tree topology only) by introducing additional projected gradient ascent (3.103) for the dual variables. Moreover, additional penalty terms are proposed allowing an extension to non-strictly convex objective functions. [HZB$^+$17] uses projected gradient flow in conjunction with a penalty-augmented Lagrangian for a centralized voltage regulation with congestion management. The main reason for the popularity of primal-dual saddle-point flow approaches is their ease of implementation in contrast to the projected gradient method, especially if the underlying plant system exhibits a sparsity pattern. As discussed e.g. in [FP10, p. 1979], primal-dual saddle-point flow lends itself to a distributed implementation, which is highly desirable in view of the individual stakeholders defined in Section 3.3. Example 3 also stated and highlighted the "warm start" problem of projected gradient flow: If the controller variables are initialized as infeasible or if individual variables become temporarily infeasible due to external events, the closed-loop system tends to invoke false equilibria. By contrast, primal-dual saddle-point flow can always be initialized even if the current state $(\boldsymbol{u}, \boldsymbol{y})$ does not satisfy all of the constraints. Moreover, the resulting controller equations (3.100) (and (3.103), if desired) always exist in explicit form, thus encouraging an automated controller design.

Remark 3.27. *There already exists a variety of* hybrid *forms between penalty, projected gradient flow or saddle-point flow methods (see Table 3.2). For instance, [Pol09a] suggests that further augmentation of* $\mathscr{L}(\boldsymbol{u}, \boldsymbol{\lambda}, \boldsymbol{\mu})$ *by another penalty term (resulting in the definition of* augmented saddle-point flow*) may help in damping oscillations, yet at the price of losing optimality with respect to the original problem. [HBHD21a, p. 14] proposes a* mixed saddle-point flow *where some of the inequality constraints are enforced directly by (gradient) projection and some are enforced by projected gradient ascent of the dual variables as in* (3.103). ◇

Remark 3.28. *In addition, the right-hand sides of the controller equations* (3.92), (3.96), *or* (3.100) *can be premultiplied with any positive semidefinite matrix* $\boldsymbol{A}(\boldsymbol{u})$ *without changing the equilibrium. The prominent choice* $\boldsymbol{A}(\boldsymbol{u}) = (\nabla^2 \widetilde{J}(\boldsymbol{u}))^{-1}$*, yielding a Newton-type gradient descent with improved convergence properties has thereby given rise to the formulation of* Newton gradient flows, projected Newton flow, *and* Newton saddle-point flow *[HBD21, HBHD21a].* ◇

With respect to practical application to electricity markets, a downside of augmentation methods and variations of saddle-point flow is that it usually sacrifices the distributed implementability mentioned above. In the light of this reasoning, particularly since temporary constraint violations in the optimization problems (3.84) and (3.90) are tolerable according to the definition of power system stability presented at the beginning of Section 2.2, we will use projected saddle-point flow (with classical saddle-point flow for the equality constraints and gradient-projected dualization of the inequality constraints) as the basic building block for the local controllers, though we will discuss possible extensions where advisable.

3.4.2 Resulting Control Laws

Before deriving specific control laws for WPs and CCs by means of projected saddle-point flow, we give two general statements with reference to the optimization problem in standard form from Definition A.18.

Lemma 3.29. *Denote by* $(\boldsymbol{u}^\star, \boldsymbol{\lambda}^\star, \boldsymbol{\mu}^\star)$ *an equlibrium of* (3.100), (3.103), *and suppose that the optimization problem* (A.19) *has a strictly convex objective function* $f_0(\boldsymbol{x})$ *and fulfills Slater's condition* (A.21). *Then* $\boldsymbol{u}^\star$ *is a solution of* (A.19).

Proof. Trivially, each $(\boldsymbol{u}^\star, \boldsymbol{\lambda}^\star, \boldsymbol{\mu}^\star)$ fulfills the KKT conditions (A.20) for (A.19). Since $f_0(\boldsymbol{u})$ is strictly convex, the KKT conditions are also sufficient for optimality if Slater's condition holds (cf. [BV15, p. 244] and Lemma A.22). Thus, each $\boldsymbol{u}^\star$ is a solution of (A.19) . □

Lemma 3.30. *Let* $f_0(\boldsymbol{u})$ *be strictly convex. Then the projected saddle-point flow* (3.100), (3.103) *asymptotically converges to the solution of* (A.19).

Proof. See [CMC16] for a proof using an extended version of LaSalle's invariance principle for Carathéodory solutions of discontinuous dynamical systems or [Goe17] for an alternative proof using maximal monotone mappings. □

Remark 3.31. *If the strict convexity assumption of* $f_0(\boldsymbol{u})$ *from Lemma 3.30 does not hold, the application of saddle-point flow may cause the closed-loop system to exhibit limit-cycle behavior [HOBD20]. However, as shown in [HOBD20], the asymptotic convergence results of Lemma 3.30 can be regained by a suitable augmentation of* $\mathscr{L}(\boldsymbol{u}, \boldsymbol{\lambda}, \boldsymbol{\mu})$. ◇

Using Lemma 3.29 and Lemma 3.30, we now consider the individual optimization problems of WPs and CCs.

For WP $\pi \in \mathcal{W}$, the Lagrangian of (3.84) equals

$$\begin{aligned}\mathscr{L}_\pi = &-\mathrm{P}_\pi + \boldsymbol{\mu}_{\mathrm{g}-,\pi}^\mathsf{T}(\underline{\boldsymbol{p}}_{\mathrm{g},\pi} - \boldsymbol{p}_{\mathrm{g},\pi}) + \boldsymbol{\mu}_{\mathrm{g}+,\pi}^\mathsf{T}(\boldsymbol{p}_{\mathrm{g},\pi} - \overline{\boldsymbol{p}}_{\mathrm{g},\pi})\\ &+ \boldsymbol{\mu}_{\mathcal{S}-,\pi}^\mathsf{T}(\underline{\boldsymbol{\Psi}}_\pi - \boldsymbol{U}_{\mathrm{f},\pi}) + \boldsymbol{\mu}_{\mathcal{S}+,\pi}^\mathsf{T}(\boldsymbol{U}_{\mathrm{f},\pi} - \overline{\boldsymbol{\Psi}}_\pi)\\ &+ \boldsymbol{\mu}_{\mathcal{I}-,\pi}^\mathsf{T}(\underline{\boldsymbol{U}}_{\mathcal{I},\pi} - \boldsymbol{U}_{\mathcal{I},\pi}) + \boldsymbol{\mu}_{\mathcal{I}+,\pi}^\mathsf{T}(\boldsymbol{U}_{\mathcal{I},\pi} - \underline{\boldsymbol{U}}_{\mathcal{I},\pi}),\end{aligned} \tag{3.104}$$

where $\boldsymbol{\mu}_{(\cdot),\pi}$ denote the vectors of Lagrange multipliers for the inequality constraints (3.84b)–(3.84d) and $\underline{\boldsymbol{\Psi}}_\pi := \mathrm{col}_{i\in\mathcal{V}_{\mathcal{W},\pi}\cup\mathcal{V}'_{\mathcal{S}}}\{\underline{\Psi}_i\}$, $\overline{\boldsymbol{\Psi}}_\pi := \mathrm{col}_{i\in\mathcal{V}_{\mathcal{W},\pi}\cup\mathcal{V}'_{\mathcal{S}}}\{\overline{\Psi}_i\}$. Since (3.84) is convex and Slater's condition is fulfilled, the KKT conditions specifying a saddle point of $\mathscr{L}_\pi$ can be applied to derive a necessary and sufficient condition for an optimizer of (3.84)[46]:

$$
\begin{aligned}
\mathbb{0} &= \nabla \mathrm{C}_\pi(\boldsymbol{p}^\star_{\mathrm{g},\pi}) - \boldsymbol{\lambda}_\pi + \boldsymbol{\omega}_\pi - \boldsymbol{\mu}^\star_{\mathrm{g}-,\pi} + \boldsymbol{\mu}^\star_{\mathrm{g}+,\pi}, && (3.105a)\\
\mathbb{0} &= -\boldsymbol{\mu}^\star_{\mathcal{S}-,\pi} + \boldsymbol{\mu}^\star_{\mathcal{S}+,\pi}, && (3.105b)\\
\mathbb{0} &= -\boldsymbol{\mu}^\star_{\mathcal{I}-,\pi} + \boldsymbol{\mu}^\star_{\mathcal{I}+,\pi} - (\nabla\underline{\boldsymbol{\Psi}}_\pi(\boldsymbol{U}_{\mathcal{I},\pi}))\circ\boldsymbol{\mu}_{\mathcal{S}-,\pi} + (\nabla\overline{\boldsymbol{\Psi}}_\pi(\boldsymbol{U}_{\mathcal{I},\pi}))\circ\boldsymbol{\mu}_{\mathcal{S}+,\pi}, && (3.105c)\\
\mathbb{0} &= (\boldsymbol{\mu}^\star_{\mathrm{g}-,\pi})\circ(\underline{\boldsymbol{p}}_{\mathrm{g},\pi} - \boldsymbol{p}_{\mathrm{g},\pi}), && (3.105d)\\
\mathbb{0} &= (\boldsymbol{\mu}^\star_{\mathrm{g}+,\pi})\circ(\boldsymbol{p}_{\mathrm{g},\pi} - \overline{\boldsymbol{p}}_{\mathrm{g},\pi}), && (3.105e)\\
\mathbb{0} &= (\boldsymbol{\mu}^\star_{\mathcal{S}-,\pi})\circ(\underline{\boldsymbol{U}}_{\mathrm{f},\pi} - \boldsymbol{U}_{\mathrm{f},\pi}), && (3.105f)\\
\mathbb{0} &= (\boldsymbol{\mu}^\star_{\mathcal{S}+,\pi})\circ(\boldsymbol{U}_{\mathrm{f},\pi} - \overline{\boldsymbol{U}}_{\mathrm{f},\pi}), && (3.105g)\\
\mathbb{0} &= (\boldsymbol{\mu}^\star_{\mathcal{I}-,\pi})\circ(\underline{\boldsymbol{U}}_{\mathcal{I},\pi} - \boldsymbol{U}_{\mathcal{I},\pi}), && (3.105h)\\
\mathbb{0} &= (\boldsymbol{\mu}^\star_{\mathcal{I}+,\pi})\circ(\boldsymbol{U}_{\mathcal{I},\pi} - \overline{\boldsymbol{U}}_{\mathcal{I},\pi}), && (3.105i)\\
\mathbb{0} &\le \boldsymbol{\mu}^\star_{\mathcal{S}-,\pi}, \boldsymbol{\mu}^\star_{\mathcal{S}+,\pi}, \boldsymbol{\mu}^\star_{\mathcal{I}-,\pi}, \boldsymbol{\mu}^\star_{\mathcal{I}+,\pi}, \boldsymbol{\mu}^\star_{\mathrm{g}-,\pi}, \boldsymbol{\mu}^\star_{\mathrm{g}+,\pi}, && (3.105j)
\end{aligned}
$$

where $\nabla \mathrm{C}_\pi(\boldsymbol{p}_{\mathrm{g},\pi}) = \mathrm{col}_{i\in\mathcal{V}_{\mathcal{W},\pi}}\{\nabla \mathrm{C}_{\mathrm{g},i}(p_{\mathrm{g},i})\}$. Since no $\mathcal{S}'$ node is adjacent to any $\mathcal{I}$ node (cf. Subsection 3.2.5), it holds that $\nabla\underline{\boldsymbol{\Psi}}_\pi(\boldsymbol{U}_{\mathcal{I},\pi}) = \nabla\overline{\boldsymbol{\Psi}}_\pi(\boldsymbol{U}_{\mathcal{I},\pi}) = \mathbb{0}$, thus application of the projected saddle-point flow (3.100), (3.103) yields the controller equations

$$
\begin{aligned}
\boldsymbol{\tau}_{\mathrm{g},\pi}\dot{\boldsymbol{p}}_{\mathrm{g},\pi} &= -\nabla \mathrm{C}_\pi(\boldsymbol{p}^\star_{\mathrm{g},\pi}) + \boldsymbol{\lambda}_\pi - \boldsymbol{\omega}_\pi + \boldsymbol{\mu}_{\mathrm{g}-,\pi} - \boldsymbol{\mu}_{\mathrm{g}+,\pi}, && (3.106a)\\
\boldsymbol{\tau}_{U_{\mathrm{f},\pi}}\dot{\boldsymbol{U}}_{\mathrm{f},\pi} &= \boldsymbol{\mu}_{\mathcal{S}-,\pi} - \boldsymbol{\mu}_{\mathcal{S}+,\pi}, && (3.106b)\\
\boldsymbol{\tau}_{U_{\mathcal{I},\pi}}\dot{\boldsymbol{U}}_{\mathcal{I},\pi} &= \boldsymbol{\mu}_{\mathcal{I}-,\pi} - \boldsymbol{\mu}_{\mathcal{I}+,\pi}, && (3.106c)\\
\boldsymbol{\tau}_{\mu_{\mathrm{g}-,\pi}}\dot{\boldsymbol{\mu}}_{\mathrm{g}-,\pi} &= \langle\underline{\boldsymbol{p}}_{\mathrm{g},\pi} - \boldsymbol{p}_{\mathrm{g},\pi}\rangle^+_{\mu_{\mathrm{g}-,\pi}}, && (3.106d)\\
\boldsymbol{\tau}_{\mu_{\mathrm{g}+,\pi}}\dot{\boldsymbol{\mu}}_{\mathrm{g}+,\pi} &= \langle\boldsymbol{p}_{\mathrm{g},\pi} - \overline{\boldsymbol{p}}_{\mathrm{g},\pi}\rangle^+_{\mu_{\mathrm{g}+,\pi}}, && (3.106e)\\
\boldsymbol{\tau}_{\mu_{\mathcal{S}-,\pi}}\dot{\boldsymbol{\mu}}_{\mathcal{S}-,\pi} &= \langle\underline{\boldsymbol{U}}_{\mathrm{f},\pi} - \boldsymbol{U}_{\mathrm{f},\pi}\rangle^+_{\mu_{\mathcal{S}-,\pi}}, && (3.106f)\\
\boldsymbol{\tau}_{\mu_{\mathcal{S}+,\pi}}\dot{\boldsymbol{\mu}}_{\mathcal{S}+,\pi} &= \langle\boldsymbol{U}_{\mathrm{f},\pi} - \overline{\boldsymbol{U}}_{\mathrm{f},\pi}\rangle^+_{\mu_{\mathcal{S}+,\pi}}, && (3.106g)\\
\boldsymbol{\tau}_{\mu_{\mathcal{I}-,\pi}}\dot{\boldsymbol{\mu}}_{\mathcal{I}-,\pi} &= \langle\underline{\boldsymbol{U}}_{\mathcal{I},\pi} - \boldsymbol{U}_{\mathcal{I},\pi}\rangle^+_{\mu_{\mathcal{I}-,\pi}}, && (3.106h)\\
\boldsymbol{\tau}_{\mu_{\mathcal{I}+,\pi}}\dot{\boldsymbol{\mu}}_{\mathcal{I}+,\pi} &= \langle\boldsymbol{U}_{\mathcal{I},\pi} - \overline{\boldsymbol{U}}_{\mathcal{I},\pi}\rangle^+_{\mu_{\mathcal{I}+,\pi}} && (3.106i)
\end{aligned}
$$

with suitable diagonal matrices $\boldsymbol{\tau}_{(\cdot)} \succ 0$. The individual strictly positive diagonal elements in $\boldsymbol{\tau}_{(\cdot)}$ are design parameters to be chosen by the individual WP. As outlined in Remark 3.28, premultiplication of the basic controller equations (3.100), (3.103) with an appropriate metric is helpful to adjust the convergence speed towards the equilibrium. The diagonal structure of $\boldsymbol{\tau}_{(\cdot)}$ helps maintain a specific transient behavior for each primal or dual variable in (3.106) separately. In particular, the speed of the controller (along with possible overshoots) increases as $\|\boldsymbol{\tau}\| \to 0$. With regard to the practical application, each individual diagonal entry in $\boldsymbol{\tau}$ must thus be sufficiently high to ensure that the respective actuator is capable of performing the

46 cf. Definition A.21 and Lemma A.22 in Appendix A.5.

required work[47]. At the same time, it is reasonable to assume that no WP has an incentive to set $\boldsymbol{\tau}_{(\cdot)}$ too high, since this would lead to an unnecessarily slow attainment of the own profit maximum. Besides these practical considerations, which are further discussed in Chapter 4, it is obvious that the specific choice of $\boldsymbol{\tau}_{(\cdot)}$ does not alter the equilibrium of (3.106), cf. Remark 3.28.

The CCs' objective $\mathsf{P}_k(\Lambda_k^\star) = 0$ from (3.90) can be accessed by means of the gradient flow $\dot{\Lambda}_k = -\nabla \mathsf{P}_k(\Lambda_k)$, yielding

$$\dot{\Lambda}_k = -\hat{\Phi}_k - \sum_{i\in\mathcal{V}_{\mathcal{Z},k}} p_{\ell,i} + \sum_{\substack{i\in\mathcal{V}_{\mathcal{Z},k}\\ i\notin\mathcal{V}_{\mathcal{L}}}} p_{\mathrm{g},i}. \tag{3.107}$$

The dynamics of the zonal price Λ_k arise from the momentary active power balance of cell k through aggregation of all net power flows from cell k to neighboring cells. Illustratively, the price for power injection in cell k increases if the cell's momentary production is lower than its own demand caused by RPs and transmission losses, and vice versa. Yet, instead of aggregating all required information at a central point, (3.107) can be set up in a fully distributed manner by defining a *consensus protocol* via a cell-specific, weakly connected communication graph $\mathcal{G}_{\mathrm{c},k} = (\mathcal{V}_{\mathcal{Z},k}, \mathcal{E}_{\mathrm{c},k})$[48]. In particular, let $\boldsymbol{D}_{\mathrm{c},k}$ denote the incidence matrix of $\mathcal{G}_{\mathrm{c},k}$. Then with Lemma B.1 from Appendix B.2, (3.107) can be implemented in a distributed fashion by

$$\boldsymbol{\tau}_{\lambda,k}\dot{\boldsymbol{\lambda}}_k = -\boldsymbol{p}_{\mathrm{g},k} + \boldsymbol{\rho}_k + \boldsymbol{p}_{\ell,k} - \boldsymbol{D}_{\mathrm{c},k}\boldsymbol{\nu}_k, \tag{3.108a}$$

$$\boldsymbol{\tau}_{\nu,k}\dot{\boldsymbol{\nu}}_k = (\boldsymbol{D}_{\mathrm{c},k})^{\mathsf{T}}\boldsymbol{\lambda}_k, \tag{3.108b}$$

where $\boldsymbol{\lambda}_k = \mathrm{col}_{i\in\mathcal{V}_{\mathcal{Z},k}}\{\lambda_i\}$ and $\boldsymbol{\rho}_k = \mathrm{col}_{i\in\mathcal{V}_{\mathcal{Z},k}}\{\rho_i\}$ with

$$\rho_i = G_{ii}U_i^2 + \sum_{j\in\mathcal{N}_i} G_{ij}U_iU_j\cos(\vartheta_{ij}). \tag{3.109}$$

As above, $\boldsymbol{\tau}_{\lambda,k}$ and $\boldsymbol{\tau}_{\nu,k}$ are real diagonal matrices of appropriate sizes with positive diagonal elements containing design parameters of the corresponding local controller. $\boldsymbol{\tau}_{\lambda,k}$ allows to adjust how fast-responding momentary power imbalances reflect in the prices, while the convergence speed of nodal prices $\boldsymbol{\lambda}_k$ towards a single cell-specific price Λ_k can be adjusted via $\boldsymbol{\tau}_{\nu,k}$.

3.4.3 Analysis of the Closed-Loop System

By virtue of using projected saddle-point flow, the initial state of the closed-loop system is not required to be located within the feasible set. However, in order to be able to attain an equilibrium, the following feasibility assumption must be made for the collection of optimization problems of WPs:

[47] Within the controller equations (3.106), the controller parameters $\boldsymbol{\tau}_{\mathrm{g},\pi}$ for active power injection are set to a comparatively high (i.e. "slow") value, whereby further differentiation of $\boldsymbol{\tau}_{\mathrm{g},\pi}$ in (3.106a) depends of the type of governing systems: While steam turbines of bulk CPPs require long ramp-up times and thus a very large $\tau_{\mathrm{g},(\cdot)}$, the respective parameters for inverter-interfaced DERs may be set to rather moderate values. By contrast, the parameters $\boldsymbol{\tau}_{\boldsymbol{\mu}_{(\cdot)},\pi}$ for the gradient ascent of dual variables can always be adjusted to a very fast value, as the dual variables correspond to "virtual" quantities that do not have to be generated physically.

[48] see Section B.2 in the appendix for some basic derivations.

Assumption 3.32. *The upper and lower bounds in* (3.83b)–(3.83c) *are chosen sufficiently loose such that there always exists a feasible solution that satisfies the constraints for all WPs* $\pi \in \mathcal{W}$.

Let $\boldsymbol{d}^\star = \operatorname{col}\{\boldsymbol{q}_\ell^\star, \boldsymbol{p}_\ell^\star\}$ be a given, constant disturbance input vector. We assume that the state vector $\boldsymbol{x}_\mathrm{p}$ of the physical plant system (3.49) fulfills the following regularity condition:

Assumption 3.33 ([TBD16, SDv17b]). *The Hessian of* $H_\mathrm{p}(\boldsymbol{x}_\mathrm{p})$ *is positive definite at steady state* $\boldsymbol{x}_\mathrm{p}^\star$.

Assumption 3.33, which is commonly known in energy-related literature as *existence of high-voltage solution* [DMSP17], can be ensured by satisfying e.g. the (relatively mild) condition presented in [SDv17a, Proposition 9] and originally derived in [DM16, Proposition 1]. The condition is satisfied if all voltage angle differences ϑ_{ij} are small and if all generator reactances $X_{\mathrm{d},i} - X'_{\mathrm{d},i}$ are small compared to the shunt susceptances b_i.

The simultaneous execution of the optimization schemes (3.106) for WPs $\pi \in \mathcal{W}$ and (3.108) for CCs $k \in \mathcal{Z}$ leads to a superposition of the derived control equations (3.106) and (3.108). Consequently, the resulting closed-loop system consists of the physical system (3.39) along with the controller equations

$$
\begin{aligned}
\boldsymbol{\tau}_\mathrm{g}\dot{\boldsymbol{p}}_\mathrm{g} &= -\nabla \mathsf{C}(\boldsymbol{p}_\mathrm{g}) + \boldsymbol{\lambda} - \boldsymbol{\omega}_\mathbf{g} + \boldsymbol{\mu}_{\mathrm{g}-} - \boldsymbol{\mu}_{\mathrm{g}+}, &\text{(3.110a)}\\
\boldsymbol{\tau}_{U_\mathrm{f}}\dot{\boldsymbol{U}}_\mathrm{f} &= \boldsymbol{\mu}_{\mathcal{S}-} - \boldsymbol{\mu}_{\mathcal{S}+}, &\text{(3.110b)}\\
\boldsymbol{\tau}_{U_\mathcal{I}}\dot{\boldsymbol{U}}_\mathcal{I} &= \boldsymbol{\mu}_{\mathcal{I}-} - \boldsymbol{\mu}_{\mathcal{I}+}, &\text{(3.110c)}\\
\boldsymbol{\tau}_{\mu_{\mathrm{g}-}}\dot{\boldsymbol{\mu}}_{\mathrm{g}-} &= \langle \underline{\boldsymbol{p}}_\mathrm{g} - \boldsymbol{p}_\mathrm{g}\rangle^+_{\boldsymbol{\mu}_{\mathrm{g}-}}, &\text{(3.110d)}\\
\boldsymbol{\tau}_{\mu_{\mathrm{g}+}}\dot{\boldsymbol{\mu}}_{\mathrm{g}+} &= \langle \boldsymbol{p}_\mathrm{g} - \overline{\boldsymbol{p}}_\mathrm{g}\rangle^+_{\boldsymbol{\mu}_{\mathrm{g}+}}, &\text{(3.110e)}\\
\boldsymbol{\tau}_{\mu_{\mathcal{S}-}}\dot{\boldsymbol{\mu}}_{\mathcal{S}-} &= \langle \underline{\boldsymbol{U}}_\mathrm{f} - \boldsymbol{U}_\mathrm{f}\rangle^+_{\boldsymbol{\mu}_{\mathcal{S}-}}, &\text{(3.110f)}\\
\boldsymbol{\tau}_{\mu_{\mathcal{S}+}}\dot{\boldsymbol{\mu}}_{\mathcal{S}+} &= \langle \boldsymbol{U}_\mathrm{f} - \overline{\boldsymbol{U}}_\mathrm{f}\rangle^+_{\boldsymbol{\mu}_{\mathcal{S}+}}, &\text{(3.110g)}\\
\boldsymbol{\tau}_{\mu_{\mathcal{I}-}}\dot{\boldsymbol{\mu}}_{\mathcal{I}-} &= \langle \underline{\boldsymbol{U}}_\mathcal{I} - \boldsymbol{U}_\mathcal{I}\rangle^+_{\boldsymbol{\mu}_{\mathcal{I}-}}, &\text{(3.110h)}\\
\boldsymbol{\tau}_{\mu_{\mathcal{I}+}}\dot{\boldsymbol{\mu}}_{\mathcal{I}+} &= \langle \boldsymbol{U}_\mathcal{I} - \overline{\boldsymbol{U}}_\mathcal{I}\rangle^+_{\boldsymbol{\mu}_{\mathcal{I}+}}, &\text{(3.110i)}\\
\boldsymbol{\tau}_\lambda\dot{\boldsymbol{\lambda}} &= -\boldsymbol{p}_\mathrm{g} + \boldsymbol{\rho} + \boldsymbol{p}_\ell - \boldsymbol{D}_\mathrm{c}\boldsymbol{\nu}, &\text{(3.110j)}\\
\boldsymbol{\tau}_\nu\dot{\boldsymbol{\nu}} &= \boldsymbol{D}_\mathrm{c}^\mathsf{T}\boldsymbol{\lambda}, &\text{(3.110k)}
\end{aligned}
$$

where the parameters and variables in (3.110) are stacked vectors or diagonal matrices of appropriate sizes for all $\pi \in \mathcal{W}$ and $k \in \mathcal{Z}$, e.g. $\boldsymbol{\tau}_\mathrm{g} = \operatorname{diag}_{\pi\in\mathcal{W}}\{\boldsymbol{\tau}_{\mathrm{g},\pi}\}$, $\boldsymbol{D}_\mathrm{c} = \operatorname{col}_{k\in\mathcal{Z}}\{\boldsymbol{D}_{\mathrm{c},k}\}$, and $\mathsf{C}(\boldsymbol{p}_\mathrm{g})$ equals the sum of all cost functions $\mathsf{C}_\pi(\boldsymbol{p}_{\mathrm{g},\pi})$ in (3.79).

Remark 3.34. *Note that* $\boldsymbol{D}_\mathrm{c}$ *has a block diagonal structure. Thus, the corresponding communication graph* $\mathcal{G}_\mathrm{c}$ *is disconnected and contains* $n_\mathcal{Z}$ *connected subgraphs.* ◊

Analysis of the Closed-Loop Equilibrium: Optimality and Zero-Frequency Deviation

At first, we point out a key result concerning the zero-frequency deviation of the closed-loop system (3.39), (3.110):

> **Lemma 3.35 (Zero Frequency Deviation).** *At each equilibrium of* (3.39), (3.110), *it holds that* $\omega_i^\star = 0$ *for all* $i \in \mathcal{V}$, *i.e. each node is operating at the nominal frequency.*

Proof. Eq. (3.39a) can be written in vector-matrix notation as $\dot{\boldsymbol{\vartheta}} = \boldsymbol{D}_p^\top \boldsymbol{\omega}$, where $\boldsymbol{D}_\mathrm{p}$ denotes the incidence matrix of the physical network $\mathscr{G}_\mathrm{p}$. Since $\mathscr{G}_\mathrm{p}$ is a weakly connected graph, each equilibrium $\boldsymbol{\omega}^\star$ with $\mathbb{0} = \boldsymbol{D}_\mathrm{p}^\top \boldsymbol{\omega}^\star$ fulfills $\boldsymbol{\omega}^\star = \omega^\star \cdot \mathbb{1}$, i.e. the nodal frequencies are synchronized at steady state. Since $\mathbb{1}^\top \boldsymbol{\rho} = \Phi$ and $\mathbb{1}^\top \boldsymbol{D}_\mathrm{c} = \mathbb{0}^\top$, left-multiplying (3.110j) by $\mathbb{1}^\top$ and inserting the equilibrium values implies

$$0 = -\sum_{i \in \mathcal{V}_\mathrm{g}} p_{\mathrm{g},i}^\star + \sum_{i \in \mathcal{V}} p_{\ell,i} + \Phi. \tag{3.111}$$

Due to the fact that $\Phi = \sum_{i \in \mathcal{V}} p_i$, a summation of all equations in (3.39b) and (3.39d) and insertion of the equilibrium values leads to

$$0 = -\sum_{i \in \mathcal{V}_\mathrm{g}} A_i \omega_i^\star + \sum_{i \in \mathcal{V}_\mathrm{g}} p_{\mathrm{g},i}^\star - \sum_{i \in \mathcal{V}} p_{\ell,i} - \Phi. \tag{3.112}$$

Comparison between (3.111) and (3.112) yields

$$0 = -\sum_{i \in \mathcal{V}_\mathrm{g}} A_i \omega_i^\star = -\omega^\star \cdot \sum_{i \in \mathcal{V}_\mathrm{g}} A_i. \tag{3.113}$$

Finally, since $A_i > 0$ holds by definition, it follows that $\omega^\star$ is zero. □

Remark 3.36. *As the control equations of WPs are absent in the proof of Lemma 3.35, zero frequency deviation of the closed-loop system can still be maintained in the presence of* clock drifts *[Sch15, Remark 5.2.11] of the frequency measurement units of WPs (where the real nodal frequency* $\omega_i(t)$ *in* (3.79) *is pertubed by an additive disturbance* $+\Delta\omega_i$*).* ◇

The next lemma states that in each equilibrium of (3.110) marginal costs are equal at each node in a cell whenever the power injection constraints (3.80) are not binding.

> **Lemma 3.37 (Equal Marginal Costs).** *Let* $\boldsymbol{x}^\star$ *denote an equilibrium of* (3.110) *and let* $k \in \mathcal{Z}$. *Then, at any node* $i, j \in \mathcal{V}_{\mathrm{g},k}$ *with* $p_{\mathrm{g},i}^\star \in (\underline{p}_{\mathrm{g},i}, \overline{p}_{\mathrm{g},i})$ *and* $p_{\mathrm{g},j}^\star \in (\underline{p}_{\mathrm{g},j}, \overline{p}_{\mathrm{g},j})$, *the marginal costs are equal, i.e.* $\nabla \mathrm{c}_{\mathrm{g},i}(p_{\mathrm{g},i}^\star) = \nabla \mathrm{c}_{\mathrm{g},j}(p_{\mathrm{g},j}^\star)$.

Proof. With (3.110k), it holds that $\mathbb{0} = \boldsymbol{D}_{\mathrm{c}}^{\mathsf{T}} \boldsymbol{\lambda}^{\star}$ at steady state. Since $\boldsymbol{D}_{\mathrm{c}} = \operatorname{diag}\{\boldsymbol{D}_{\mathrm{c},k}\}$ and since each $\boldsymbol{D}_{\mathrm{c},k}$ is the incidence matrix of a weakly connected graph, this implies that each row of $\boldsymbol{\lambda}_k^{\star}$ has the same value, i.e. $\boldsymbol{\lambda}_k^{\star} =: \Lambda_k \cdot \mathbb{1}$. Moreover, from (3.110d) and (3.110e), we get

$$\langle \underline{p}_{\mathrm{g},i} - p_{\mathrm{g},i}^{\star} \rangle_{\mu_{\mathrm{g}-,i}}^{+} = \langle p_{\mathrm{g},i}^{\star} - \overline{p}_{\mathrm{g},i} \rangle_{\mu_{\mathrm{g}+,i}}^{+} = 0. \tag{3.114}$$

If $p_{\mathrm{g},i}^{\star} \in (\underline{p}_{\mathrm{g},i}, \overline{p}_{\mathrm{g},i})$, then both brackets in (3.114) contain negative arguments, which implies that $\mu_{\mathrm{g}-,i} = \mu_{\mathrm{g}+,i} = 0$ (cf (A.1)). Now, with $\boldsymbol{\omega}^{\star} = \mathbb{0}$ from Lemma 3.35, it follows via (3.110a) that $\nabla \mathrm{C}_{\mathrm{g},i}(p_{\mathrm{g},i}^{\star}) = \lambda_i^{\star} = \Lambda_k$. Hence, all marginal costs are equal to the cell-specific value Λ_k. □

A trivial implication of Lemma 3.37 is that the special case of *quadratic cost functions*

$$\mathrm{C}_{\mathrm{g},i}(p_{\mathrm{g},i}) = \frac{1}{2} \frac{p_{\mathrm{g},i}^2}{\varpi_i}, \qquad \varpi_i > 0, \quad i \in \mathcal{V}_{\mathrm{g},k}, \tag{3.115}$$

yields *active power sharing* in cell $k \in \mathcal{Z}$, which is defined as follows:

Definition 3.38 (Active Power Sharing [SOA+14, p. 2465]). *Let* $\boldsymbol{\varpi}_k = \operatorname{col}_{l \in \mathcal{V}_{\mathrm{g},k}}\{\varpi_{k,l}\} > \mathbb{0}$ *denote a vector of constant weighting factors. Then* active power sharing *applies in cell* $k \in \mathcal{Z}$ *if*

$$\frac{p_{\mathrm{g},i}^{\star}}{\varpi_{k,i}} = \frac{p_{\mathrm{g},j}^{\star}}{\varpi_{k,j}}, \qquad i, j \in \mathcal{V}_{\mathrm{g},k} \tag{3.116}$$

holds.

Shifted Passivity and Stability

Since the physical plant system is amenable to a port-Hamiltonian representation (3.49), it is convenient to evaluate the closed-loop stability based on passivity arguments. As discussed in [van17, Section 4.4], PHSs can be decomposed into an interconnection of smaller subsystems to simplify passivity analysis. For the sake of clarity, we thus conceptually segregate (3.110) into a *frequency controller* (3.110a), (3.110d)–(3.110e), (3.110j)–(3.110k) and a *voltage controller* (3.110b)–(3.110c), (3.110f)–(3.110i) and analyze each of these components separately.

Lemma 3.39. *The physical plant system* (3.49) *with output* $\boldsymbol{y}_{\mathrm{p}} = \boldsymbol{G}_{\mathrm{p}}^{\mathsf{T}} \boldsymbol{z}_{\mathrm{p}}$ *is (locally) shifted passive with respect to* $\boldsymbol{x}_{\mathrm{p}}^{\star}$, *if*

$$\left[\boldsymbol{z}_{\mathrm{p}} - \boldsymbol{z}_{\mathrm{p}}^{\star}\right]^{\mathsf{T}} \left[\boldsymbol{\mathcal{R}}(\boldsymbol{x}_{\mathrm{p}}) - \boldsymbol{\mathcal{R}}(\boldsymbol{x}_{\mathrm{p}}^{\star})\right] \geq 0 \tag{3.117}$$

with $\boldsymbol{\mathcal{R}}(\boldsymbol{x}_{\mathrm{p}}) = \boldsymbol{R}_{\mathrm{p}} \boldsymbol{z}_{\mathrm{p}} + \boldsymbol{r}_{\mathrm{p}}$ *holds.*

Proof. The physical plant system is (locally) shifted passive with respect to $\boldsymbol{x}_{\mathrm{p}}^{\star}$, if the shifted plant Hamiltonian [van17, p. 136] $\widetilde{H}_{\mathrm{p}}(\widetilde{\boldsymbol{x}}_{\mathrm{p}}) := H_{\mathrm{p}}(\boldsymbol{x}_{\mathrm{p}}) - (\widetilde{\boldsymbol{x}}_{\mathrm{p}})^{\mathsf{T}} \nabla H_{\mathrm{p}}(\boldsymbol{x}_{\mathrm{p}}^{\star}) - H_{\mathrm{p}}(\boldsymbol{x}_{\mathrm{p}}^{\star})$ with $\widetilde{\boldsymbol{x}}_{\mathrm{p}} = \boldsymbol{x}_{\mathrm{p}} - \boldsymbol{x}_{\mathrm{p}}^{\star}$ satisfies $\widetilde{H}_{\mathrm{p}}(\widetilde{\boldsymbol{x}}_{\mathrm{p}}) \succ 0$ and

$$\dot{\widetilde{H}}_{\mathrm{p}}(\widetilde{\boldsymbol{x}}_{\mathrm{p}}) \leq \widetilde{\boldsymbol{y}}_{\mathrm{p}}^{\mathsf{T}} \widetilde{\boldsymbol{u}}_{\mathrm{p}} \tag{3.118}$$

with $\boldsymbol{y}_{\mathrm{p}} = \boldsymbol{G}_{\mathrm{p}}^{\mathsf{T}} \boldsymbol{z}_{\mathrm{p}}$.

The positive definiteness condition is satisfied locally due to Assumption 3.33. To investigate (3.118), let the disturbance input $\boldsymbol{d}^{\star}$ be constant in time and define $\boldsymbol{\mathcal{R}}(\boldsymbol{x}_{\mathrm{p}}) = \boldsymbol{R}_{\mathrm{p}} \nabla H_{\mathrm{p}}(\boldsymbol{x}_{\mathrm{p}}) + \boldsymbol{r}_{\mathrm{p}}$. Then (3.49) can be written as

$$\boldsymbol{E}\dot{\boldsymbol{x}}_{\mathrm{p}} = \boldsymbol{J}_{\mathrm{p}} \nabla H_{\mathrm{p}}(\boldsymbol{x}_{\mathrm{p}}) - \boldsymbol{\mathcal{R}}(\boldsymbol{x}_{\mathrm{p}}) + \boldsymbol{G}_{\mathrm{p}} \boldsymbol{u}_{\mathrm{p}} + \boldsymbol{G}_{\mathrm{d}} \boldsymbol{d}^{\star} \tag{3.119}$$

with each equilibrium $\boldsymbol{x}_{\mathrm{p}}^{\star}$ fulfilling

$$\mathbb{0} = \boldsymbol{J}_{\mathrm{p}} \nabla H_{\mathrm{p}}(\boldsymbol{x}_{\mathrm{p}}^{\star}) - \boldsymbol{\mathcal{R}}(\boldsymbol{x}_{\mathrm{p}}^{\star}) + \boldsymbol{G}_{\mathrm{p}} \boldsymbol{u}_{\mathrm{p}}^{\star} + \boldsymbol{G}_{\mathrm{d}} \boldsymbol{d}^{\star} \tag{3.120}$$

for a constant input vector $\boldsymbol{u}_{\mathrm{p}}^{\star}$. Since the shifted Hamiltonian $\widetilde{H}_{\mathrm{p}}(\widetilde{\boldsymbol{x}}_{\mathrm{p}})$ is positive definite with minimum $\widetilde{H}_{\mathrm{p}}(\mathbb{0}) = 0$, the shifted plant dynamics, i.e. (3.119) minus (3.120), can be expressed in terms of $\widetilde{H}_{\mathrm{p}}(\widetilde{\boldsymbol{x}}_{\mathrm{p}})$ as follows:

$$\boldsymbol{E}\dot{\widetilde{\boldsymbol{x}}}_{\mathrm{p}} = \boldsymbol{J}_{\mathrm{p}} \nabla \widetilde{H}_{\mathrm{p}}(\boldsymbol{x}_{\mathrm{p}}) - \left[\boldsymbol{\mathcal{R}}(\boldsymbol{x}_{\mathrm{p}}) - \boldsymbol{\mathcal{R}}(\boldsymbol{x}_{\mathrm{p}}^{\star})\right] + \boldsymbol{G}_{\mathrm{p}} \widetilde{\boldsymbol{u}}_{\mathrm{p}}. \tag{3.121}$$

With $\dot{\widetilde{H}}_{\mathrm{p}}(\widetilde{\boldsymbol{x}}_{\mathrm{p}}) = \widetilde{\boldsymbol{z}}_{\mathrm{p}}^{\mathsf{T}} \boldsymbol{E} \dot{\widetilde{\boldsymbol{x}}}_{\mathrm{p}}$ and bearing in mind that $\boldsymbol{J}_{\mathrm{p}}$ is skew-symmetric, the passivity condition (3.118) is equivalent to (3.117). □

Corollary 3.40. *If the physical plant system* (3.49) *with output* $\boldsymbol{y}_{\mathrm{p}} = \boldsymbol{G}_{\mathrm{p}}^{\mathsf{T}} \boldsymbol{z}_{\mathrm{p}}$ *is lossless, then it is (locally) shifted passive with respect to* $\boldsymbol{x}_{\mathrm{p}}^{\star}$.

Proof. For lossless grids, it holds that $\boldsymbol{r}_{\mathrm{p}} = \mathbb{0}$. Hence, (3.117) is always fulfilled due to the fact that $\boldsymbol{R}_{\mathrm{p}} \succcurlyeq 0$. Accordingly, (local) shifted passivity of (3.49) follows from Lemma 3.39. □

Next, we examine the shifted passivity of the frequency controller (3.110a), (3.110d)–(3.110e), (3.110j)–(3.110k):

Lemma 3.41. *The frequency controller* (3.110a), (3.110d)–(3.110e), (3.110j)–(3.110k) *with input* $\boldsymbol{u}_{\mathrm{c}} = -\boldsymbol{\omega}_{\mathrm{g}}$ *and output* $\boldsymbol{y}_{u1} = \boldsymbol{p}_{\mathrm{g}}$ *is (locally) shifted passive, if*

$$(\boldsymbol{p}_{\mathrm{g}} - \boldsymbol{p}_{\mathrm{g}}^{\star})^{\mathsf{T}} (\nabla \mathrm{c}(\boldsymbol{p}_{\mathrm{g}}) - \nabla \mathrm{c}(\boldsymbol{p}_{\mathrm{g}}^{\star})) \geq (\boldsymbol{\lambda} - \boldsymbol{\lambda}^{\star})^{\mathsf{T}} (\boldsymbol{\rho} - \boldsymbol{\rho}^{\star}). \tag{3.122}$$

Proof. By defining the frequency controller state $\boldsymbol{\xi}_1 = \mathrm{col}\{\boldsymbol{\tau}_{\mathrm{g}}\boldsymbol{p}_{\mathrm{g}}, \boldsymbol{\tau}_{\mu_{\mathrm{g}-}}\boldsymbol{\mu}_{\mathrm{g}-}, \boldsymbol{\tau}_{\mu_{\mathrm{g}+}}\boldsymbol{\mu}_{\mathrm{g}+}, \boldsymbol{\tau}_{\lambda}\boldsymbol{\lambda}, \boldsymbol{\tau}_{\nu}\boldsymbol{\nu}\}$ and the frequency controller Hamiltonian $H_1(\boldsymbol{\xi}_1) = \frac{1}{2}\boldsymbol{\xi}_1^{\mathsf{T}}\boldsymbol{\tau}_c^{-1}\boldsymbol{\xi}_1$ with $\boldsymbol{\tau}_c = \mathrm{diag}\{\boldsymbol{\tau}_{\mathrm{g}}, \boldsymbol{\tau}_{\mu_{\mathrm{g}-}}, \boldsymbol{\tau}_{\mu_{\mathrm{g}+}}, \boldsymbol{\tau}_{\lambda}, \boldsymbol{\tau}_{\nu}\} \succ 0$, (3.110a), (3.110d)–(3.110e), (3.110j)–(3.110k) can be written in port-Hamiltonian form

$$\dot{\boldsymbol{\xi}}_1 = \boldsymbol{J}_1 \nabla H_1(\boldsymbol{\xi}_1) - \boldsymbol{r}_1 + \boldsymbol{G}_{u1}\boldsymbol{u}_{\mathrm{c}} + \boldsymbol{G}_{d1}\boldsymbol{p}_{\ell}, \tag{3.123}$$

$$\boldsymbol{y}_{u1} = \boldsymbol{G}_{u1}^{\mathsf{T}}\boldsymbol{\zeta}_1, \tag{3.124}$$

$$\boldsymbol{y}_{d1} = \boldsymbol{G}_{d1}^{\mathsf{T}}\boldsymbol{\zeta}_1, \tag{3.125}$$

where $\boldsymbol{\zeta}_1 = \nabla H_1(\boldsymbol{\xi}_1) = \mathrm{col}\{\boldsymbol{p}_{\mathrm{g}}, \boldsymbol{\mu}_{\mathrm{g}-}, \boldsymbol{\mu}_{\mathrm{g}+}, \boldsymbol{\lambda}, \boldsymbol{\nu}\}$ and

$$\boldsymbol{J}_1 = \begin{bmatrix} \boldsymbol{0} & \boldsymbol{0} & \boldsymbol{0} & \hat{\boldsymbol{I}}_g & \boldsymbol{0} \\ \boldsymbol{0} & \boldsymbol{0} & \boldsymbol{0} & \boldsymbol{0} & \boldsymbol{0} \\ \boldsymbol{0} & \boldsymbol{0} & \boldsymbol{0} & \boldsymbol{0} & \boldsymbol{0} \\ -\hat{\boldsymbol{I}}_g^{\mathsf{T}} & \boldsymbol{0} & \boldsymbol{0} & \boldsymbol{0} & \boldsymbol{D}_{\mathrm{c}} \\ \boldsymbol{0} & \boldsymbol{0} & \boldsymbol{0} & -\boldsymbol{D}_{\mathrm{c}}^{\mathsf{T}} & \boldsymbol{0} \end{bmatrix}, \qquad \boldsymbol{r}_1 = \begin{bmatrix} \nabla \mathsf{C}(\boldsymbol{p}_{\mathrm{g}}) \\ -\langle \underline{\boldsymbol{p}}_{\mathrm{g}} - \boldsymbol{p}_{\mathrm{g}} \rangle_{\boldsymbol{\mu}_{\mathrm{g}-}}^{+} \\ -\langle \boldsymbol{p}_{\mathrm{g}} - \overline{\boldsymbol{p}}_{\mathrm{g}} \rangle_{\boldsymbol{\mu}_{\mathrm{g}+}}^{+} \\ -\boldsymbol{\rho} \\ \boldsymbol{0} \end{bmatrix}, \tag{3.126}$$

$$\boldsymbol{G}_{u1} = \begin{bmatrix} \boldsymbol{I} \\ \boldsymbol{0} \\ \boldsymbol{0} \\ \boldsymbol{0} \\ \boldsymbol{0} \end{bmatrix}, \qquad \boldsymbol{G}_{d1} = \begin{bmatrix} \boldsymbol{0} \\ \boldsymbol{0} \\ \boldsymbol{0} \\ \boldsymbol{I} \\ \boldsymbol{0} \end{bmatrix}. \tag{3.127}$$

With the shifted controller Hamiltonian

$$\widetilde{H}_1(\widetilde{\boldsymbol{\xi}}_1) := H_1(\boldsymbol{\xi}_1) - (\widetilde{\boldsymbol{\xi}}_1)^{\mathsf{T}}\nabla H_1(\boldsymbol{\xi}_1^{\star}) - H_1(\boldsymbol{\xi}_1^{\star}), \tag{3.128}$$

the shifted controller co-state $\widetilde{\boldsymbol{\zeta}}_1$ equals

$$\widetilde{\boldsymbol{\zeta}}_1 = \nabla \widetilde{H}_1(\widetilde{\boldsymbol{\xi}}_1) = \nabla H_1(\boldsymbol{\xi}_1) - \nabla H_1(\boldsymbol{\xi}_1^{\star}) = \boldsymbol{\zeta}_1 - \boldsymbol{\zeta}_1^{\star}. \tag{3.129}$$

Since the disturbance input $\boldsymbol{p}_{\ell}$ is assumed to be constant, (3.123) can be expressed in shifted coordinates as

$$\dot{\widetilde{\boldsymbol{\xi}}}_1 = \boldsymbol{J}_1 \nabla \widetilde{H}_1(\widetilde{\boldsymbol{\xi}}_1) - [\boldsymbol{r}_1(\boldsymbol{\xi}_1) - \boldsymbol{r}_1(\boldsymbol{\xi}_1^{\star})] + \boldsymbol{G}_{u1}\widetilde{\boldsymbol{u}}_{c}, \tag{3.130}$$

$$\widetilde{\boldsymbol{y}}_{u1} = \boldsymbol{G}_{u1}^{\mathsf{T}}\widetilde{\boldsymbol{\zeta}}_1. \tag{3.131}$$

Due to strict convexity of $H_1(\boldsymbol{\xi}_1)$, positive definiteness $\widetilde{H}_1(\widetilde{\boldsymbol{\xi}}_1) \succ 0$ is always satisfied. Hence the frequency controller is (locally) shifted passive if

$$\dot{\widetilde{H}}_1(\widetilde{\boldsymbol{\xi}}_1) = (\nabla \widetilde{H}_1(\widetilde{\boldsymbol{\xi}}_1))^{\mathsf{T}}\dot{\widetilde{\boldsymbol{\xi}}}_1 = \widetilde{\boldsymbol{\zeta}}_1^{\mathsf{T}}[\boldsymbol{r}_1(\boldsymbol{\xi}_1) - \boldsymbol{r}_1(\boldsymbol{\xi}_1^{\star})] + \widetilde{\boldsymbol{\zeta}}_1^{\mathsf{T}}\boldsymbol{G}_{u1}\widetilde{\boldsymbol{u}}_c \leq \widetilde{\boldsymbol{y}}_{u1}^{\mathsf{T}}\widetilde{\boldsymbol{u}}_c. \tag{3.132}$$

By inserting (3.124), we get

$$\begin{aligned} 0 \overset{!}{\geq} \widetilde{\boldsymbol{\zeta}}_1^{\mathsf{T}}[\boldsymbol{r}_1(\boldsymbol{\xi}_1) - \boldsymbol{r}_1(\boldsymbol{\xi}_1^{\star})] &= -(\boldsymbol{p}_{\mathrm{g}} - \boldsymbol{p}_{\mathrm{g}}^{\star})^{\mathsf{T}}(\nabla \mathsf{C}(\boldsymbol{p}_{\mathrm{g}}) - \nabla \mathsf{C}(\boldsymbol{p}_{\mathrm{g}}^{\star})) \\ &+ \widetilde{\boldsymbol{p}}_{\mathrm{g}}^{\mathsf{T}}\widetilde{\boldsymbol{\mu}}_{\mathrm{g}-} - \widetilde{\boldsymbol{p}}_{\mathrm{g}}^{\mathsf{T}}\widetilde{\boldsymbol{\mu}}_{\mathrm{g}+} + \widetilde{\boldsymbol{\mu}}_{\mathrm{g}-}^{\mathsf{T}}\langle \underline{\boldsymbol{p}}_{\mathrm{g}} - \boldsymbol{p}_{\mathrm{g}} \rangle_{\boldsymbol{\mu}_{\mathrm{g}-}}^{+} + \widetilde{\boldsymbol{\mu}}_{\mathrm{g}+}^{\mathsf{T}}\langle \boldsymbol{p}_{\mathrm{g}} - \overline{\boldsymbol{p}}_{\mathrm{g}} \rangle_{\boldsymbol{\mu}_{\mathrm{g}+}}^{+} \\ &+ (\boldsymbol{\lambda} - \boldsymbol{\lambda}^{\star})^{\mathsf{T}}(\boldsymbol{\rho} - \boldsymbol{\rho}^{\star}). \end{aligned} \tag{3.133}$$

In [SDv15, Proposition 3], it is shown that $\widetilde{\boldsymbol{\mu}}^{\mathsf{T}}\langle \boldsymbol{g}\rangle^{+}_{\boldsymbol{\mu}} \leq \widetilde{\boldsymbol{\mu}}^{\mathsf{T}}\boldsymbol{g}$ and $\widetilde{\boldsymbol{\mu}}^{\mathsf{T}}\boldsymbol{g}^{\star} \leq \mathbb{0}$ hold for each convex function $\boldsymbol{g} = \mathrm{col}_i\{g_i\}$. Hence $\widetilde{\boldsymbol{\mu}}_{\mathrm{g}-}^{\mathsf{T}}\langle \underline{\boldsymbol{p}}_{\mathrm{g}} - \boldsymbol{p}_{\mathrm{g}}\rangle^{+}_{\boldsymbol{\mu}_{\mathrm{g}-}} \leq \widetilde{\boldsymbol{\mu}}_{\mathrm{g}-}^{\mathsf{T}}(\underline{\boldsymbol{p}}_{\mathrm{g}} - \boldsymbol{p}_{\mathrm{g}}) = \widetilde{\boldsymbol{\mu}}_{\mathrm{g}-}^{\mathsf{T}}(\underline{\boldsymbol{p}}_{\mathrm{g}} - \boldsymbol{p}_{\mathrm{g}}^{\star} - \widetilde{\boldsymbol{p}}_{\mathrm{g}}) \leq -\widetilde{\boldsymbol{\mu}}_{\mathrm{g}-}^{\mathsf{T}}\widetilde{\boldsymbol{p}}_{\mathrm{g}}$. With the same procedure it can be calculated that $\widetilde{\boldsymbol{\mu}}_{\mathrm{g}+}^{\mathsf{T}}\langle \boldsymbol{p}_{\mathrm{g}} - \overline{\boldsymbol{p}}_{\mathrm{g}}\rangle^{+}_{\boldsymbol{\mu}_{\mathrm{g}+}} \leq +\widetilde{\boldsymbol{\mu}}_{\mathrm{g}+}^{\mathsf{T}}\widetilde{\boldsymbol{p}}_{\mathrm{g}}$. Thus, the second line of (3.133) is less than or equal to zero, which leads to (3.122). □

Corollary 3.42. *If the plant system* (3.49) *is lossless, then the frequency controller with input* $\boldsymbol{u}_{\mathrm{c}} = -\boldsymbol{\omega}_{\mathrm{g}}$ *and output* $\boldsymbol{y}_{u1} = \boldsymbol{p}_{\mathrm{g}}$ *is (locally) shifted passive.*

Proof. For lossless networks, it holds that $\boldsymbol{\rho} = \mathbb{0}$, hence (3.122) in Lemma 3.41 is always fulfilled due to the fact that $\mathtt{C}(\boldsymbol{p}_{\mathrm{g}})$ is strictly convex by Assumption 3.12. □

Remark 3.43. *Since the interconnection between frequency controller and physical plant system*

$$\boldsymbol{u}_{c1} = -\boldsymbol{y}_{p1} = -\boldsymbol{G}_{p1}^{\mathsf{T}}\nabla H_{\mathrm{p}}(\boldsymbol{x}_{\mathrm{p}}) = -\boldsymbol{\omega}_{\mathrm{g}}, \tag{3.134}$$

$$\boldsymbol{u}_{p1} = \boldsymbol{y}_{c1} = \boldsymbol{G}_{u1}^{\mathsf{T}}\nabla H_1(\boldsymbol{\xi}_1) = \boldsymbol{p}_g \tag{3.135}$$

is power-preserving, i.e. $\boldsymbol{u}_{c1}^{\mathsf{T}}\boldsymbol{y}_{c1} + \boldsymbol{u}_{p1}^{\mathsf{T}}\boldsymbol{y}_{p1} = 0$, *shifted passivity of the subsystems in terms of Lemmas 3.39 and 3.41 implies shifted passivity of the closed-loop system* (3.49), (3.110a), (3.110d)–(3.110e), (3.110j)–(3.110k). *In fact, the conditions stated in Lemmas 3.39 and 3.41 are only sufficient and not necessary, as an excess of passivity in one subsystem can compensate for the lack of passivity in an other subsystem, cf. [van17, p. 23].* ◇

Based on the previous conditions we now formulate a stability criterion for the lossy, physical plant system along with frequency and voltage controller:

Theorem 3.44. *Assume that the conditions of Lemmas 3.39 and 3.41 hold. For a constant input* $\boldsymbol{d}^{\star}$, *let* $(\boldsymbol{x}_{\mathrm{p}}^{\star}, \boldsymbol{\xi}_1^{\star}, \boldsymbol{\xi}_2^{\star})$ *denote an equilibrium of* (3.49), (3.110). *Then there exists a neighborhood* $\mathcal{B}$ *around* $(\boldsymbol{x}_{\mathrm{p}}^{\star}, \boldsymbol{\xi}_1^{\star}, \boldsymbol{\xi}_2^{\star})$ *such that if* $(\boldsymbol{x}_{\mathrm{p}}, \boldsymbol{\xi}_1, \boldsymbol{\xi}_2) \in \mathcal{B}$, *then the state asymptotically converges to* $(\boldsymbol{x}_{\mathrm{p}}^{\star}, \boldsymbol{\xi}_1^{\star}, \boldsymbol{\xi}_2^{\star})$.

Proof. With the voltage controller state $\boldsymbol{\xi}_2 = \mathrm{col}\{\boldsymbol{\tau}_{\mu_{\mathcal{S}-}}\boldsymbol{\mu}_{\mathcal{S}-}, \boldsymbol{\tau}_{\mu_{\mathcal{S}+}}\boldsymbol{\mu}_{\mathcal{S}+}, \boldsymbol{\tau}_{U_{\mathrm{f}}}\boldsymbol{U}_{\mathrm{f}}, \boldsymbol{\tau}_{\mu_{\mathcal{I}-}}\boldsymbol{\mu}_{\mathcal{I}-}, \boldsymbol{\tau}_{\mu_{\mathcal{I}+}}\boldsymbol{\mu}_{\mathcal{I}+}, \boldsymbol{\tau}_{U_{\mathcal{I}}}\boldsymbol{U}_{\mathcal{I}}\}$, let $\boldsymbol{\xi}_2^{\star}$ denote an equilibrium of (3.110b)–(3.110c), (3.110f)–(3.110i) and define the Lyapunov function candidate $\widetilde{V}_2(\widetilde{\boldsymbol{\xi}}_2) = \frac{1}{2}\widetilde{\boldsymbol{\xi}}_2^{\mathsf{T}}\boldsymbol{\tau}_2^{-1}\widetilde{\boldsymbol{\xi}}_2^{\mathsf{T}}$, where $\boldsymbol{\tau}_2 = \mathrm{diag}\{\boldsymbol{\tau}_{\mu_{\mathcal{S}-}}, \boldsymbol{\tau}_{\mu_{\mathcal{S}+}}, \boldsymbol{\tau}_{U_{\mathrm{f}}}, \boldsymbol{\tau}_{\mu_{\mathcal{I}-}}, \boldsymbol{\tau}_{\mu_{\mathcal{I}+}}, \boldsymbol{\tau}_{U_{\mathcal{I}}}\} \succ 0$ and $\widetilde{\boldsymbol{\xi}}_2 = \boldsymbol{\xi}_2 - \boldsymbol{\xi}_2^{\star}$. This allows to write out an overall Lyapunov function candidate $\widetilde{V}(\widetilde{\boldsymbol{x}}_{\mathrm{p}}, \widetilde{\boldsymbol{\xi}}_1, \widetilde{\boldsymbol{\xi}}_2)$ for the overall closed-loop system as the sum of $\widetilde{H}_{\mathrm{p}}$, $\widetilde{H}_1$, and $\widetilde{V}_2$, i.e.

$$\widetilde{V}(\widetilde{\boldsymbol{x}}_{\mathrm{p}}, \widetilde{\boldsymbol{\xi}}_1, \widetilde{\boldsymbol{\xi}}_2) = \widetilde{H}_{\mathrm{p}}(\widetilde{\boldsymbol{x}}_{\mathrm{p}}) + \widetilde{H}_1(\widetilde{\boldsymbol{\xi}}_1) + \widetilde{V}_2(\widetilde{\boldsymbol{\xi}}_2) \succ 0. \tag{3.136}$$

With respect to $\dot{\widetilde{V}}$, we observe that

$$\dot{\widetilde{V}}(\widetilde{\boldsymbol{x}}_\mathrm{p},\widetilde{\boldsymbol{\xi}}_1,\widetilde{\boldsymbol{\xi}}_2) = \dot{\widetilde{H}}_\mathrm{p}(\widetilde{\boldsymbol{x}}_\mathrm{p}) + \dot{\widetilde{H}}_1(\widetilde{\boldsymbol{\xi}}_1) + \dot{\widetilde{V}}_2(\widetilde{\boldsymbol{\xi}}_2), \tag{3.137}$$

where the first summand of (3.137) equals

$$\begin{aligned}\dot{\widetilde{H}}_\mathrm{p}(\widetilde{\boldsymbol{x}}_\mathrm{p}) &= (\nabla\widetilde{H}_\mathrm{p}(\widetilde{\boldsymbol{x}}_\mathrm{p})^\mathsf{T}\boldsymbol{J}_\mathrm{p}(\nabla\widetilde{H}_\mathrm{p}(\widetilde{\boldsymbol{x}}_\mathrm{p})) - (\nabla\widetilde{H}_\mathrm{p}(\widetilde{\boldsymbol{x}}_\mathrm{p})^\mathsf{T}\left[\boldsymbol{\mathcal{R}}(\boldsymbol{x}_\mathrm{p}) - \boldsymbol{\mathcal{R}}(\boldsymbol{x}_\mathrm{p}^\star)\right] \\ &= -(\boldsymbol{z}_\mathrm{p} - \boldsymbol{z}_\mathrm{p}^\star)^\mathsf{T}\left[\boldsymbol{\mathcal{R}}(\boldsymbol{x}_\mathrm{p}) - \boldsymbol{\mathcal{R}}(\boldsymbol{x}_\mathrm{p}^\star)\right]\end{aligned} \tag{3.138}$$

due to the skew-symmetry of $\boldsymbol{J}_\mathrm{p}$. The second summand of (3.137) is

$$\dot{\widetilde{H}}_1(\widetilde{\boldsymbol{\xi}}_1) = -\widetilde{\boldsymbol{p}}_\mathrm{g}^\mathsf{T}(\nabla\mathtt{C}(\boldsymbol{p}_\mathrm{g}) - \nabla\mathtt{C}(\boldsymbol{p}_\mathrm{g}^\star)) + (\boldsymbol{\lambda}-\boldsymbol{\lambda}^\star)^\mathsf{T}(\boldsymbol{\rho}-\boldsymbol{\rho}^\star). \tag{3.139}$$

The third summand of (3.137) equals

$$\begin{aligned}\dot{\widetilde{V}}_2(\widetilde{\boldsymbol{\xi}}_2) = {}&\widetilde{\boldsymbol{\mu}}_{\mathcal{S}-}^\mathsf{T}\langle\underline{\boldsymbol{\Psi}} - \boldsymbol{U}_\mathrm{f}\rangle^+_{\boldsymbol{\mu}_{\mathcal{S}-}} + \widetilde{\boldsymbol{\mu}}_{\mathcal{S}+}^\mathsf{T}\langle\boldsymbol{U}_\mathrm{f} - \overline{\boldsymbol{\Psi}}\rangle^+_{\boldsymbol{\mu}_{\mathcal{S}+}} + \widetilde{\boldsymbol{U}}_\mathrm{f}^\mathsf{T}\widetilde{\boldsymbol{\mu}}_{\mathcal{S}-} - \widetilde{\boldsymbol{U}}_\mathrm{f}^\mathsf{T}\widetilde{\boldsymbol{\mu}}_{\mathcal{S}+} \\ &+ \widetilde{\boldsymbol{\mu}}_{\mathcal{I}-}^\mathsf{T}\langle\underline{\boldsymbol{U}}_\mathcal{I} - \boldsymbol{U}_\mathcal{I}\rangle^+_{\boldsymbol{\mu}_{\mathcal{I}-}} + \widetilde{\boldsymbol{\mu}}_{\mathcal{I}+}^\mathsf{T}\langle\boldsymbol{U}_\mathcal{I} - \overline{\boldsymbol{U}}_\mathcal{I}\rangle^+_{\boldsymbol{\mu}_{\mathcal{I}+}}.\end{aligned} \tag{3.140}$$

In [SDv15, Proposition 3], it was shown that $\widetilde{\boldsymbol{\mu}}^\mathsf{T}\langle\boldsymbol{g}\rangle^+_{\boldsymbol{\mu}} \leq \widetilde{\boldsymbol{\mu}}^\mathsf{T}\boldsymbol{g}$ and $\widetilde{\boldsymbol{\mu}}^\mathsf{T}\boldsymbol{g}^\star \leq \boldsymbol{0}$ hold for each convex function $\boldsymbol{g} = \mathrm{col}_i\{g_i\}$. Hence $\widetilde{\boldsymbol{\mu}}_{\mathcal{S}-}^\mathsf{T}\langle\underline{\boldsymbol{\Psi}} - \boldsymbol{U}_\mathrm{f}\rangle^+_{\boldsymbol{\mu}_{\mathcal{S}-}} \leq \widetilde{\boldsymbol{\mu}}_{\mathcal{S}-}^\mathsf{T}(\underline{\boldsymbol{\Psi}} - \boldsymbol{U}_\mathrm{f}) = \widetilde{\boldsymbol{\mu}}_{\mathcal{S}-}^\mathsf{T}(\underline{\boldsymbol{\Psi}} - \boldsymbol{U}_\mathrm{f}^\star - \widetilde{\boldsymbol{U}}_\mathrm{f}) \leq -\widetilde{\boldsymbol{\mu}}_{\mathcal{S}-}^\mathsf{T}\widetilde{\boldsymbol{U}}_\mathrm{f}$. With the same procedure it can be calculated for $\mathcal{S}+, \mathcal{I}-$, and $\mathcal{I}+$ that the first two rows of (3.140) are less than or equal to zero and thus $\dot{\widetilde{V}}_2(\widetilde{\boldsymbol{\xi}}_2) \leq 0$. With the assumption from Lemma 3.39, the first row of (3.138) is ≤ 0 and with the assumption from Lemma 3.41, it also holds that $\dot{\widetilde{H}}_1(\widetilde{\boldsymbol{\xi}}_1) \leq 0$ in (3.139). Hence $\dot{\widetilde{V}}(\widetilde{\boldsymbol{x}}_\mathrm{p},\widetilde{\boldsymbol{\xi}}_1,\widetilde{\boldsymbol{\xi}}_2) \leq 0$ in (3.137). Since $\widetilde{V} \succ 0$ and $\dot{\widetilde{V}} \leq 0$, the discontinuous closed-loop system (3.49), (3.110) fulfills the condition of [CMC16, Lemma 4.3]. According to [CMC16, Theorem 4.5], the limit set of each solution of (3.49), (3.110) starting in $\mathcal{B}$ is a singleton, hence $(\boldsymbol{x}_\mathrm{p}^\star, \boldsymbol{\xi}_1^\star, \boldsymbol{\xi}_2^\star)$ is (locally) asymptotically stable [Kha02, Corollary 4.1]. □

3.4.4 Summary and Illustrative Example

In this section, we have introduced an optimization-based control scheme for WPs and CCs via projected saddle-point flow ensuring frequency and voltage regulation. While the latter can be ensured directly via local constraints (3.84b) and (3.84c), frequency regulation is achieved indirectly via the real-time price $\boldsymbol{\lambda}$ by incorporating $\boldsymbol{\omega}$ in the profit functions of each WP[49]. The derived price-based control scheme provides a rigorous compliance with the unbundling paradigm: WPs as competitive players require only the nodal prices λ_i transmitted by the local CC individually for each node. Apart from this communicated quantity, the decision-making of WPs is fully decentralized. Namely, there is no need for the WPs to reveal their own cost

49 In contrast to this *native* implementation of the frequency control task, the optimization-based frequency control schemes presented e.g. in [TBD16, SDv17a, ZMLB18] artificially impose a frequency-dependent term as extra control input in addition to a saddle-point-based controller.

function C_π to competitors or CCs, where dishonesty by the WPs would be likely anyway (see e.g. [TJ13, p. 562]). In addition, a distributed realization can be adopted for the control actions of CCs as an outcome of the reformulation (3.108). The resulting local controllers then only need to have knowledge about the local values $p_{\mathrm{g},i} - p_{\ell,i}$ and $\vec{U}_i$, voltages $\vec{U}_j$ of direct neighbors $j \in \mathcal{N}_i$, as well as line parameters and communication variables ν_{ij} of adjacent lines. Moreover, since no WP $\pi \in \mathcal{W}$ can unilaterally increase its profit[50] P_π whenever at steady-state $\boldsymbol{p}_{\mathrm{g},\pi} = \boldsymbol{p}^\star_{\mathrm{g},\pi}$, each $\boldsymbol{p}^\star_{\mathrm{g}}$ represents a Nash equilibrium of the multi-player game (3.83), cf. [BLMH$^+$21].

Due to the port-Hamiltonian model of the physical plant system (3.49) and frequency controller (3.110a), (3.110d)–(3.110e), (3.110j)–(3.110k) as well as the nonnegativity property [SDv15, Proposition 3] which is exploited in the proofs of Lemma 3.41 and Theorem 3.44, closed-loop stability can easily be assessed by passivity arguments. Even if the control scheme is certainly intended for lossy networks, a strict mathematical proof that provides an a priori statement on closed-loop stability is only available for lossless networks. This is due to the fact that the conditions (3.117) and (3.122) can only be evaluated online, yet are always fulfilled in the lossless case. This result is in line with the concern that there exists no analytical energy-like function for "generic" power grids with nonzero transmission losses, which has been addressed by many authors over the last eight decades [Mag47, Nar84, KBP85, ASB01, SALB05, VT15]. In fact, energy functions which enable a Lyapunov-based stability analysis are only available for very specific academic test systems [BA03, GSAB05, OGA$^+$05]. However, even for highly low-dimensional systems, there are counterexamples where all Lyapunov-like functions are not applicable [Nar84]. Yet an advantageous property of conditions (3.117) and (3.122) is their *modular* structure: If passivity could be verified for two smaller subsystems (3.49), their power-preserving interconnection[51] always leads to a passive overall system, which potentially gives rise to a *plug-and-play* based approach to stability analysis, see [APM16, SP19, SMKH20] for more in-depth discussions. We illustrate some of the theoretical insights of Section 3.4 by the following academic example:

Example 4

In this example, we investigate the frequency stability and active power sharing of the closed-loop system resulting from the exemplary physical plant system from Fig. 3.8 along with the controller (3.110)*. For the sake of brevity, we choose* $\boldsymbol{G} = (R/X) \cdot \boldsymbol{B}$ *with* $(R/X) \in \mathbb{N} \cup \{0\}$ *yet to be determined and set all power and voltage constraints equal to infinity. The units of all parameters are given in p.u. with* $S_{\mathrm{base}} = 1\,\mathrm{MVA}$ *and* $U_{\mathrm{base}} = 10\,\mathrm{kV}$*. The parameter values of the plant system can be found in Appendix B.4. The controller parameters in* (3.110) *are all set to* $\boldsymbol{\tau}_{\mathrm{c}} = 0.1 \cdot \boldsymbol{I}$ *and the cost functions are chosen to*

$$\mathrm{C}_{\mathrm{g},i}(p_{\mathrm{g},i}) = \frac{1}{2}\frac{1}{\varpi_i} \cdot p_{\mathrm{g},i}^2 \tag{3.141}$$

50 Figuratively, this is the case since at equilibrium, the profit margins of all players are zero. Indeed, if they deviate from their strategy $(\boldsymbol{p}^\star_{\mathrm{g},\pi}, \boldsymbol{U}^\star_{\mathrm{f},\pi}, \boldsymbol{U}^\star_{\mathcal{I},\pi})$, the new strategy would either be infeasible with respect to the constraints (3.84b)–(3.84d) or they would be worse off as the additional cost $\nabla\mathrm{C}_\pi(\boldsymbol{p}_{\mathrm{g},\pi})$ for power production exceeds the additional revenue $\nabla\mathrm{R}_\pi(\boldsymbol{p}_{\mathrm{g},\pi})$.

51 e.g. by coupling all boundary voltage phasors $\vec{\boldsymbol{U}}^{\mathrm{b}}_1, \vec{\boldsymbol{U}}^{\mathrm{b}}_2$ and power injections $\boldsymbol{p}^{\mathrm{b}}_{\ell,1}, \boldsymbol{p}^{\mathrm{b}}_{\ell,2}, \boldsymbol{q}^{\mathrm{b}}_{\ell,1}, \boldsymbol{q}^{\mathrm{b}}_{\ell,2}$ via $\vec{\boldsymbol{U}}^{\mathrm{b}}_1 = \vec{\boldsymbol{U}}^{\mathrm{b}}_2$, $\boldsymbol{p}^{\mathrm{b}}_{\ell,1} = -\boldsymbol{p}^{\mathrm{b}}_{\ell,2}$, and $\boldsymbol{q}^{\mathrm{b}}_{\ell,1} = -\boldsymbol{q}^{\mathrm{b}}_{\ell,2}$, respectively.

with weighting factors $\varpi_1 = 1$, $\varpi_2 = 1.1$, $\varpi_3 = 1.2$, $\varpi_4 = 1.3$, *and* $\varpi_5 = 1.4$. *The network is initially in steady state with nominal frequency* $\Omega_{n,i} = 50\,\text{Hz}$ *and constant loads* $p_{\ell,i}$, $q_{\ell,i}$. *At* $t = 10\,\text{s}$ *and* $t = 40\,\text{s}$, *a step load change of* $\Delta p_{\ell,i} = +0.1p.u.$ *occurs at load nodes* 6 *and* 7, *respectively.*

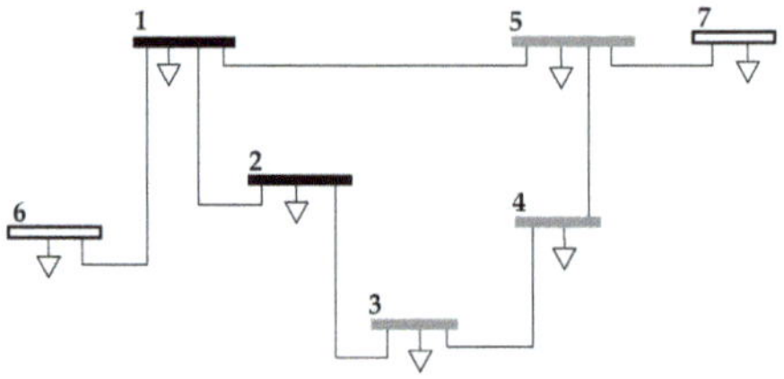

Figure 3.8: Stylized representation of the exemplary physical plant system with $n_{\mathcal{S}} = 2$, $n_{\mathcal{I}} = 3$, and $n_{\mathcal{L}} = 2$. $\mathcal{S}$ nodes are depicted in black, $\mathcal{I}$ nodes are depicted in gray, and $\mathcal{L}$ nodes are depicted in white. Each bus is equipped with an uncontrollable load modeled by (uncontrollable) active and reactive power consumption.

Fig. 3.9 shows that for $R/X \in \{0, 1, 2\}$, *all nodal frequencies converge to the nominal frequency within* $10\,\text{s}$ *after the load jumps. For* $R/X = 3$, *however, there is no synchronization of frequencies.*

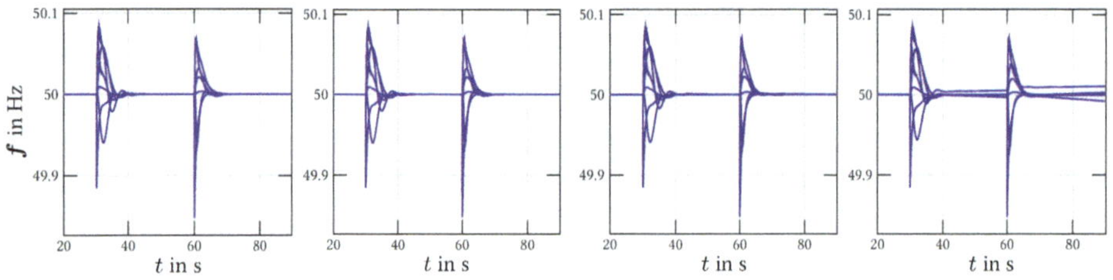

(a) $R/X = 0$ (b) $R/X = 1$ (c) $R/X = 2$ (d) $R/X = 3$

Figure 3.9: Nodal frequencies for Example 4.

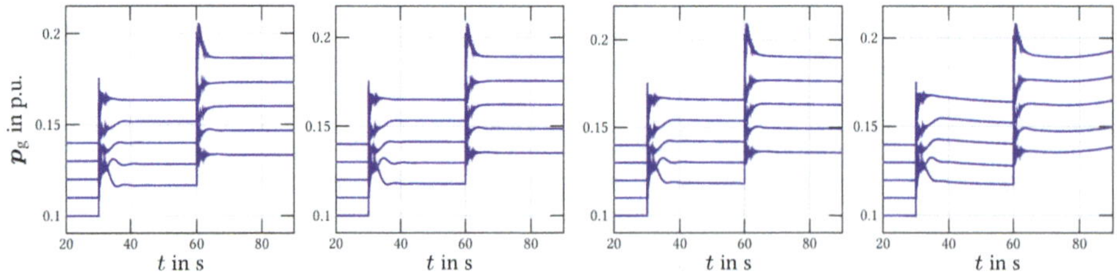

(a) $R/X = 0$ (b) $R/X = 1$ (c) $R/X = 2$ (d) $R/X = 3$

Figure 3.10: Active power injections for Example 4.

For $R/X \in \{0, 1, 2\}$, *the active power generations* $p_{g,i}$ *are equidistant for each post-fault equilibrium, i.e. active power sharing is maintained (see Fig. 3.10). As the* R/X *ratio increases,* *also the final values of $\boldsymbol{p}_g$ increase slightly. This can be traced back to the fact that the WPs have to compensate the increasingly high transmission losses. For $R/X = 3$, as already seen for the frequencies, there is no convergence to a final value $\boldsymbol{p}_g^\star$.*

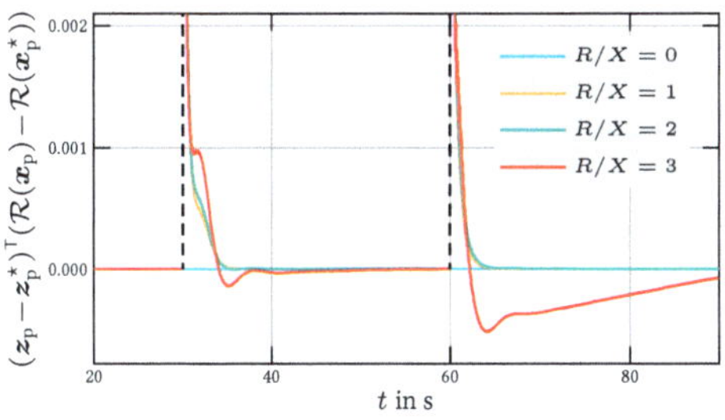

Figure 3.11: Dissipations in the system for ascending R/X ratios. The system becomes unstable for $R/X = 3$.

As apparent in Fig. 3.11, a negative (shifted) dissipation (3.138) *can be observed for $R/X = 3$, while* (3.138) *is nonnegative for $R/X \in \{0, 1, 2\}$. The curve of* (3.138) *along with Figs. 3.9 and 3.10, suggests that the momentary dissipation rate is a useful tool to qualitatively assess the stability margin of the closed-loop system.*
Now we reset R/X equal to 1 *and assume that the measurement of ω_1 at node 1 constantly deviates from the actual frequency by* $+0.1$ Hz *to simulate a clock drift.*

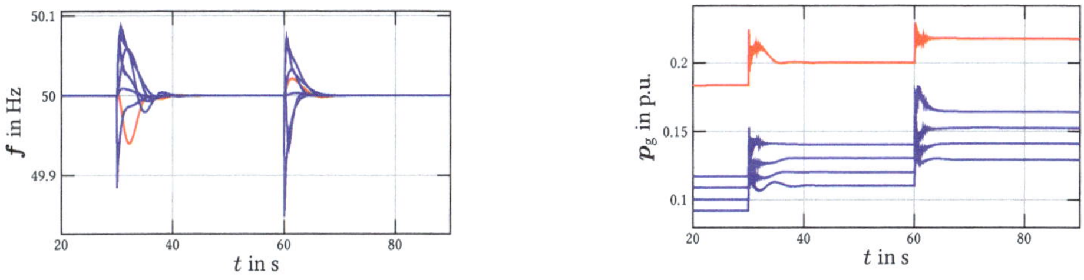

(a) Nodal frequencies f_i

(b) Active power injections $p_{g,i}$

Figure 3.12: Nodal Frequencies and active power injections in the presence of clock drifts. The red lines indicate the nodal frequency and active power injection at node 1.

As depicted in Fig. 3.12, simulations show that the controller is robust in terms of clock drifts and is able to restore the nominal frequency $\Omega_{n,i}$ (cf. Remark 3.36). However, active power sharing can no longer be achieved by node 1*, which is in line with the findings from [Sch15, Remark 5.2.11] that clock drifts affect the ratios* (3.116) *of active power sharing. This can easily be verified by the fact that Lemma 3.37 concerning equal marginal costs is violated in the presence of clock drifts. However, as argued in [Sch15], the introduced quantitative errors in power sharing are negligible for most practical applications.*

Remark 3.45. *The fact that variations of $U_{\mathcal{S},i}$ or $U_{\mathcal{I},i}$ typically come at no additional cost for the respective WP (cf. Remark 3.22) can be exploited by imposing additional responsibilities on* *the voltage controller* (3.110b)–(3.110c). *For instance, the augmentation*

$$\boldsymbol{\tau}_{U_\mathrm{f}} \dot{\boldsymbol{U}}_\mathrm{f} = -\boldsymbol{U}_\mathrm{f} + \boldsymbol{U}_\mathrm{f}^\mathrm{set} + \boldsymbol{\mu}_{\mathcal{S}-} - \boldsymbol{\mu}_{\mathcal{S}+}, \tag{3.142a}$$

$$\boldsymbol{\tau}_{U_\mathcal{I}} \dot{\boldsymbol{U}}_\mathcal{I} = -\boldsymbol{U}_\mathcal{I} + \boldsymbol{U}_\mathcal{I}^\mathrm{set} + \boldsymbol{\mu}_{\mathcal{I}-} - \boldsymbol{\mu}_{\mathcal{I}+} \tag{3.142b}$$

can be employed to follow a specific setpoint $\boldsymbol{U}_\mathrm{f}^\mathrm{set} = \mathrm{col}_{i\in\mathcal{V}'_\mathcal{S}}\{\Psi_i(U_i^\mathrm{set})\}$, $\boldsymbol{U}_\mathcal{I}^\mathrm{set} = \mathrm{col}_{i\in\mathcal{V}_\mathcal{I}}\{U_i^\mathrm{set}\}$. *Alternatively, to also achieve* reactive power sharing *[SFO15, p. 60]*

$$\frac{q_{\mathrm{g},i}^\star}{\chi_i} = \frac{q_{\mathrm{g},j}^\star}{\chi_j}, \qquad i,j \in \mathcal{V}_\mathrm{g} \tag{3.143}$$

with $q_{\mathrm{g},i} = q_{\mathrm{inj},i}$, $q_{\mathrm{g},j} = q_{\mathrm{inj},j}$ *and* $\chi_i, \chi_j > 0$, *the distributed consensus-based control scheme from [Sch15, Section 5.3] can be applied*[52] *if the network* $\mathscr{G}_\mathrm{p}$ *is mainly inductive* ($\boldsymbol{G} \approx \boldsymbol{0}$), *yielding*

$$\tau_{U_{\mathrm{f},i}} \dot{U}_{\mathrm{f},i} = -k_i \sum_{j\in\mathcal{N}_i} \left(\frac{q_{\mathrm{g},i}}{\chi_i} - \frac{q_{\mathrm{g},j}}{\chi_j}\right) + \mu_{\mathcal{S}-,i} - \mu_{\mathcal{S}+,i}, \qquad i \in \mathcal{V}'_\mathcal{S}, \quad k_i > 0, \tag{3.144a}$$

$$\tau_{U_{\mathcal{I},i}} \dot{U}_{\mathcal{I},i} = -k_i \sum_{j\in\mathcal{N}_i} \left(\frac{q_{\mathrm{g},i}}{\chi_i} - \frac{q_{\mathrm{g},j}}{\chi_j}\right) + \mu_{\mathcal{I}-,i} - \mu_{\mathcal{I}+,i}, \qquad i \in \mathcal{V}_\mathcal{I}, \quad k_i > 0. \tag{3.144b}$$

Note that even with one of these augmentations, steady-state compliance with the voltage amplitude constraints (3.84c)–(3.84d) *is still preserved due to the projected gradient flow of the dual variables* (3.110f)–(3.110i). ◇

3.5 Interconnection of Zonal Prices

Lemma 3.37 has revealed that at each equilibrium of (3.110), nodal prices $\lambda_i, \lambda_j \in \mathcal{V}_{\mathcal{Z},k}$ are equal to a common zonal price Λ_k for each cell $k \in \mathcal{Z}$. Incentives for increased or decreased power generation are thus imposed implicitly by means of differences in zonal prices. However, Remark 3.34 implies that there is no direct relationship between each individual Λ_k to be adjusted. Therefore, in the following Subsection 3.5.1, specific couplings between CCs are introduced that enable such relationships. Subsequently, Subsection 3.5.2 reveals the connection of the resulting closed-loop equilibrium with a (notional) *centralized* optimization problem. On this basis, Subsection 3.5.3, which contains the main result of Section 3.5, evaluates the economic efficiency of this closed-loop equilibrium as a first milestone towards Contribution 2.

3.5.1 Coupling of Zonal Prices

To provide interdependencies between the zonal prices, additional constraints of the form

$$\lambda_i \stackrel{!}{=} \eta_{ij} \cdot \lambda_j, \tag{3.145}$$

[52] For the sake of simplicity, we neglect the filter elements for measured active and reactive power injections $p_{\mathrm{g},i}$, $q_{\mathrm{g},i}$ originally included in the controller equations of [Sch15, Section 5.3].

with an appropriate multiplier $\eta_{ij} > 0$ can be imposed for pairs of nodes $i, j \in \mathcal{V}$ which are located in *different* cells. For this purpose, the equilibrium condition of (3.110k) along with the linear dependencies (3.145) can be written compactly as

$$\mathbb{0} = (\boldsymbol{D}_{\mathrm{c}}^{+})^{\mathsf{T}} \boldsymbol{\lambda}. \tag{3.146}$$

Hence, the resulting controller equations (3.110j)–(3.110k) are given by

$$\boldsymbol{\tau}_{\lambda} \dot{\boldsymbol{\lambda}} = -\boldsymbol{p}_{\mathrm{g}} + \boldsymbol{\rho} + \boldsymbol{p}_{\ell} - \boldsymbol{D}_{\mathrm{c}}^{+} \boldsymbol{\nu}, \tag{3.147a}$$

$$\boldsymbol{\tau}_{\nu} \dot{\boldsymbol{\nu}} = (\boldsymbol{D}_{\mathrm{c}}^{+})^{\mathsf{T}} \boldsymbol{\lambda}. \tag{3.147b}$$

With this notation, $\boldsymbol{D}_{\mathrm{c}}^{+}$ can be interpreted as the incidence matrix of an extended, weighted communication graph $\mathcal{G}_{\mathrm{c}}^{+} = (\mathcal{V}, (\mathcal{E}_{\mathrm{c}}, \mathcal{E}_{\mathrm{c}}^{\mathrm{b}}))$, where $\mathcal{E}_{\mathrm{c}}^{\mathrm{b}}$ represents communication across cell boundaries. The weights of the edges of $\mathcal{G}_{\mathrm{c}}^{+}$ are equal to η_{ij}, if the two adjacent nodes belong to different cells, and equal to 1, if i and j are located in the same cell.

Assumption 3.46. *The multipliers η_{ij} are chosen in a* feasible *sense such that* $\dim \ker(\boldsymbol{D}_{\mathrm{c}}^{+})^{\mathsf{T}} > 0$. *Practically speaking, we assume that there exists at least one $\boldsymbol{\lambda} > \mathbb{0}$ fulfilling* (3.146).

Remark 3.47. *Assumption 3.46 is fullfilled if and only if* $\log\{\boldsymbol{\eta}\} \in \mathrm{im}(\hat{\boldsymbol{D}}_{\mathrm{c}})$. *In this case, each cell $k \in \mathcal{Z}$ can be characterized with a specific* participation factor $\kappa_k > 0$ *such that each η_{ij} is calculated by $\eta_{ij} = \kappa_{k_1} / \kappa_{k_2}$ if i is located in cell k_1 and j is located in cell k_2.* ◊

Lemma 3.48 (Connectivity of Zonal Prices). *If $\mathcal{G}_{\mathrm{c}}^{+}$ is weakly connected, then*

$$\frac{\lambda_1}{\kappa_{k_1}} = \frac{\lambda_2}{\kappa_{k_2}} = \cdots = \frac{\lambda_n}{\kappa_{k_n}} =: \Lambda^{\circ}, \tag{3.148}$$

where κ_{k_i} denotes the corresponding participation factor belonging to the cell k where node $i \in \mathcal{V}$ is located.

Proof. Define the auxiliary matrices $\boldsymbol{K}_1 = \mathrm{diag}_{ij \in (\mathcal{E}_{\mathrm{c}}, \mathcal{E}_{\mathrm{c}}^{\mathrm{b}})}\{\eta_{ij}\} \succ 0$ and $\boldsymbol{K}_2 = \mathrm{diag}_{i \in \mathcal{V}}\{\kappa_{k_i}\} \succ 0$. Then, $(\boldsymbol{D}_{\mathrm{c}}^{+})^{\mathsf{T}} \boldsymbol{\lambda} = \mathbb{0}$ is equivalent to

$$\underbrace{\boldsymbol{K}_1^{-1} \mathbb{0}}_{\mathbb{0}} = \boldsymbol{K}_1^{-1} (\boldsymbol{D}_{\mathrm{c}}^{+})^{\mathsf{T}} \boldsymbol{\lambda} = \underbrace{\boldsymbol{K}_1^{-1} (\boldsymbol{D}_{\mathrm{c}}^{+})^{\mathsf{T}} \boldsymbol{K}_2}_{=:(\boldsymbol{D}_{\mathrm{c}}^{\circ})^{\mathsf{T}}} \underbrace{\boldsymbol{K}_2^{-1} \boldsymbol{\lambda}}_{=:\boldsymbol{\lambda}^{\circ}}. \tag{3.149}$$

Inserting the definition of $\boldsymbol{K}_1$ into (3.149) with $\eta_{ij} = \kappa_{k_1} / \kappa_{k_2}$ reveals that $\boldsymbol{D}_{\mathrm{c}}^{\circ}$ is the incidence matrix of a new communication graph $\mathcal{G}_{\mathrm{c}}^{\circ}$, which is equivalent to $\mathcal{G}_{\mathrm{c}}^{+}$ with all edge weights reset to 1. Thus each solution $\boldsymbol{\lambda}^{\circ}$ of the resulting equation $\mathbb{0} = (\boldsymbol{D}_{\mathrm{c}}^{\circ})^{\mathsf{T}} \boldsymbol{\lambda}^{\circ}$ is of the form $\boldsymbol{\lambda}^{\circ} = \mathbb{1} \cdot \mathrm{const}$, i.e. each component of $\boldsymbol{\lambda}^{\circ}$ has the same value. Finally, since $\boldsymbol{\lambda}^{\circ} = \boldsymbol{K}_2^{-1} \boldsymbol{\lambda} = \mathrm{col}_{i \in \mathcal{V}}\{\lambda_i / \kappa_{k_i}\}$, this leads to (3.148). □

Remark 3.49. *If $\mathscr{G}_c^+$ is not weakly connected, then* (3.148) *holds separately for all nodes in each weakly connected subgraph in $\mathscr{G}_c^+$.* ♢

Remark 3.50. *The uniform price Λ° resulting when $\kappa_k = 1$ for all $k \in \mathcal{Z}$ refers to the social optimum (cf. footnote 35), yielding a minimization of the overall costs of WPs for power generation as well as equal revenues for the same amount of active power generation irrespective of where it is located. Accordingly, κ_k describes the multiplicity of the zonal price Λ_k compared to the uniform price Λ°.* ♢

In the following, we will discuss to what extent $\boldsymbol{\kappa} = \mathrm{col}_{k\in\mathcal{Z}}\{\kappa_k\}$ can be applied to modify the impact of specific cells on the overall network.

3.5.2 Comparison with Centralized Optimization

Conceptions of an *overall utility* within competitive scenarios are strongly related to the idea of global efficiency, which was alluded to in Subsection 3.3.1. The information about which solution is considered "socially preferable" to which other solution can thus be expressed by means of welfare functions (cf. Appendix A.6). In this context, the next theorem reveals an interesting connection between the multiple optimization problems of WPs and CCs and a notional centralized optimization problem with a modified utilitarian welfare function which is a *weighted sum* of the individual WPs' profits:

Theorem 3.51 (Equivalence to Centralized Optimization). *Define the centralized optimization problem*

$$\max_{\boldsymbol{p}_\mathrm{g},\boldsymbol{U}_\mathrm{f},\boldsymbol{U}_\mathcal{I}} \quad \mathrm{P}^\kappa \tag{3.150a}$$

$$\text{subject to} \quad \boldsymbol{\Phi} = \sum_{i\in\mathcal{V}_\mathrm{g}} p_{\mathrm{g},i} - \sum_{i\in\mathcal{V}} p_{\ell,i}, \tag{3.150b}$$

$$\underline{\boldsymbol{p}}_\mathrm{g} \leq \boldsymbol{p}_\mathrm{g} \leq \overline{\boldsymbol{p}}_\mathrm{g}, \tag{3.150c}$$

$$\underline{\boldsymbol{U}}_\mathrm{f} \leq \boldsymbol{U}_\mathrm{f} \leq \overline{\boldsymbol{U}}_\mathrm{f}, \tag{3.150d}$$

$$\underline{\boldsymbol{U}}_\mathcal{I} \leq \boldsymbol{U}_\mathcal{I} \leq \overline{\boldsymbol{U}}_\mathcal{I}, \tag{3.150e}$$

where

$$\mathrm{P}^\kappa = -\sum_{k\in\mathcal{Z}}\sum_{i\in\mathcal{V}_{\mathcal{Z},k}} \frac{1}{\kappa_k}\cdot \mathrm{C}_{\mathrm{g},i}(p_{\mathrm{g},i}) - \sum_{i\in\mathcal{V}_\mathrm{g}} \omega_i \cdot p_{\mathrm{g},i}. \tag{3.151}$$

Then each optimizer $(\boldsymbol{p}_\mathrm{g}^\star, \boldsymbol{U}_\mathrm{f}^\star, \boldsymbol{U}_\mathcal{I}^\star)$ of (3.150) *is an equilibrium of* (3.110a)–(3.110i), (3.147a)–(3.147b) *and vice versa. If all $\mathrm{C}_{\mathrm{g},i}(p_{\mathrm{g},i})$ in* (3.151) *are strictly convex, then $(\boldsymbol{p}_\mathrm{g}^\star, \boldsymbol{U}_\mathrm{f}^\star, \boldsymbol{U}_\mathcal{I}^\star)$ is unique.*

Proof. Constraint (3.150b) is equivalent to (cf. [SDv17b, p. 2615])

$$\hat{\boldsymbol{D}}_{\mathrm{c}}\hat{\boldsymbol{\nu}} = \boldsymbol{p}_{\mathrm{g}} - \boldsymbol{p}_{\ell} - \boldsymbol{\rho}, \tag{3.152}$$

where $\boldsymbol{\rho} = \mathrm{col}_i\{\rho_i\}$ as in (3.55)–(3.56) and $\hat{\boldsymbol{D}}_{\mathrm{c}}$ is the incidence matrix of a weakly connected communication graph. If we choose $\hat{\boldsymbol{D}}_{\mathrm{c}} = \boldsymbol{D}_{\mathrm{c}}^{\circ}$ and define $\mathtt{C}^{\kappa}(\boldsymbol{p}_{\mathrm{g}}) = \sum_{k\in\mathcal{Z}}\sum_{i\in\mathcal{V}_k^{\mathcal{Z}}}\frac{1}{\kappa_k}\cdot \mathtt{C}_i(p_{\mathrm{g},i})$, then the Lagrangian of (3.150) becomes

$$\begin{aligned}
\mathscr{L}^{\kappa} &= \mathtt{C}^{\kappa}(\boldsymbol{p}_{\mathrm{g}}) + \left(\sum_{i\in\mathcal{V}_{\mathrm{g}}}\omega_i\cdot p_{\mathrm{g},i}\right) + \hat{\boldsymbol{\lambda}}^{\mathsf{T}}(\boldsymbol{D}_{\mathrm{c}}^{\circ}\hat{\boldsymbol{\nu}} - \boldsymbol{p}_{\mathrm{g}} + \boldsymbol{p}_{\ell} + \boldsymbol{\rho}) \\
&\quad + \hat{\boldsymbol{\mu}}_{\mathrm{g}-}^{\mathsf{T}}(\underline{\boldsymbol{p}}_{\mathrm{g}} - \boldsymbol{p}_{\mathrm{g}}) + \hat{\boldsymbol{\mu}}_{\mathrm{g}+}^{\mathsf{T}}(\boldsymbol{p}_{\mathrm{g}} - \overline{\boldsymbol{p}}_{\mathrm{g}}) + \hat{\boldsymbol{\mu}}_{\mathcal{S}-}^{\mathsf{T}}(\underline{\boldsymbol{U}}_{\mathrm{f}} - \boldsymbol{U}_{\mathrm{f}}) + \hat{\boldsymbol{\mu}}_{\mathcal{S}+}^{\mathsf{T}}(\boldsymbol{U}_{\mathrm{f}} - \overline{\boldsymbol{U}}_{\mathrm{f}}) \\
&\quad + \hat{\boldsymbol{\mu}}_{\mathcal{I}-}^{\mathsf{T}}(\underline{\boldsymbol{U}}_{\mathcal{I}} - \boldsymbol{U}_{\mathcal{I}}) + \hat{\boldsymbol{\mu}}_{\mathcal{I}+}^{\mathsf{T}}(\boldsymbol{U}_{\mathcal{I}} - \overline{\boldsymbol{U}}_{\mathcal{I}}),
\end{aligned} \tag{3.153}$$

where $\hat{\boldsymbol{\lambda}}$ denotes the Lagrange multiplier for equality constraint (3.152) and $\hat{\boldsymbol{\mu}}_{(\cdot)}$ are the Lagrange multipliers for the inequality constraints (3.150c)–(3.150e). Since (3.150) is convex and Slater's condition is fulfilled, the primal-dual optimizer of (3.150) is given by the KKT point $(\cdot)^{\sharp}$ defined by

$$\mathbb{0} = -\nabla\mathtt{C}^{\kappa}(\boldsymbol{p}_{\mathrm{g}}^{\sharp}) - \omega_i + \hat{\boldsymbol{\lambda}}^{\sharp} + \hat{\boldsymbol{\mu}}_{\mathrm{g}-}^{\sharp} - \hat{\boldsymbol{\mu}}_{\mathrm{g}+}^{\sharp}, \tag{3.154a}$$

$$\mathbb{0} = \hat{\boldsymbol{\mu}}_{\mathcal{S}-}^{\sharp} - \hat{\boldsymbol{\mu}}_{\mathcal{S}+}^{\sharp}, \tag{3.154b}$$

$$\mathbb{0} = -\hat{\boldsymbol{\mu}}_{\mathcal{I}-}^{\sharp} + \hat{\boldsymbol{\mu}}_{\mathcal{I}+}^{\sharp}, \tag{3.154c}$$

$$\mathbb{0} = (\hat{\boldsymbol{\mu}}_{\mathrm{g}-}^{\sharp}) \circ (\underline{\boldsymbol{p}}_{\mathrm{g}} - \boldsymbol{p}_{\mathrm{g}}^{\sharp}), \tag{3.154d}$$

$$\mathbb{0} = (\hat{\boldsymbol{\mu}}_{\mathrm{g}+}^{\sharp}) \circ (\boldsymbol{p}_{\mathrm{g}}^{\sharp} - \overline{\boldsymbol{p}}_{\mathrm{g}}), \tag{3.154e}$$

$$\mathbb{0} = (\hat{\boldsymbol{\mu}}_{\mathcal{S}-}^{\sharp}) \circ (\underline{\boldsymbol{U}}_{\mathrm{f}} - \boldsymbol{U}_{\mathrm{f}}^{\sharp}), \tag{3.154f}$$

$$\mathbb{0} = (\hat{\boldsymbol{\mu}}_{\mathcal{S}+}^{\sharp}) \circ (\boldsymbol{U}_{\mathrm{f}}^{\sharp} - \overline{\boldsymbol{U}}_{\mathrm{f}}), \tag{3.154g}$$

$$\mathbb{0} = (\hat{\boldsymbol{\mu}}_{\mathcal{I}-}^{\sharp}) \circ (\underline{\boldsymbol{U}}_{\mathcal{I}} - \boldsymbol{U}_{\mathcal{I}}^{\sharp}), \tag{3.154h}$$

$$\mathbb{0} = (\hat{\boldsymbol{\mu}}_{\mathcal{I}+}^{\sharp}) \circ (\boldsymbol{U}_{\mathcal{I}}^{\sharp} - \overline{\boldsymbol{U}}_{\mathcal{I}}), \tag{3.154i}$$

$$\mathbb{0} = -\boldsymbol{p}_{\mathrm{g}}^{\sharp} + \boldsymbol{\rho} + \boldsymbol{p}_{\ell} - \boldsymbol{D}_{\mathrm{c}}^{\circ}\hat{\boldsymbol{\nu}}^{\sharp}, \tag{3.154j}$$

$$\mathbb{0} = (\boldsymbol{D}_{\mathrm{c}}^{\circ})^{\mathsf{T}}\hat{\boldsymbol{\lambda}}^{\sharp}, \tag{3.154k}$$

$$\mathbb{0} \leq \hat{\boldsymbol{\mu}}_{\mathcal{S}-}^{\sharp}, \hat{\boldsymbol{\mu}}_{\mathcal{S}+}^{\sharp}, \hat{\boldsymbol{\mu}}_{\mathcal{I}-}^{\sharp}, \hat{\boldsymbol{\mu}}_{\mathcal{I}+}^{\sharp}, \hat{\boldsymbol{\mu}}_{\mathrm{g}-}^{\sharp}, \hat{\boldsymbol{\mu}}_{\mathrm{g}+}^{\sharp}. \tag{3.154l}$$

Inserting the definition (3.149) in (3.154k) and comparison with the right-hand side of (3.110k) yields $\hat{\boldsymbol{\lambda}}^{\sharp} = \boldsymbol{K}_2^{-1}\boldsymbol{\lambda}^{\star}$. Hence with $\nabla\mathtt{C}^{\kappa}(\boldsymbol{p}_{\mathrm{g}}^{\sharp}) = \boldsymbol{K}_2^{-1}\nabla\mathtt{C}(\boldsymbol{p}_{\mathrm{g}}^{\sharp})$ in (3.154a) and by comparison between (3.154a)–(3.154l) and (3.110a)–(3.110k) we get the equivalences $\boldsymbol{p}_{\mathrm{g}}^{\sharp} = \boldsymbol{p}_{\mathrm{g}}^{\star}$, $\boldsymbol{U}_{\mathrm{f}}^{\sharp} = \boldsymbol{U}_{\mathrm{f}}^{\star}$, $\boldsymbol{U}_{\mathcal{I}}^{\sharp} = \boldsymbol{U}_{\mathcal{I}}^{\star}$, $\hat{\boldsymbol{\mu}}_{\mathrm{g}-}^{\sharp} = \boldsymbol{K}_2^{-1}\boldsymbol{\mu}_{\mathrm{g}-}^{\star}$, $\hat{\boldsymbol{\mu}}_{\mathrm{g}+}^{\sharp} = \boldsymbol{K}_2^{-1}\boldsymbol{\mu}_{\mathrm{g}+}^{\star}$, $\hat{\boldsymbol{\mu}}_{\mathcal{S}-}^{\sharp} = \boldsymbol{\mu}_{\mathcal{S}-}^{\star}$, $\hat{\boldsymbol{\mu}}_{\mathcal{S}+}^{\sharp} = \boldsymbol{\mu}_{\mathcal{S}+}^{\star}$, $\hat{\boldsymbol{\mu}}_{\mathcal{I}-}^{\sharp} = \boldsymbol{\mu}_{\mathcal{I}-}^{\star}$, $\hat{\boldsymbol{\mu}}_{\mathcal{I}+}^{\sharp} = \boldsymbol{\mu}_{\mathcal{I}+}^{\star}$. Moreover, the dual optimizer $\boldsymbol{\nu}^{\star}$ from (3.110j) (using the notation (3.147a)) and the primal optimizer $\hat{\boldsymbol{\nu}}^{\sharp}$ of (3.152) are connected via the relationship $-\boldsymbol{D}_{\mathrm{c}}^{\circ}\hat{\boldsymbol{\nu}}^{\sharp} + \boldsymbol{D}_{\mathrm{c}}^{+}\boldsymbol{\nu}^{\star} = \mathbb{0}$.

From (3.149) it follows that $\boldsymbol{D}_{\mathrm{c}}^{\circ} = \boldsymbol{K}_2 \boldsymbol{D}_{\mathrm{c}}^{+} \boldsymbol{K}_1^{-1}$ with $\boldsymbol{K}_1, \boldsymbol{K}_2 \succ 0$. Accordingly, for each equilibrium $\boldsymbol{\lambda}^{\star}$ of (3.110k) there exists a $\hat{\boldsymbol{\lambda}}^{\sharp}$ fulfilling (3.154k), and vice versa.

In summary, for each primal-dual optimizer of (3.150) there exists exactly one corresponding equilibrium of (3.110a)–(3.110i), (3.147a)–(3.147b). In particular, $(\boldsymbol{p}_{\mathrm{g}}^{\sharp}, \boldsymbol{U}_{\mathrm{f}}^{\sharp}, \boldsymbol{U}_{\mathcal{I}}^{\sharp}) = (\boldsymbol{p}_{\mathrm{g}}^{\star}, \boldsymbol{U}_{\mathrm{f}}^{\star}, \boldsymbol{U}_{\mathcal{I}}^{\star})$. Since (3.150) is a convex optimization problem, convergence of the trajectory $(\boldsymbol{p}_{\mathrm{g}}(t), \boldsymbol{U}_{\mathrm{f}}(t), \boldsymbol{U}_{\mathcal{I}}(t))$ to $(\boldsymbol{p}_{\mathrm{g}}^{\star}, \boldsymbol{U}_{\mathrm{f}}^{\star}, \boldsymbol{U}_{\mathcal{I}}^{\star})$ is guaranteed (cf. [Ant94, Theorem 2]).

If the cost functions $\mathrm{C}_{\mathrm{g},i}(p_{\mathrm{g},i})$ are strictly convex, then P^{κ} is strictly concave. Hence, the equilibrium $(\boldsymbol{p}_{\mathrm{g}}^{\sharp}, \boldsymbol{U}_{\mathrm{f}}^{\sharp}, \boldsymbol{U}_{\mathcal{I}}^{\sharp})$ from centralized optimization and therewith the equilibrium $(\boldsymbol{p}_{\mathrm{g}}^{\star}, \boldsymbol{U}_{\mathrm{f}}^{\star}, \boldsymbol{U}_{\mathcal{I}}^{\star})$ from distributed optimization are unique. □

Remark 3.52. *If $\boldsymbol{\kappa} = \mathbb{1}$, then the objective function of* (3.150) *is*

$$\mathrm{P}^{\kappa} = \sum_{\pi \in \mathcal{P}} \mathrm{P}_{\pi} - \sum_{k \in \mathcal{Z}} \mathrm{C}_k^{\ell} + \sum_{k \in \mathcal{Z}} \mathrm{P}_k = -\sum_{i \in \mathcal{V}} \mathrm{C}_i(p_{\mathrm{g},i}), \tag{3.155}$$

and thus equal to the sum of all payoffs of WPs (3.79), *RPs* (3.88), *and CCs* (3.90). *Accordingly, the distributed controller* (3.110a)–(3.110i), (3.147a)–(3.147b) *leads to a constrained minimization of the overall costs of power production.* ◇

3.5.3 Analysis of Pareto Efficiency

As stated in Theorem 3.51, the interaction of distributed WPs and CCs, subject to cell-specific pricing via κ_k, leads to the same equilibrium $(\boldsymbol{p}_{\mathrm{g}}^{\star}, \boldsymbol{U}_{\mathrm{f}}^{\star}, \boldsymbol{U}_{\mathcal{I}}^{\star})$ as if a centralized authority with full knowledge of the whole plant system would solve the optimization problem (3.150) with modified cost $\mathrm{C}^{\kappa}(\boldsymbol{p}_{\mathrm{g}})$, where the cumulative cost functions of each cell $k \in \mathcal{Z}$ are divided by κ_k. This equivalence reveals some further key properties of the developed distributed controller scheme that allows the evaluation of the Pareto efficiency of the equilibrium from a multi-objective perspective.

Remark 3.53. *With the same reasoning as in the proof of Lemma 3.35, it can be proven that for each equilibrium solution of* (3.39), (3.154), *it holds that $\boldsymbol{\omega} = \mathbb{0}$, i.e. zero frequency deviation is preserved.* ◇

In particular, it can be shown that each $\boldsymbol{\kappa} > \mathbb{0}$ achieves an *efficient* allocation in the sense that for a given $\boldsymbol{\kappa}' > 0$, there is no possibility to find a "better" $\boldsymbol{\kappa}'' \neq \boldsymbol{\kappa}'$, $\boldsymbol{\kappa}'' > 0$ such that at least one cell is at a lower cost and no cell is at a higher cost. This is formalized in the next theorem.

Theorem 3.54 (Pareto Efficiency of κ). *For each $k \in \mathcal{Z}$, consider the optimization problems*

$$\max_{\boldsymbol{p}_{\mathrm{g},k}, \boldsymbol{U}_{\mathrm{f},k}, \boldsymbol{U}_{\mathcal{I},k}} \quad \mathrm{P}_k^{\mathrm{total}} \tag{3.156a}$$

$$\text{subject to} \quad (\boldsymbol{D}_{\mathrm{c},k}^{+})^{\mathsf{T}} \boldsymbol{\lambda}_k = \mathbb{0}, \tag{3.156b}$$

$$\underline{\boldsymbol{p}}_{\mathrm{g},k} \leq \boldsymbol{p}_{\mathrm{g},k} \leq \overline{\boldsymbol{p}}_{\mathrm{g},k}, \tag{3.156c}$$

$$\underline{\boldsymbol{U}}_{\mathrm{f},k} \leq \boldsymbol{U}_{\mathrm{f},k} \leq \overline{\boldsymbol{U}}_{\mathrm{f},k}, \tag{3.156d}$$

$$\underline{\boldsymbol{U}}_{\mathcal{I},k} \leq \boldsymbol{U}_{\mathcal{I},k} \leq \overline{\boldsymbol{U}}_{\mathcal{I},k}, \tag{3.156e}$$

where

$$\mathrm{P}_k^{\mathrm{total}} = \left(\sum_{i \in \mathcal{V}_{\mathcal{Z},k}} \sum_{\pi \in \mathcal{P}} \mathrm{P}_{\pi,i} \right) - \mathrm{C}_k^{\ell} + \mathrm{P}_k \tag{3.157}$$

denotes the net profits of all network participants in cell $k \in \mathcal{Z}$ and where $\boldsymbol{D}_{\mathrm{c},k}^{+}$ in (3.156b) *contains all elements of* (3.146) *that belong to the nodes in cell k. Let $\kappa_1 > \mathbb{0}$ be fixed and let the corresponding value of $\mathrm{P}_k^{\mathrm{total}}$ for an equilibrium of* (3.156) *with parameter κ set to κ_1 be denoted by $\mathrm{P}_k^{\mathrm{total}\star}(\kappa_1)$. Then, there exists no other* dominating *$\kappa_2 > \mathbb{0}$ with $\kappa_2 \neq \kappa_1$ such that the following two conditions hold:*

$$\forall\, k \in \mathcal{Z}: \quad \mathrm{P}_k^{\mathrm{total}\star}(\kappa_2) \geq \mathrm{P}_k^{\mathrm{total}\star}(\kappa_1), \tag{3.158a}$$

$$\exists\, k \in \mathcal{Z}: \quad \mathrm{P}_k^{\mathrm{total}\star}(\kappa_2) > \mathrm{P}_k^{\mathrm{total}\star}(\kappa_1). \tag{3.158b}$$

Proof. Taking into account that (3.156) are concave optimization problems and following the same lines as in the proof of Theorem 3.51, it can be shown that each KKT point of (3.156) is given by (3.154a)–(3.154l) and vice versa. For each $\kappa > \mathbb{0}$, the equivalent problem (3.150) is a linear scalarization of (3.156) with positive weights $1/\kappa_k > 0$. With [ED18, Proposition 9], it is implied that for each $\kappa' > \mathbb{0}$, the solution of (3.156) is a part of the Pareto front, hence there exists no $\kappa' > \mathbb{0}$ which is dominated by another $\kappa'' > 0$ in terms of (3.158). □

The observation of Theorem 3.54 that any market equilibrium resulting from a given $\kappa > \mathbb{0}$ yields a Pareto efficient allocation for players $\pi \in \mathcal{W}$ is fully consistent with the first fundamental theorem of welfare theory (see Theorem A.25 in Appendix A.6). However, as opposed to the "copper-plate based" statements of Appendix A.6, Theorem 3.54 provides a statement for *real-world* electricity markets which are faced with operational constraints and a lossy power network. We illustrate the effects of different values of κ by means of Example 4.

Example 4 ***(cont'd)***

Recall the illustrative 7-node example from Subsection 3.4.4. We now assume that the network is divided into two cells as depicted in Fig. 3.13a, with κ_1 set to 1.

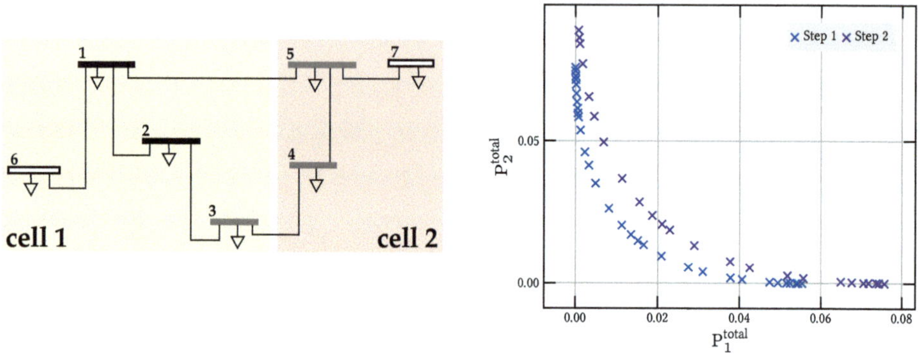

(a) Schematic diagram of cell partitioning

(b) Scatter plot of $(\mathrm{P}_1^{\mathrm{total}}, \mathrm{P}_2^{\mathrm{total}})$

Figure 3.13: Exemplary 7-node system divided into two cells.

Fig. 3.13b shows a scatter plot of the resulting steady-state $(\mathrm{P}_1^{\mathrm{total}}, \mathrm{P}_2^{\mathrm{total}})$ after each of both load steps, and under different choices of $\kappa_2 \in [0.2, 20]$. It can be seen that in both cases, each point $(\mathrm{P}_1^{\mathrm{total}}, \mathrm{P}_2^{\mathrm{total}})$ is part of the Pareto front, thus no κ_2 yields a solution $(\mathrm{P}_1^{\mathrm{total}}, \mathrm{P}_2^{\mathrm{total}})$ which is dominated by another κ_2'.

3.6 Discussion

In this chapter, we have elaborated a distributed intra-cell mechanism for optimization-based control in zonal electricity markets. The proposed mechanism allows a mixture of price-taking WPs, price-setting CCs, and inelastic RPs. The interplay of different optimization problems of WPs results in the overall control system (3.49), (3.110) for which frequency and voltage regulation is preserved. The entailed constraints for WPs are either technically motivated or used to incorporate other obligations of WPs, e.g. from day-ahead clearing or long-term contracts. In this regard, the feasible set of each WP's optimization problem (3.83) contains the capacity that is currently available to the real-time market and not contracted otherwise. In line with the brownfield paradigm addressed in Chapter 1, WPs are not obliged to disclose their own profit function or their momentary flexibility to local CCs or to competitors. The proposed control scheme thereby provides a valuable alternative and improvement to other concepts currently under discussion, which are always characterized by an increasingly strong normative interventionism (see Subsection 2.1.3).

Apart from the cell-based communication structure, all WPs compete in a common market, which helps reconcile individual profit and social welfare, thus limiting the *moral hazard* phenomenon (cf. [ZPE17, p. 180]). By pushing the market price to marginal cost level, the allocation $(\boldsymbol{p}^{\star}_{\mathrm{g},\pi}, \boldsymbol{U}^{\star}_{\mathrm{f},\pi}, \boldsymbol{U}^{\star}_{\mathcal{I},\pi})$ resulting as a closed-loop equilibrium is individually rational from the perspective of each WP $\pi \in \mathcal{W}$, since any other feasible allocation would lead to lower individual profits. Moreover, Theorem 3.54 guarantees that by permitting cell-specific couplings via some $\boldsymbol{\kappa} > \mathbb{0}$, any equilibrium is also Pareto efficient.

Despite these promising findings, however, Pareto efficiency is unsatisfactory if used as the sole solution concept, since no concrete statement has yet been attempted on the proper choice of particularly suitable or unsuitable $\boldsymbol{\kappa} \in \mathbb{R}^{n_{\mathcal{Z}}}_{>0}$, which could theoretically lead to allocations with arbitrarily poor social welfare. In the next chapter, we shed light on such unique allocations by proposing different control schemes for the dynamic and distributed settlement of $\boldsymbol{\kappa}$ throughout the interconnected power system.

4 Real-Time Incentives by Zonal Pricing

In this chapter, distributed control schemes for inter-cell pricing are developed, working towards Contribution 2 and 4 (see Fig. 2.3). Section 4.1 defines the requirements for the control framework to be developed. Section 4.2 gives an overview of the developed overall control structure along with the dependencies of the individual sub-controllers. Section 4.3 provides real-time control strategies for the automatic regulation of zonal price differences by means of κ in order to fulfill Contribution 4. Section 4.4 provides the main results for Contribution 2 by proposing a *dynamic balancing* controller which guarantees fair pricing from the perspective of RPs. Section 4.5 analyzes the overall closed-loop system resulting from the interconnection of the proposed controllers. Section 4.6 gives a conclusive statement about the main findings of this chapter[53].

4.1 Introduction

The market-based controller developed in Chapter 3 is able to achieve a solution that is *technically feasible* in the sense that each market-clearing solution provides compliance with all technical constraints as well as zero deviation from nominal frequency caused by local supply-demand matching. However, the resulting power injections and the generated wholesale and retail prices Λ_k, $\Lambda_{\mathrm{R},k}$ are usually not "monetarily feasible" in the sense that the overall (net) payments throughout the network balance to zero, i.e. that all revenues of WPs are covered by respective payments from RPs. Moreover, despite the global supply-demand matching, the prices generated by CCs are in general unable to reflect *local* scarcity, since geo-spatial network effects such as power flow limits cannot be internalized [TBW15]. In particular, the methods discussed so far do not prevent from after-market redispatch by the system operator due to transmission congestion. Therefore, in this dissertation the free parameter κ is used to design regional short-term incentives for the activation of additional power generation in undersupplied locations and the (market-based) cutback of power generation in oversupplied locations.

The aim of balancing the overall payments throughout the network can be formalized as follows:

---

[53] A preliminary version of the results of this chapter has been published in the journal paper [KZV+22].

Definition 4.1 (Balance of Payments (BoP)). *Let*

$$\sigma(t) = \int_0^t \left(\sum_{\pi \in \mathcal{W}} \mathtt{R}_\pi(t') - \sum_{k \in \mathcal{Z}} \sum_{i \in \mathcal{V}_{\mathcal{Z},k}} \mathtt{C}_k^\ell(t') \right) \mathrm{d}t \tag{4.1}$$

denote the time-integrated sum of revenues of WPs and RPs. Then the payments of the interconnected power system are said to be balanced *if*

$$\lim_{t \to \infty} \sigma(t) = 0, \tag{4.2}$$

i.e. if the time-integrated revenues of all WPs are equal to the time-integrated costs of all RPs as $t \to \infty$.

A weaker requirement is that not the integrated, but only the momentary payments must tend towards zero:

Definition 4.2 (BoP Containment)). *The BoP of the interconnected power system is said to be* contained *if the momentary revenues of all WPs are equal to the momentary costs of all RPs as* $t \to \infty$*:*

$$\lim_{t \to \infty} \sum_{\pi \in \mathcal{W}} \mathtt{R}_\pi(t) - \sum_{k \in \mathcal{Z}} \sum_{i \in \mathcal{V}_{\mathcal{Z},k}} \mathtt{C}_k^\ell(t) = 0. \tag{4.3}$$

Trivially, Definition 4.2 is fulfilled, if Definition 4.1 is fulfilled, but not vice versa.

Remark 4.3. *If BoP is satisfied, then it immediately follows that the only net (i.e. external) payments by network participants are given by the sum of WPs' cost for power generation.* ◊

Remark 4.4. *If BoP containment is satisfied, then the BoP error is constant with respect to time.* ◊

While real-time zonal pricing is considered to reduce structural and regional mismatches in generation and consumption by signaling local grid conditions and thus providing investment signals for WPs to deploy generation capacity in load centers [Sta19], it should be noted that RPs in particular are also inflexible with regard to geographical displacements of their own power production or consumption[54]. Therefore, as a third measure towards fairness among network participants, we impose the following normative requirement with regard to the retail prices $\Lambda_{\mathrm{R},k}$:

[54] Namely, forcing a hypothetical "competition on locations" among RPs would not only be meaningless, but also arbitrarily discriminating.

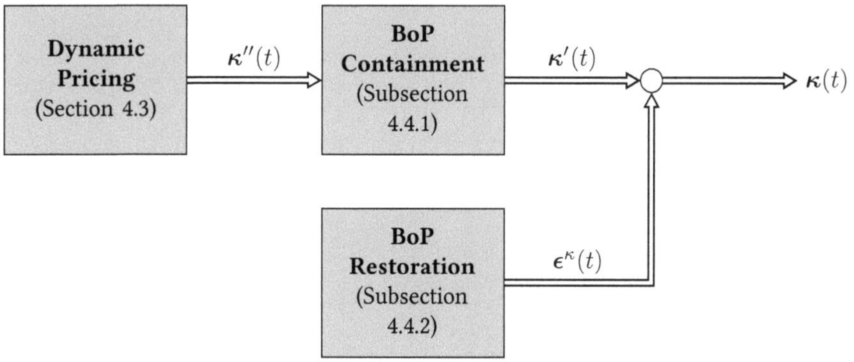

Figure 4.1: Schematic diagram of overall procedure for dynamic pricing.

Claim 4.5. *RPs should pay a* uniform *retail price* Λ_{R} *per amount of power, independent of in which price zone they are located.*

As a consequence of Claim 4.5, we choose the uniform retail price as $\Lambda_{\mathrm{R}} = \Lambda^0$ (as opposed to the cell-specific wholesale prices $\Lambda_k = \kappa_k \cdot \Lambda^0$, cf. Lemma 3.48.)

4.2 Overall Control Structure

Fig. 4.1 shows the overall procedure towards a purposeful and economically feasible $\boldsymbol{\kappa}$ as addressed in the following sections. At first, several approaches for the automatic generation of spatially different participation factors $\boldsymbol{\kappa}''$ (*dynamic pricing*) are proposed in Section 4.3. Main focus is given to the development of a real-time congestion management strategy, which is derived and elaborated in detail in Section 4.3.1. Subsequently, Section 4.4 discusses the generation of an "adjusted" participation factor $\boldsymbol{\kappa}$, which satisfies BoP. Inspired by the terminology used in frequency control (cf. footnote 23), we distinguish in the following between a "primary" controller for BoP containment, which has the goal of limiting the BoP error to a finite value (cf. Remark 4.4), and a "secondary/tertiary" controller for *BoP restoration*, which has the goal of optimally restoring the BoP error to zero, in order to also comply with (4.1). Hence, in Subsection 4.4.1, a *regularized* $\boldsymbol{\kappa}'$ is obtained, which already satisfies BoP containment (4.3), but not (4.1). Ultimately, in Subsection 4.4.2 the constant BoP error (cf. Remark 4.4) is compensated by a suitable *restoration signal* ϵ^{κ}. All methodologies necessary to resolve this last component are then derived in Chapter 5. In accordance with our main principle adopted so far, all of the components considered in this chapter aim for a distributed communication structure. Fig. 4.2 provides an aggregate overview of the controller components to be developed in Sections 4.3 and 4.4 as well as their relation to the distributed frequency and voltage controller from Chapter 3. We can see a clear resemblance between the light gray colored controller components in Fig. 4.2 and the three layers in Fig. 2.3.

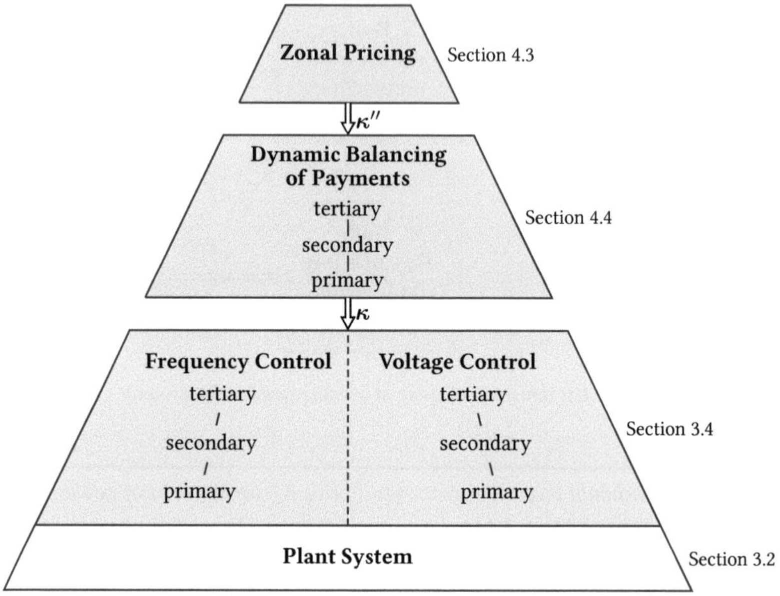

Figure 4.2: Outline of overall controller structure for real-time zonal pricing.

A throughout intent of the resulting overall control mechanism is to exercise and implement different notions of fairness among network participants: For the interaction between CCs, fairness means that regional price differences (imposed via non-uniform entries in κ'') are supposed to help reward power generation at locations that are favorable from a network perspective and to incentivize grid-supportive behavior in both short and long term. At individual cell level, fairness is achieved by using market balancing to prevent accumulations of positive or negative amounts of capital at any of the CCs.

4.3 Dynamic Locational Pricing in Real-Time

Section 3.5 has revealed that each strictly positive κ'' is economically efficient in the sense of Theorem 3.54. By adjusting κ'' appropriately, specific desired relationships between zonal prices are imposed, which can serve as a real-time control mechanism for the reallocation of electricity supply from one cell to another.

In the following Subsection 4.3.1, we apply a feedback control strategy for κ'' to obtain continuous-time congestion management, which is conducted among neighboring CCs by adjusting the cell-specific prices. As discussed in the literature review in Subsection 2.2.3 (see also Table 2.2), current price-based congestion management techniques are either based on an offline calculation of linear sensitivity factors or rely on simplified, network-reduced power system models. By contrast, the presented control scheme puts special emphasis on deploying a network-preserving model (cf. footnote 27) of inter-cell interaction that considers both active and reactive power flows and accounts for the potential overloads on each individual boundary line. In order to further extend this principle to the nodal voltages, Subsection 4.3.2 introduces an additional feedback control strategy for adjusting the voltage setpoints $\boldsymbol{U}_{\mathcal{S}}^{\text{set}}$ and $\boldsymbol{U}_{\mathcal{I}}^{\text{set}}$. Finally, Subsection 4.3.3 discusses various approaches for influencing $\boldsymbol{\kappa}''$ in the sense of Contribution 4 in addition to congestion management.

4.3.1 Real-Time Congestion Management with Zonal Pricing

We describe the cell topology by the condensed graph $\hat{\mathscr{G}}_\text{p}$ (as defined in Section 3.1). Let $\hat{P}_{ij}$ and $\hat{Q}_{ij}$ denote the sending-end active and reactive power flows respectively from node $i \in \mathcal{V}_{\mathcal{Z},k}$ to node $j \in \mathcal{V}_{\text{z},l}$ with $k \neq l$. As it generally holds that $\hat{P}_{ij} \neq -\hat{P}_{ji}$ and $\hat{Q}_{ij} \neq -\hat{Q}_{ji}$, we define

$$\hat{P}_m = \begin{cases} \hat{P}_{ij}, & |\hat{P}_{ji}| \geq |\hat{P}_{ji}|, \\ -\hat{P}_{ji}, & \text{otherwise}, \end{cases} \qquad \hat{Q}_m = \begin{cases} \hat{Q}_{ij}, & |\hat{Q}_{ji}| \geq |\hat{Q}_{ij}|, \\ -\hat{Q}_{ji}, & \text{otherwise}, \end{cases} \tag{4.4}$$

and $\vec{S}_m := \hat{P}_m + \jmath\hat{Q}_m$ for each line $m \in \hat{\mathcal{E}}_\text{p}$. All complex power flows $\vec{S}_m$ are concatenated in the vector $\vec{\boldsymbol{S}} = \text{col}_{m\in\{1,\dots,|\hat{\mathcal{E}}_\text{p}|\}}\{\vec{S}_m\}$. With $\boldsymbol{S}^\text{max}$ specifying the maximum permissible apparent power flow, the *congestion factor* $\mathcal{C}_m \in \mathbb{C}$ of line $m \in \hat{\mathcal{E}}_\text{p}$ is thus defined as

$$\mathcal{C}_m := \frac{\vec{S}_m}{S_m^\text{max}}, \qquad m \in \hat{\mathcal{E}}_\text{p}. \tag{4.5}$$

Definition 4.6 (Line Congestion). *Line $m \in \hat{\mathcal{E}}_\text{p}$ is said to be* congested, *if*

$$|\mathcal{C}_m| > 1. \tag{4.6}$$

For the sake of argument, the next example paves the way to a resonable choice of $\boldsymbol{\kappa}''$ in case of line congestion.

Example 5 (Congestion Management by Zonal Pricing)
Consider a WoC scenario with two cells interconnected by a single power line as depicted in Fig. 4.3a.

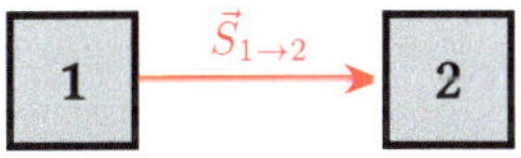

(a) Network with two cells

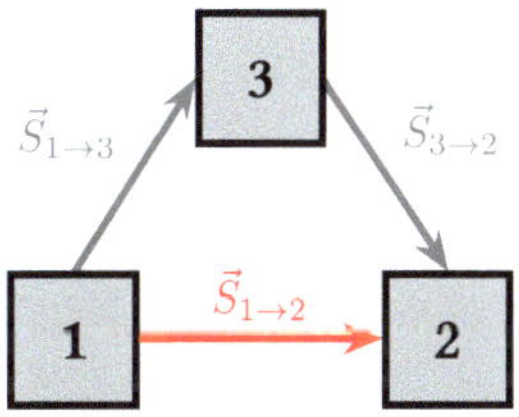

(b) Network with three cells

Figure 4.3: Example networks to illustrate the influence of $\boldsymbol{\kappa}''$ in congestion management.

We assume that the power line is congested (i.e. $|\vec{S}_{1\to 2}| > S_{1\to 2}^{\max}$) with $\hat{P}_{1\to 2} > 0$. Obviously, the active power flow from cell 1 to cell 2 can be mitigated by raising κ_2'' (to stimulate active power injection in cell 1) and/or by lowering κ_1'' (to restrain active power injection in cell 2). Conversely, if $\hat{P}_{1\to 2} < 0$, then raising κ_1'' or lowering κ_2'' is advisable.

Now consider a scenario of three *interconnected cells as depicted in Fig. 4.3b and assume again that the power line from cell 1 to cell 2 is congested with $|\vec{S}_{1\to 2}| > S_{1\to 2}^{\max}$ and $\hat{P}_{1\to 2} > 0$. While the qualitative statements on modifying κ_1'' and κ_2'' stay the same as in the two-cell scenario, it is not obvious whether increasing or decreasing κ_3'' is more favorable in order to support the congestion management. Indeed, an increased power flow through parallel lines $1 \to 3$ and $3 \to 2$ can contribute to further relieving congestion on line $1 \to 2$, hence it is not reasonable to keep κ_3'' completely unchanged. This fact applies even more to scenarios with several parallel branches or more than one adjacent cells whereby a purely heuristic determination of appropriate κ_k'' certainly becomes increasingly complex.*

Apparently, if $|\vec{S}_{3\to 1}| = |\vec{S}_{3\to 2}|$, i.e. if an equal amount of power is exchanged between cells 3 and 1 and between cells 3 and 2, then an increased or decreased power injection in cell 3 (triggered by a changed participation factor κ_3'') has no effect on the power flow over the congested line $1 \to 2$. On the contrary, if $|\vec{S}_{3\to 1}| \gg |\vec{S}_{3\to 2}|$, then the influence of cell 3 on the congested line is quite similar to the influence of cell 1, thus cells 1 and 3 should have similar participation factors. Analogously, if $|\vec{S}_{3\to 1}| \ll |\vec{S}_{3\to 2}|$, then the influence of cell 3 is highly similar to the influence of cell 2. Thus, the participation factors κ_2'' and κ_3'' should be similar. In summary, it emerges that if two cells exchange a high amount of power, then the respective participation factors should be tighly coupled. Conversely, the participation factors of two cells exchanging little power should be decoupled.

In order to keep all of the following participation factors limited to strictly positive values in $\mathbb{R}_{>0}$, we apply the transformation $\phi_i := \ln(\kappa_i'')$. Then, based on the above discussion, the (transformed) participation factor of cell 3 can be calculated out of ϕ_1 and ϕ_2 as the weighted sum

$$\phi_3 = \frac{|\vec{S}_{3\to 1}|\phi_1 + |\vec{S}_{3\to 2}|\phi_2}{|\vec{S}_{3\to 1}| + |\vec{S}_{3\to 2}|}. \tag{4.7}$$

Following the same principle, the participation factors for cells 1 and 2 are calculated as

$$\phi_1 = \frac{|\vec{S}_{1\to 2}|\phi_2 + |\vec{S}_{1\to 3}|\phi_3 - \psi_1}{|\vec{S}_{1\to 2}| + |\vec{S}_{1\to 3}|}, \qquad \phi_2 = \frac{|\vec{S}_{2\to 1}|\phi_1 + |\vec{S}_{2\to 3}|\phi_3 - \psi_2}{|\vec{S}_{2\to 1}| + |\vec{S}_{2\to 3}|}, \tag{4.8}$$

where ψ_1 and ψ_2 are suitable activation functions (yet to be determined) which indicate the congestion status of the boundary lines incident to the respective cell. In particular, choosing $\psi_1 > 0$ signals that increased power injection in cell 1 has an undesirable *impact on congestion of the incident line $1 \to 2$ (i.e. the participation factor of cell 1 should be reduced), and $\psi_2 < 0$ signals that increased power injection in cell 2 has a* desirable *impact on congestion of the incident line $1 \to 2$, thus the participation factor of cell 2 should be increased.*

Combining (4.7) *and* (4.8) *finally yields*

$$\boldsymbol{\psi} = \underbrace{\begin{bmatrix} -|\vec{S}_{1\to2}| - |\vec{S}_{1\to3}| & |\vec{S}_{1\to2}| & |\vec{S}_{1\to3}| \\ |\vec{S}_{1\to2}| & -|\vec{S}_{1\to2}| - |\vec{S}_{2\to3}| & |\vec{S}_{2\to3}| \\ |\vec{S}_{1\to3}| & |\vec{S}_{2\to3}| & -|\vec{S}_{1\to3}| - |\vec{S}_{2\to3}| \end{bmatrix}}_{=:\boldsymbol{\mathcal{A}}} \boldsymbol{\phi}. \tag{4.9}$$

It can be seen in (4.9) *that* $\boldsymbol{\mathcal{A}}$ *is the Laplacian matrix of the graph in Fig. 4.3b weighted by the apparent power flows between cells.*

Based on the insights gained from Example 5, we can thus set up

$$\mathbb{0} = \underbrace{-\hat{\boldsymbol{D}}_{\mathrm{p}}\mathrm{diag}\{|\vec{\boldsymbol{S}}|\}\hat{\boldsymbol{D}}_{\mathrm{p}}^{\mathsf{T}}}_{\boldsymbol{\mathcal{A}}}\boldsymbol{\phi} - \boldsymbol{\psi} \tag{4.10}$$

for the determination of participation factors $\boldsymbol{\kappa}'' = \mathbf{exp}(\boldsymbol{\phi})$. As discussed in Subsection 3.4.1, constraint (4.10) can be obtained in feedback form via the gradient flow

$$\boldsymbol{\tau}_{\phi}\dot{\boldsymbol{\phi}} = -\boldsymbol{\mathcal{A}}\boldsymbol{\phi} - \boldsymbol{\psi}. \tag{4.11}$$

However, while in Example 5 only line $1 \to 2$ was assumed to be congested, in a general setup potentially all lines can be more or less affected. In order to map the individual contribution of each cell to (possible) congestion on one or more lines, we choose the activation functions $\boldsymbol{\psi}$ to

$$\boldsymbol{\psi} = \hat{\boldsymbol{D}}_{\mathrm{p}}\boldsymbol{\mho}, \tag{4.12}$$

where $\boldsymbol{\mho} \in \mathbb{R}^{|\hat{\mathcal{E}}_{\mathrm{p}}|} = \mathrm{col}_{m\in\hat{\mathcal{E}}_{\mathrm{p}}}\{\mho_m\}$ is an appropriate barrier function penalizing lines $m \in \hat{\mathcal{E}}_{\mathrm{p}}$ which are close to congestion. The signal flow diagram in Fig. 4.4 briefly summarizes the influence of line congestion on the prices $\boldsymbol{\lambda}(t)$ and thus the active power generation $\boldsymbol{p}_{\mathrm{g}}(t)$.

Before discussing two possible options how to properly select $\mho_m$, we conclude with three additional remarks:

Remark 4.7. *Using the logarithmic quantities* $\boldsymbol{\phi}$ *instead of* $\boldsymbol{\kappa}''$ *also has the practical benefit that arithmetic averaging* (4.7) *or* (4.8) *clusters around the social optimum with* $\boldsymbol{\kappa}'' = \mathbb{1}$ *and thus no upward outliers can occur.* ◇

Remark 4.8. *For high voltage networks where it is justified to assume that power lines are purely inductive, voltage profiles are flat (i.e.* $U_i \approx 1$ *for* $i \in \mathcal{V}$*), and voltage angle differences* ϑ_{ij} *are small, power line susceptances can be used to approximate power flows throughout the power system [vDD14]. Accordingly, it is then valid to use the approximation*

$$\boldsymbol{\mathcal{A}} = \hat{\boldsymbol{D}}_{\mathrm{p}}\mathrm{diag}\{|\vec{\boldsymbol{S}}|\}\hat{\boldsymbol{D}}_{\mathrm{p}}^{\mathsf{T}} \approx -\beta\hat{\boldsymbol{D}}_{\mathrm{p}}\mathrm{diag}\{\hat{\boldsymbol{B}}\}\hat{\boldsymbol{D}}_{\mathrm{p}}^{\mathsf{T}}, \tag{4.13}$$

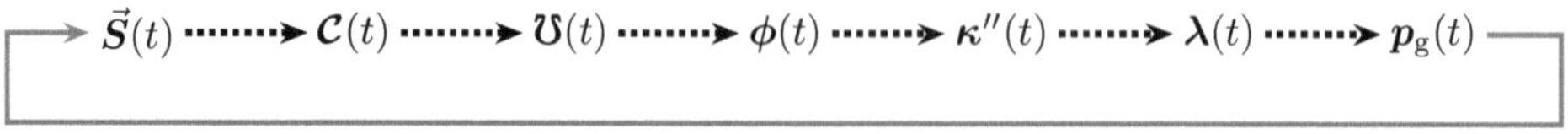

Figure 4.4: Signal flow diagram illustrating the relationship between line congestion $\boldsymbol{C}(t)$ and prices $\boldsymbol{\lambda}(t)$ created by the proposed real-time congestion controller.

where $\beta > 0$ *and* $\hat{\boldsymbol{B}} \in \mathbb{R}_{<0}^{|\hat{\mathcal{E}}_\mathrm{p}|}$ *is the vector of susceptances of inter-cell power lines (with the same sorting as the edges in* $\hat{\boldsymbol{D}}_\mathrm{p}$*). The approximation in* (4.13) *has the major advantage that* $\boldsymbol{\mathcal{A}}$ *is now independent of the momentary power flows over inter-cell lines and thus can be calculated offline.* ♢

Remark 4.9. *An insightful interpretation of* (4.10) *can be made via generalized coordinates: In this regard,* ϕ_k *(and thus* κ''_k*) can be interpreted as the* potential *(i.e. generalized effort) variable of cell* k*, while the apparent power flows* $\vec{S}_{k\to l}$*, which indicate how closely the corresponding values of* $\boldsymbol{\kappa}'$ *should be coupled (cf. Example 5), act as* generalized conductances. *Accordingly,* ψ_k *can be interpreted as local injection of generalized currents (i.e. as generalized* flow *variable), inducing* differences of potential *(i.e. generalized efforts)—in particular, an increase of the local potential (if* $\psi_k > 0$*) or decrease of local potential (if* $\psi_k < 0$*). Formulating the node-voltage method for the resulting* generalized circuit *with potentials* $\boldsymbol{\phi}$*, conductances* $|\vec{\boldsymbol{S}}|$*, and flows* $\boldsymbol{\psi}$ *also leads to* (4.10)*, where* $\boldsymbol{\mathcal{A}}$ *acts as nodal admittance matrix.* ♢

Congestion Controller without Integral Action

As indicated in (4.10), $\boldsymbol{\psi}$ valuates the impact of each cell on overall congestion. By using (4.12), we can assemble this impact from line-wise "static" barrier functions to increasingly penalize those lines that are close to congestion. With regard to Example 5 where $|\vec{S}_{1\to 2}| > \vec{S}^{\max}_{1\to 2}$ and $\hat{P}_{1\to 2} = \Re\{\vec{S}_{1\to 2}\} > 0$, the participation factor κ''_1 has to be decreased in order to alleviate congestion between cells 1 and 2, while κ''_2 should be increased. Conversely, if $\hat{P}_{1\to 2} = \Re\{\vec{S}_{1\to 2}\} < 0$, then κ''_1 has to be decreased and κ''_2 has to be increased.

In this light, we propose a static barrier function for each line as follows:

$$\mho_m = \begin{cases} \Re\{\mathcal{C}_m\} \cdot \dfrac{|\mathcal{C}_m| - \mathcal{C}_m^{\min}}{(1 - |\mathcal{C}_m|)(1 - \mathcal{C}_m^{\min})}, & |\mathcal{C}_m| > \mathcal{C}_m^{\min}, \\ 0, & \text{otherwise}, \end{cases} \tag{4.14}$$

where $\mathcal{C}_m^{\min} \in (0, 1)$ is a (user-defined) threshold for control actions. As long as $|\mathcal{C}_m|$ is below this threshold, it holds that $\mho_m = 0$. Moreover, $|\mho_m| \to \infty$ as $|\mathcal{C}_m| \to 1$[55] (see Fig. 4.5).

Remark 4.10. *Note that the individual values of* ϕ_k *and* ψ_k *in* (4.11) *can be calculated locally by the respective CC* $k \in \mathcal{Z}$*, since the CC only requires information about neighboring* κ *values as well as power flows across its own cell boundary.* ♢

---

55 Analogously, it can be stated that $|\mho_m| \to \pm\infty$ as $|\vec{S}_m| \to \vec{S}_m^{\max}$.

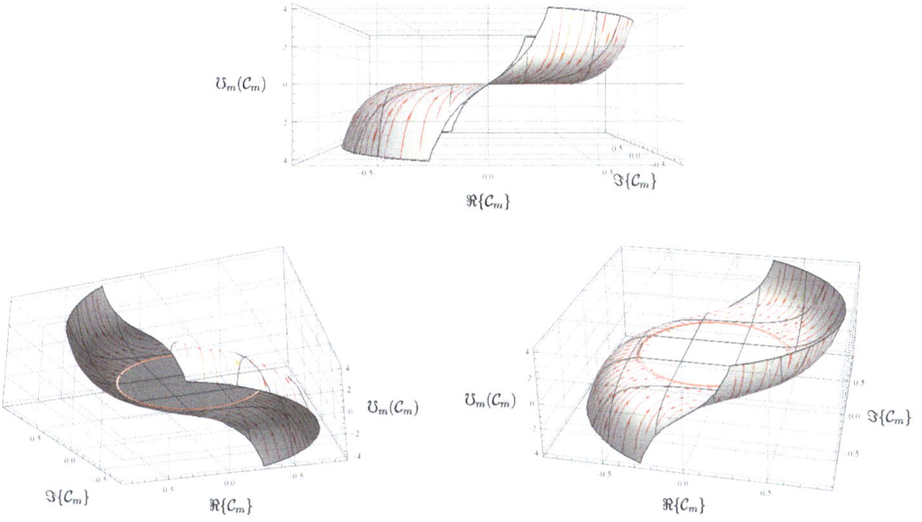

Figure 4.5: Exemplary plot of barrier function $\mho_m(\mathcal{C}_m)$ with $\mathcal{C}_m^{\min} = 0.5$ from different viewing angles. The red circle highlights all points where $|\mathcal{C}_m| = \mathcal{C}_m^{\min}$ holds.

Remark 4.11. *The proposed barrier function* (4.14) *represents only one possible choice of feedback for* $\psi = \hat{\boldsymbol{D}}_\mathrm{p}\boldsymbol{\mho}$. *Basically, the only requirement which* $\boldsymbol{\mho}$ *must certainly satisfy is* $\operatorname{sgn}(\mho_m) = \operatorname{sgn}(\Re\{\mathcal{C}_m\})$ *in order to ensure that* ψ *has the "correct" sign which helps to alleviate congestion. Furthermore,* $|\mho_m|$ *should be at least monotonically increasing with respect to* $|\mathcal{C}_m|$. ♢

Congestion Controller with Integral Action

Typically, the maximum rated power $\vec{S}_m^{\max}$ can be exceeded for a certain amount of time before a secure operation of the power system can no longer be guaranteed [CVV17]. Thus, the momentary apparent power flow should generally be permitted to temporarily exceed the maximum rated power. In order to capture this fact, we propose a more sophisticated feedback law for ψ by incorporating a simple model for line overheating to replace the static feedback (4.14), namely

$$\dot{\mho}_m = \begin{cases} -\frac{1}{\tau_{\mho_m}^-}(1-|\mathcal{C}_m|)\mho_m, & |\mathcal{C}_m| < \mathcal{C}_m^{\min}, \\ -\frac{1}{\tau_{\mho_m}^-}(1-|\mathcal{C}_m|)\mho_m + \frac{1}{\tau_{\mho_m}^+}(|\mathcal{C}_m| - \mathcal{C}_m^{\min})\Re\{\mathcal{C}_m\}, & \mathcal{C}_m^{\min} \le |\mathcal{C}_m| \le 1, \\ \frac{1}{\tau_{\mho_m}^+}\Re\{\mathcal{C}_m\}, & |\mathcal{C}_m| > 1. \end{cases} \tag{4.15}$$

Here, $\mathcal{C}_m^{\min}$ denotes the threshold below which the line is defined as *cooling*. Accordingly, $|\mho_m|$ can be interpreted as a measure of line overheating. In particular, if the line is congested ($|\mathcal{C}_m| > 1$), then it heats up and $|\mho_m|$ in (4.15) increases. If the line is not congested and $|\mathcal{C}_m| < \mathcal{C}_m^{\min}$, then it cools down (such that $|\mho_m|$ in (4.15) decreases) and the cooling rate increases proportionally with less power flowing over the line. If $\mathcal{C}_m^{\min} \leq |\mathcal{C}_m| \leq 1$, then both effects are overlaid. The design parameters $\tau_{\mho_m}^-, \tau_{\mho_m}^+ > 0$ model the specific heating and cooling rates, which are dependent on e.g. the conductor type of the power line or local weather conditions [Dou88].

Both Remark 4.10 and Remark 4.11 analogously apply to the dynamic barrier function in (4.15).

4.3.2 Adjustment of Voltage Setpoints

If the voltage controller (3.110b)–(3.110c) is equipped with setpoints $\boldsymbol{U}_{\mathcal{S}}^{\text{set}}, \boldsymbol{U}_{\mathcal{I}}^{\text{set}}$ as proposed in Remark 3.45, additional control of the *reactive* power flows can be implemented as a second measure for congestion management. For this purpose, the setpoints U_i^{set} can be used to manipulate voltage differences along congested lines and thus mitigate congestion by reducing reactive power flows where necessary.

Reconsider the exemplary scenario from Example 5 with line $1 \to 2$ assumed to be congested with $|\vec{S}_{1\to 2}| > \vec{S}_{1\to 2}^{\max}$ and $\hat{Q}_{1\to 2} = \Im\{\vec{S}_{1\to 2}\} > 0$. Here, the reactive power flow can be decreased by increasing the voltage amplitudes in cell 2 and/or by decreasing the voltage amplitudes in cell 1, while the voltage amplitudes in cell 3 do not need to be modified[56].

In general, cell-specific setpoints U_k^{set} for each $k \in \mathcal{Z}$ can be controlled by the dynamic setpoint correction scheme

$$\tau_{U_k^{\text{set}}} \dot{U}_k^{\text{set}} = -U_k^{\text{set}} + U_k^{\text{nom}} + \overline{U}^{\Delta} \langle U_k^{\Delta} \rangle_{-1}^{1}, \tag{4.16}$$

where U_k^{nom} denotes the nominal voltage of cell $k \in \mathcal{Z}$, $\overline{U}^{\Delta} > 0$ is the maximum admissible deviation between the steady-state voltage setpoint and the nominal voltage, U_k^{Δ} is the "desired" voltage deviation between U_k^{set} and U_k^{nom}, and

$$\langle U_k^{\Delta} \rangle_{-1}^{1} := \begin{cases} 1, & U_k^{\Delta} > 1, \\ U_k^{\Delta}, & -1 \leq U_k^{\Delta} \leq 1, \\ -1, & U_k^{\Delta} < 1 \end{cases} \tag{4.17}$$

is a saturation operator restricting U_k^{Δ} onto the domain $[-1, 1]$. With the summand $\overline{U}^{\Delta} \langle U_k^{\Delta} \rangle_{-1}^{1}$ in (4.16), it is guaranteed that the voltage setpoint U_k^{set} always remains in a sufficiently small neighborhood around the nominal voltage U_k^{nom}.

A simple yet effective feedback control scheme for the desired voltage deviations $\boldsymbol{U}^{\Delta}$ can then be applied by choosing for example

$$U_k^{\Delta} = \sum_{\substack{m \in \mathcal{E}_{\text{p}}^{\text{b}} \\ m = i \to j,\, i \in \mathcal{Z}_k}} \operatorname{sgn}(\Im\{\mathcal{C}_m\}) \cdot |\mathcal{C}_m|. \tag{4.18}$$

56 An essential difference between active and reactive power flows is that reactive power flows are mainly local phenomena and thus should also be compensated by local interventions [vDD14, LAPG17].

Here, the desired voltage deviation U_k^{Δ} is proportional to the absolute value $|\mathcal{C}_m|$ of the congestion factor. The multiplier $\mathrm{sgn}(\Im\{\mathcal{C}_m\})$ reflects the direction of reactive power flow, which is required in order to yield the correct orientation for U_k^{Δ}. With respect to all cells, we thus get the local setpoint correction dynamics

$$\boldsymbol{\tau}_{U^{\mathrm{set}}}\dot{\hat{\boldsymbol{U}}}^{\mathrm{set}} = -\hat{\boldsymbol{U}}^{\mathrm{set}} + \hat{\boldsymbol{U}}^{\mathrm{nom}} + \overline{U}^{\Delta}\langle \boldsymbol{U}^{\Delta}\rangle_{-1}^{1}, \tag{4.19a}$$

$$\boldsymbol{U}^{\Delta} = \hat{\boldsymbol{D}}_{\mathrm{p}}\left(\mathbf{sgn}(\Im\{\boldsymbol{\mathcal{C}}\}) \circ |\boldsymbol{\mathcal{C}}|\right), \tag{4.19b}$$

where $\boldsymbol{\tau}_{U^{\mathrm{set}}} = \mathrm{diag}_{k\in\mathcal{Z}}\{\tau_{U_k^{\mathrm{set}}}\}$, $\hat{\boldsymbol{U}}^{\mathrm{set}} = \mathrm{col}_{k\in\mathcal{Z}}\{U_k^{\mathrm{set}}\}$, and $\hat{\boldsymbol{U}}^{\mathrm{nom}} = \mathrm{col}_{k\in\mathcal{Z}}\{U_k^{\mathrm{nom}}\}$.

4.3.3 Extensions, Combinations, and Long-Term Perspective

In Subsection 4.3.1, the participation factor $\boldsymbol{\kappa}''$ was exploited to establish a real-time congestion management strategy. In the following subsection, we briefly discuss several other possible means for regulating $\boldsymbol{\kappa}''$ such that alternative notions of "grid-supportive behavior" as objected in Chapter 1 may be realized.

A somewhat classical approach is to ponder on the active power balances separately for each cell[57]. By e.g. setting

$$\kappa_k'' = \frac{\sum\limits_{i\in\mathcal{V}_{\mathcal{Z},k}} p_{\ell,i}(t) + \hat{\Phi}_k(t)}{\sum\limits_{i\in\mathcal{V}_{\mathrm{g},k}} p_{\mathrm{g},i}(t)}, \tag{4.20}$$

the participation factor is equal to 1 as soon as the active power balance is fulfilled for cell $k \in \mathcal{Z}$, while the revenues for power generation automatically increase if cell k is a load center. By this basic approach, regional incentives can be created to deploy additional generation capacity primarily where there is a momentary or permanent local undersupply. Conversely, cell-specific prices automatically drop in case there is a lot of available local generation capacity and/or few local consumption. Since all information required in (4.20) can be measured decentrally in cell k, there is actually no need for communication with neighboring CCs. An obvious disadvantage of (4.20) is that there exists no solution for κ_k'' whenever there is no momentary power generation in cell k for some reason.

A considerably simpler approach, which invariably yields a nonsingular solution while requiring no measurements from WPs' power generations, is

$$\kappa_k'' = \frac{\sum\limits_{i\in\mathcal{V}_{\mathcal{Z},k}} p_{\ell,i}(t)}{\sum\limits_{k\in\mathcal{Z}}\sum\limits_{i\in\mathcal{V}_{\mathcal{Z},k}} p_{\ell,i}(t)}. \tag{4.21}$$

57 The concept of cell-specific balancing is closely related to the MG paradigm, which suggests that local power balancing helps to reduce (global) network losses, cf. [Sch15, p. 56].

In this case, the pricing scheme only accounts for the momentary distribution of consumption among all cells. To obtain meaningful results for this approach, cells should have nominally the same amount of consumers. In contrast to (4.20), it is necessary to obtain delay-free information about the consumption in all other cells[58].

Besides the examples shown above, also a combination of pricing regimes can be applied such that the controller equations for $\kappa_k''(t)$ (and consequently the incentive signaling of the pricing strategy) are composed of several components in an additive manner. For the remainder of this subsection, denote by κ_k° the "cell-individual" calculations governed e.g. by (4.20) or (4.21). Then, by introducing a *communality factor* $\iota \in [0,1]$ and choosing

$$\kappa_k''(t) = \iota \cdot K + (1-\iota) \cdot \kappa_k^\circ(t), \qquad K > 0, \tag{4.22}$$

the *degree of excess* among cell-specific prices can be seamlessly adjusted from fully individualized prices $\kappa_k''(t) = \kappa_k^\circ(t)$ for $\iota = 0$ up to a commonly shared price $\kappa_k''(t) = K$ for $\iota = 1$.

Besides a fixed common value $K > 0$, more sophisticated methods for the "commonly shared part" of (4.22) can be employed. The following lemma exemplarily shows how dynamic consensus among CCs can be reached by by invoking distributed averaging theory [DdGP10, SDF17].

Lemma 4.12. *Let $\boldsymbol{\mathcal{A}}_\kappa$ be the adjacency matrix of a weakly connected and possibly weighted communication graph $\mathcal{G}_\kappa = (\mathcal{Z}, \mathcal{E}_\kappa)$. Then for each equilibrium $((\kappa_k'')^\star, e_k^\star)$ of the distributed-averaging-based controller*

$$\kappa_k'' = \iota \cdot e_k + (1-\iota) \cdot \kappa_k^\circ, \qquad k \in \mathcal{Z}, \tag{4.23a}$$

$$\dot{e}_k = -\sum_{l \in \mathcal{Z}} \varpi_{kl}(e_k - \kappa_l''), \qquad k \in \mathcal{Z} \tag{4.23b}$$

with convex weights $\varpi_{kl} = [\boldsymbol{\mathcal{A}}_\kappa]_{kl} / \left(\sum_{l \in \mathcal{N}_k} [\boldsymbol{\mathcal{A}}_\kappa]_{kl}\right)$, it holds that $\mathbf{e}^\star$ is a weighted average of the neighboring participation factors κ_l''.

Proof. By replacing $\varpi_{kl} = [\boldsymbol{\mathcal{A}}]_{kl} / \left(\sum_{l \in \mathcal{N}_k} [\boldsymbol{\mathcal{A}}]_{kl}\right)$ in (4.23b) and equating the right-hand side of (4.23b) to zero, we get

$$e_k^\star = \sum_{l \in \mathcal{Z}} [\boldsymbol{\mathcal{A}}]_{kl} \cdot (\kappa_l'')^\star. \tag{4.24}$$

Thus, $e_k^\star$ is a weighted average of all neighboring participation factors $(\kappa_l'')^\star$, whereby the weights are given by $\boldsymbol{\mathcal{A}}$. □

Remark 4.13. *If $\mathcal{G}_\kappa$ is complete, then $e_k^\star = e_l^\star$ holds for all $k, l \in \mathcal{Z}$, i.e. the commonly shared part of $\boldsymbol{\kappa}''(t)$ in (4.23a) has the same value throughout the whole network.* ♢

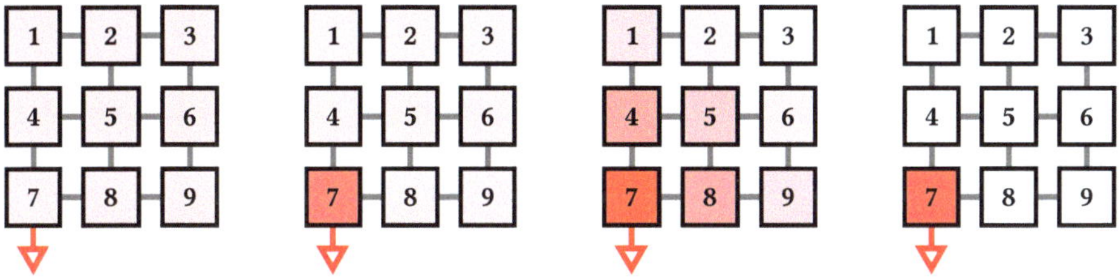

(a) $\iota = 1$ (b) $\iota = 0.5$ with complete $\mathscr{G}_\kappa$ (c) $\iota = 0.5$ with $\mathscr{G}_\kappa = \hat{\mathscr{G}}_\mathrm{p}$ (d) $\iota = 0$

Figure 4.6: Zonal prices in case of load center in cell 7 under different choice of communality factor ι. The intensity of the red color in each individual cell signals the resulting increase in local prices and thus the added incentive for supply in this respective cell.

Fig. 4.6 schematically shows the effects of combined pricing (4.23) for a load center in cell 7 for different communality factors $\iota \in \{0, 0.5, 1\}$ with $\kappa_k^\circ(t)$ being set according to (4.21). It emerges that $\iota = 1$ leads to a regionally independent (uniform) price (see Fig. 4.6a), while for $\iota = 0$ the load surplus in cell 7 leads to an increased price in the same cell only, while all other cells remain unaffected (see Fig. 4.6d). For $\iota \in (0, 1)$, on the other hand, the specific properties of the weakly connected graph $\mathscr{G}_\kappa$ become relevant: As Fig. 4.6b (complete graph $\mathscr{G}_\kappa$) shows, all cells other than cell 7 share a unique price $\Lambda_{\neg 7} < \Lambda_7$, while in Fig. 4.6c (incomplete graph $\mathscr{G}_\kappa = \hat{\mathscr{G}}_\mathrm{p}$), the prices Λ_k in cells $k \neq 7$ gradually drop with their spatial distance to cell 7.

4.4 Dynamic Balancing of Payments

While the previous section focused primarily on notions of fairness for WPs, this section is devoted to also ensuring fairness for RPs in terms of Claim 4.5 by further refinements on participation factors $\boldsymbol{\kappa}''$. In particular, Subsection 4.4.1 combines the fairness objective of Claim 4.5 with the BoP containment from Definition 4.2. Subsequently, Section 4.4.2 introduces an optimization-based framework for BoP restoration to also comply with the BoP requirement as in Definition 4.1.

4.4.1 Balance of Payments Containment

Before considering specific actions for the observance of BoP containment, it is important to note the basic finding that the resulting power flows are invariant with regard to scalar multiplications of the participation vector $\boldsymbol{\kappa}''$:

[58] Even if this approach is not decentralized, a fully distributed consensus-based control scheme can be derived by applying Lemma B.1 analogously to the control laws in Section 3.4, such that calculations of (4.21) are solely based on neighbor-to-neighbor communication among CCs.

Lemma 4.14. *Let* $(\boldsymbol{p}_{\mathrm{g}}^{\sharp}, \boldsymbol{U}_{\mathrm{f}}^{\sharp}, \boldsymbol{U}_{\mathcal{I}}^{\sharp})$ *denote an equilibrium of* (3.110a)–(3.110i), (3.147a)–(3.147b). *For a given* $\boldsymbol{\kappa}_1$, *and let* $(\boldsymbol{p}_{\mathrm{g}}^{\star}, \boldsymbol{U}_{\mathrm{f}}^{\star}, \boldsymbol{U}_{\mathcal{I}}^{\star})$ *denote the corresponding equilibrium for* $\boldsymbol{\kappa}_2 = a \cdot \boldsymbol{\kappa}_1$ *with* $a > 0$. *Then it holds that* $(\boldsymbol{p}_{\mathrm{g}}^{\sharp}, \boldsymbol{U}_{\mathrm{f}}^{\sharp}, \boldsymbol{U}_{\mathcal{I}}^{\sharp}) = (\boldsymbol{p}_{\mathrm{g}}^{\star}, \boldsymbol{U}_{\mathrm{f}}^{\star}, \boldsymbol{U}_{\mathcal{I}}^{\star})$.

Proof. With Theorem 3.51 it follows that each equilibrium $(\boldsymbol{p}_{\mathrm{g}}^{\sharp}, \boldsymbol{U}_{\mathrm{f}}^{\sharp}, \boldsymbol{U}_{\mathcal{I}}^{\sharp})$ of (3.110a)–(3.110i), (3.147a)–(3.147b) with $\kappa = \kappa_1$ is defined by (3.154) with (3.154a) equal to

$$\mathbb{0} = -\nabla \mathrm{C}^{\kappa_1}(\boldsymbol{p}_{\mathrm{g}}^{\sharp}) - \boldsymbol{\omega}^{\sharp} + \hat{\boldsymbol{\lambda}}^{\sharp} + \hat{\boldsymbol{\mu}}_{\mathrm{g}-}^{\sharp} - \hat{\boldsymbol{\mu}}_{\mathrm{g}+}^{\sharp}, \tag{4.25}$$

while each equilibrium $(\boldsymbol{p}_{\mathrm{g}}^{\star}, \boldsymbol{U}_{\mathrm{f}}^{\star}, \boldsymbol{U}_{\mathcal{I}}^{\star})$ of (3.110a)–(3.110i), (3.147a)–(3.147b) with $\kappa = \kappa_2$ fulfills

$$\mathbb{0} = -\nabla \mathrm{C}^{\kappa_2}(\boldsymbol{p}_{\mathrm{g}}^{\star}) - \boldsymbol{\omega}^{\star} + \hat{\boldsymbol{\lambda}}^{\star} + \hat{\boldsymbol{\mu}}_{\mathrm{g}-}^{\star} - \hat{\boldsymbol{\mu}}_{\mathrm{g}+}^{\star} = -\frac{1}{a} \cdot \nabla \mathrm{C}^{\kappa_1}(\boldsymbol{p}_{\mathrm{g}}^{\star}) - \boldsymbol{\omega}^{\star} + \hat{\boldsymbol{\lambda}}^{\star} + \hat{\boldsymbol{\mu}}_{\mathrm{g}-}^{\star} - \hat{\boldsymbol{\mu}}_{\mathrm{g}+}^{\star}, \tag{4.26}$$

$a > 0$, together with (3.154b), where each $(\cdot)^{\sharp}$ in (3.154b)–(3.154l) is replaced by $(\cdot)^{\star}$. With Lemma 3.35, it follows that $\boldsymbol{\omega}^{\sharp} = \boldsymbol{\omega}^{\star} = \mathbb{0}$. Multiplying (4.26) by a thus leads to

$$\begin{aligned} \mathbb{0} &= -\nabla \mathrm{C}^{\kappa_1}(\boldsymbol{p}_{\mathrm{g}}^{\sharp}) + \hat{\boldsymbol{\lambda}}^{\sharp} + \hat{\boldsymbol{\mu}}_{\mathrm{g}-}^{\sharp} - \hat{\boldsymbol{\mu}}_{\mathrm{g}+}^{\sharp}, & \text{(4.27a)} \\ &= -\nabla \mathrm{C}^{\kappa_1}(\boldsymbol{p}_{\mathrm{g}}^{\star}) + a \cdot \hat{\boldsymbol{\lambda}}^{\star} + a \cdot \hat{\boldsymbol{\mu}}_{\mathrm{g}-}^{\star} - a \cdot \hat{\boldsymbol{\mu}}_{\mathrm{g}+}^{\star}. & \text{(4.27b)} \end{aligned}$$

By comparison of coefficients[59] between (4.27a), (3.154b)–(3.154l) and (4.27b), (3.154b)–(3.154l)[60], it immediately follows that for each equilibrium $(\cdot)^{\sharp}$ to $\boldsymbol{\kappa}_1$ fulfilling (4.27a), (3.154b)–(3.154l), there exists an equilibrium $(\cdot)^{\star}$ to $\boldsymbol{\kappa}_2 = a \cdot \boldsymbol{\kappa}_1$ with $\boldsymbol{p}_{\mathrm{g}}^{\star} = \boldsymbol{p}_{\mathrm{g}}^{\sharp}$, $\hat{\boldsymbol{\lambda}}^{\star} = a \cdot \hat{\boldsymbol{\lambda}}^{\sharp}$, $\hat{\boldsymbol{\mu}}_{\mathrm{g}-}^{\star} = a \cdot \hat{\boldsymbol{\mu}}_{\mathrm{g}-}^{\sharp}$, $\hat{\boldsymbol{\mu}}_{\mathrm{g}+}^{\star} = a \cdot \hat{\boldsymbol{\mu}}_{\mathrm{g}+}$, $\hat{\boldsymbol{\mu}}_{\mathcal{S}-}^{\star} = \hat{\boldsymbol{\mu}}_{\mathcal{S}-}^{\sharp}$, $\hat{\boldsymbol{\mu}}_{\mathcal{S}+}^{\star} = \hat{\boldsymbol{\mu}}_{\mathcal{S}+}^{\sharp}$, $\hat{\boldsymbol{\mu}}_{\mathcal{I}-}^{\star} = \hat{\boldsymbol{\mu}}_{\mathcal{I}-}^{\sharp}$, $\hat{\boldsymbol{\mu}}_{\mathcal{I}+}^{\star} = \hat{\boldsymbol{\mu}}_{\mathcal{I}+}^{\sharp}$, $\boldsymbol{U}_{\mathrm{f}}^{\star} = \boldsymbol{U}_{\mathrm{f}}^{\sharp}$, $\boldsymbol{U}_{\mathcal{I}}^{\star} = \boldsymbol{U}_{\mathcal{I}}^{\sharp}$, and $\hat{\boldsymbol{\nu}}^{\star} = \hat{\boldsymbol{\nu}}^{\sharp}$. In particular, it holds that $(\boldsymbol{p}_{\mathrm{g}}^{\sharp}, \boldsymbol{U}_{\mathrm{f}}^{\sharp}, \boldsymbol{U}_{\mathcal{I}}^{\sharp}) = (\boldsymbol{p}_{\mathrm{g}}^{\star}, \boldsymbol{U}_{\mathrm{f}}^{\star}, \boldsymbol{U}_{\mathcal{I}}^{\star})$. □

Lemma 4.12 implies that there exists a remaining degree of freedom of dimension one which only affects the prices (and thus the costs and revenues of WPs and RPs), whilst leaving all physical quantities (i.e. power flows and nodal voltages) unchanged. This degree of freedom can be exploited to generate a *regularized* $\boldsymbol{\kappa}'(t) = a^{\star} \cdot \boldsymbol{\kappa}''(t)$, where $a^{\star}$ is the *unique* scalar that induces BoP containment without altering the physical actuation (and thus the incentives originally governed by $\boldsymbol{\kappa}''(t)$). With the observation that the requirement of BoP containment (4.3) can be written as $((\boldsymbol{\kappa}')^{\star})^{\mathsf{T}} \hat{\boldsymbol{p}}_{\mathrm{g}}^{\star} = \mathbb{1}^{\mathsf{T}} \hat{\boldsymbol{p}}_{\ell}^{\star}$, the defining condition for $a^{\star}$ thus equals

$$-a^{\star} \sum_{k \in \mathcal{Z}} (\kappa_k'')^{\star} \hat{p}_{\mathrm{g},k}^{\star} + \sum_{k \in \mathcal{Z}} \hat{p}_{\ell,k}^{\star} = 0. \tag{4.28}$$

[59] Note that strict convexity of $\mathrm{C}^{\kappa}(\boldsymbol{p}_{\mathrm{g}})$ implies that $\nabla \mathrm{C}^{\kappa}(\boldsymbol{p}_{\mathrm{g}})$ is invertible.

[60] Again, each $(\cdot)^{\sharp}$ in (3.154b)–(3.154l) has to be replaced by $(\cdot)^{\star}$.

Due to the separability of (4.28) with respect to $\boldsymbol{\kappa}''$, we again employ the distributed consensus-based control approach derived in Appendix B.2 to obtain the distributed controller

$$\boldsymbol{\tau}_a \dot{\boldsymbol{a}} = -\boldsymbol{a} \circ \boldsymbol{\kappa}'' \circ \hat{\boldsymbol{p}}_{\mathrm{g}} + \hat{\boldsymbol{p}}_{\ell} + \hat{\boldsymbol{D}}_{\mathrm{p}} \boldsymbol{\nu}_a, \tag{4.29a}$$

$$\boldsymbol{\tau}_{\nu_a} \dot{\boldsymbol{\nu}}_a = -\hat{\boldsymbol{D}}_{\mathrm{p}}^{\mathsf{T}} \boldsymbol{\nu}_a. \tag{4.29b}$$

Using (4.29), each individual CC can execute its control actions by relying solely on own local measurements as well as on neighbor-to-neighbor communication, thus in the same way as already implemented for the congestion controllers in Subsections 4.3.1 and 4.3.2.

Remark 4.15. *It is convenient to resort to the cell incidence matrix $\hat{\boldsymbol{D}}_{\mathrm{p}}$ as the communication matrix in* (4.29)*, as it is already encountered for the settlement of cell-specific participation factors κ_k''* (4.11) *via $\boldsymbol{\mathcal{A}}$ in* (4.10)*. Hence, existing communication infrastructure can be reused.* ◇

Remark 4.16. *If the distributed controller* (4.29) *is interpreted as primal-dual saddle-point flow (cf. Subsection 3.4.1), then each equilibrium $(\boldsymbol{a}^\star, \boldsymbol{\nu}_{a}^\star)$ of* (4.29) *is an optimizer of the convex optimization problem*

$$\min_{\boldsymbol{\kappa}'} \quad \frac{1}{2}(\boldsymbol{\kappa}')^{\mathsf{T}} \begin{bmatrix} \frac{\hat{p}_{\mathrm{g},1}}{\kappa_1''} & & \mathbb{0} \\ & \ddots & \\ \mathbb{0} & & \frac{\hat{p}_{\mathrm{g},N}}{\kappa_N''} \end{bmatrix} \boldsymbol{\kappa}' \tag{4.30a}$$

$$\text{subject to} \quad (\boldsymbol{\kappa}')^{\mathsf{T}} \hat{\boldsymbol{p}}_{\mathrm{g}} = \mathbb{1}^{\mathsf{T}} \hat{\boldsymbol{p}}_{\ell}, \tag{4.30b}$$

where $a_k^\star = (\kappa_k')^\star / \kappa_k''$ and $\boldsymbol{\nu}_{\boldsymbol{a}}^\star$ is the vector of Lagrange multipliers for the affine constraint (4.30b)*. One possible generalization of* (4.30) *is to replace* (4.30a) *by an arbitrary convex objective function $\mathrm{J}^{\kappa'}(\boldsymbol{\kappa}')$. This optimization problem then also leads to participation factors fulfilling BoP containment, without the need to compute non-regularized participation factors $\boldsymbol{\kappa}''$. However, since the relation $\boldsymbol{\kappa}' = a \cdot \boldsymbol{\kappa}''$ with $\boldsymbol{\kappa}'' > \mathbb{0}$ has been dropped, the strict positivity requirement is obeyed separately, yielding*

$$\min_{\boldsymbol{\kappa}', \boldsymbol{a}} \quad \mathrm{J}^{\kappa'}(\boldsymbol{\kappa}') \tag{4.31a}$$

$$\text{subject to} \quad \hat{\boldsymbol{D}}_{\mathrm{p}} \boldsymbol{a} - \boldsymbol{\kappa}' \circ \hat{\boldsymbol{p}}_{\mathrm{g}} + \hat{\boldsymbol{p}}_{\ell} = \mathbb{0}, \tag{4.31b}$$

$$\boldsymbol{\kappa}' - \boldsymbol{\varepsilon} \geq \mathbb{0}, \tag{4.31c}$$

where $\boldsymbol{\varepsilon} > \mathbb{0}$ is an arbitrarily small safety constant. Using (4.31)*, the distributed consensus-based controller becomes*

$$\boldsymbol{\tau}_{\kappa} \dot{\boldsymbol{\kappa}}' = -\nabla \mathrm{J}^{\kappa'}(\boldsymbol{\kappa}') + \boldsymbol{\nu}_{\kappa}^{\mathsf{T}} \hat{\boldsymbol{p}}_{\mathrm{g}} + \boldsymbol{\mu}_{\kappa}, \tag{4.32a}$$

$$\boldsymbol{\tau}_a \dot{\boldsymbol{a}} = -\hat{\boldsymbol{D}}_{\mathrm{p}}^{\mathsf{T}} \boldsymbol{\nu}_{\kappa}, \tag{4.32b}$$

$$\boldsymbol{\tau}_{\sigma} \dot{\boldsymbol{\nu}}_{\kappa} = \hat{\boldsymbol{D}}_{\mathrm{p}} \boldsymbol{a} - \boldsymbol{\kappa}' \circ \hat{\boldsymbol{p}}_{\mathrm{g}} + \hat{\boldsymbol{p}}_{\ell}, \tag{4.32c}$$

$$\boldsymbol{\tau}_{\mu_{\kappa}} \dot{\boldsymbol{\mu}}_{\kappa} = \langle -\boldsymbol{\kappa}' + \boldsymbol{\varepsilon} \rangle_{\boldsymbol{\mu}_{\kappa}}^{+}, \tag{4.32d}$$

where $\boldsymbol{\nu}_{\kappa}$ and $\boldsymbol{\mu}_{\kappa}$ denote the Lagrange multipliers for constraints (4.31b) *and* (4.31c), *respectively.* ◇

From (4.28) in conjunction with (4.29), it immediately follows that $\boldsymbol{a}^\star = a^\star \cdot \mathbb{1}$, where

$$a^\star = \left(\frac{\sum\limits_{k\in\mathcal{Z}} \sum\limits_{i\in\mathcal{V}_{\mathcal{Z},k}} (\kappa''_k \cdot p_{\mathrm{g},i})}{\sum\limits_{i\in\mathcal{V}} p_{\ell,i}} \right)^{-1} = \left(\frac{\sum\limits_{k\in\mathcal{Z}} \sum\limits_{i\in\mathcal{V}_{\mathcal{Z},k}} (\kappa''_k \cdot p_{\ell,i})}{\sum\limits_{i\in\mathcal{V}} p_{\ell,i}} + \frac{\sum\limits_{k\in\mathcal{Z}} \sum\limits_{i\in\mathcal{V}_{\mathcal{Z},k}} (\kappa''_k \cdot \hat{\Phi}_k)}{\sum\limits_{i\in\mathcal{V}} p_{\ell,i}} \right)^{-1}. \tag{4.33}$$

Equation (4.33) reveals some interesting connections between the steady-state wholesale prices $\Lambda_k^\star = \kappa_k^\star \cdot (\Lambda^0)^\star = a^\star \cdot (\kappa''_k)^\star \cdot \Lambda_{\mathrm{R}}^\star$ and the steady-state retail prices $\Lambda_{\mathrm{R}}^\star = (\Lambda^0)^\star$, which are discussed below.

Remark 4.17. *In the lossless case, the inverse of $a^\star$ is a weighted average of the individual cell-specific participation factors $(\boldsymbol{\kappa}'')^\star$. Conversely,* (4.33) *suggests that for the case of uniform prices $\boldsymbol{\kappa}' = \mathbb{1}$, BoP containment is only fulfilled for the (academic) case of lossless networks. In other words, the social optimum never produces a (monetarily) balanced solution in real-world power systems unless $\boldsymbol{\kappa}''$ is regularized by the scalar multiplication $\boldsymbol{\kappa}' = a^\star \cdot \boldsymbol{\kappa}''$ with $0 < a^\star < 1$. Without this regulation, the monetary imbalance increases as transmission losses increase.* ◇

Remark 4.18. *The regularization by means of $a^\star$ (respectively $\boldsymbol{a}^\star$) does not only account for the aggregate transmission losses*[61]*, but also ensures an automatic and fair monetary reallocation of these transmission losses to all RPs in real time*[62]*. Accordingly, since $\mathbb{1}^\top \hat{\boldsymbol{p}}_{\mathrm{g}} > \mathbb{1}^\top \hat{\boldsymbol{p}}_{\ell}$ for lossy networks, there always exists at least one cell $k \in \mathcal{Z}$ with $\kappa'_k < 1$, regardless of the choice of $\boldsymbol{\kappa}'' \in \mathbb{R}_{>0}^{|\mathcal{Z}|}$. Through the representation* (4.29)*, each CC is able to calculate the momentary value $a^\star$ in a distributed way.* ◇

4.4.2 Balance of Payments Restoration

Recall the time-integrated sum of payments from Definition 4.1 with (3.78) and (3.88):

$$\sigma(t) = \int_0^t -\sum_{k\in\mathcal{Z}} (\Lambda_k \cdot \kappa_k \cdot p_{\mathrm{g},i} - \hat{\omega}_k \cdot \hat{p}_{\mathrm{g},k}) + \sum_{k\in\mathcal{Z}} \Lambda^0 \cdot \hat{p}_{\ell,k} \,\mathrm{d}t'. \tag{4.34}$$

Applying the above "primary" control actions for BoP containment (4.3) leads to a finite, but generally nonzero end value $\sigma^\star$ (cf. Remark 4.4). If $\sigma^\star > 0$, then there exists an excess of

[61] Transmission losses are already considered by the automatic dispatch of $\boldsymbol{p}_{\mathrm{g}}$ developed in Chapter 3 and therefore do not have to be explicitly contained in (4.33).

[62] By contrast, the current electricity market always requires additional grid tariffs to rebalance the resulting mismatch. See the discussion in Section 2.1.

capital since the RPs' payments are higher than the revenues that have been received by WPs. Conversely, if $\sigma^\star < 0$, then a deficit of payments exists because the revenues of WPs exceed the payments from RPs. In the following, we will derive the "secondary/tertiary" control actions to *optimally* regulate σ towards zero as $t \to \infty$.

A reasonable choice for the additive restoration signals $\boldsymbol{\epsilon}^\kappa$ is given by the first-order lag relation

$$\dot{\epsilon}_k^\kappa = -\epsilon_k^\kappa + \hat{p}_{\mathrm{g},k} \cdot \sigma + u_k, \qquad k \in \mathcal{Z} \tag{4.35}$$

where u_k is a cell-specific control input yet to be determined, and where the positive sign of $+\hat{p}_{\mathrm{g},k} \cdot \sigma$ ensures that an excess of overall payments by RPs ($\sigma > 0$) yields a *positive* correction $\epsilon_k^\kappa > 0$ (and vice versa),

Now the overall balance (4.34) along with the cell-specific ϵ_k^κ dynamics (4.35) can be written compactly by the port-Hamiltonian representation

$$\underbrace{\begin{bmatrix} \dot{\boldsymbol{\epsilon}}^\kappa \\ \dot{\sigma} \end{bmatrix}}_{\dot{\boldsymbol{x}}_\sigma} = \Bigg(\underbrace{\begin{bmatrix} \mathbf{0} & \hat{\boldsymbol{p}}_\mathrm{g} \\ -\hat{\boldsymbol{p}}_\mathrm{g}^\mathsf{T} & 0 \end{bmatrix}}_{\boldsymbol{J}_\sigma} - \underbrace{\begin{bmatrix} \boldsymbol{\Lambda}' & \mathbb{0} \\ \mathbb{0}^\mathsf{T} & 0 \end{bmatrix}}_{\boldsymbol{R}_\sigma} \Bigg) \nabla H_\sigma(\boldsymbol{x}_\sigma) + \underbrace{\begin{bmatrix} \boldsymbol{I} \\ \mathbb{0}^\mathsf{T} \end{bmatrix}}_{\boldsymbol{G}_\sigma} \boldsymbol{u} + \begin{bmatrix} \mathbb{0} \\ d \end{bmatrix}, \tag{4.36a}$$

$$H_\sigma(\boldsymbol{x}_\sigma) = \frac{1}{2} \cdot \Lambda^0 \cdot (\boldsymbol{\epsilon}^\kappa)^\mathsf{T} \boldsymbol{\epsilon}^\kappa + \frac{1}{2}\sigma^2, \tag{4.36b}$$

where $\boldsymbol{\Lambda}' := (1/\Lambda^0) \cdot \boldsymbol{I} \succ 0$, and $d = \Lambda^0 \cdot (\mathbb{1}^\mathsf{T} \hat{\boldsymbol{p}}_\ell - \hat{\boldsymbol{p}}_\mathrm{g}^\mathsf{T} \boldsymbol{\kappa}')$ is a disturbance that decays to 0 as $t \to \infty$.

Lemma 4.19. *Let the cell-specific power injections* $\hat{\boldsymbol{p}}_\mathrm{g} \neq \mathbb{0}$ *be constant and let* (4.3) *hold. Then the autonomous system* (4.36) *exhibits a globally asymptotically stable equilibrium at the origin.*

Proof. Consider the Lyapunov function candidate $H_\sigma(\boldsymbol{x}_\sigma) \succ 0$ in (4.36b). The time derivative of $H_\sigma(\boldsymbol{x}_\sigma)$ along the trajectories of (4.36) is given by $\dot{H}_\sigma = -(\boldsymbol{\epsilon}^\kappa)^\mathsf{T} \boldsymbol{\epsilon}^\kappa$. Now define the set $E = \{\boldsymbol{x}_\sigma \in \mathbb{R}^{|\mathcal{Z}|+1} : \dot{H}_\sigma(\boldsymbol{x}_\sigma) = 0\} = \{\boldsymbol{x}_\sigma \in \mathbb{R}^{|\mathcal{Z}|+1} : \boldsymbol{\epsilon}^\kappa = \mathbb{0}\}$. Since the largest invariant set contained in E is the singleton $\{(\boldsymbol{\epsilon}^\kappa = \mathbb{0}, \sigma = 0)\}$ and $H_\sigma(\boldsymbol{x}_\sigma)$ is radially unbounded, global asymptotic stability of the origin follows by LaSalle's invariance principle [Kha02, Corollary 4.2]. □

Now the "optimal" control input $\boldsymbol{u} = \boldsymbol{u}^\star$ shall be chosen in such a way that the resulting closed-loop trajectory $((\boldsymbol{\epsilon}^\kappa)^\star(t), \sigma^\star(t))$ of (4.36) is regulated towards zero so that a user-defined quadratic Lagrange-type performance index is minimized. In particular, we specify

$$\min_{\boldsymbol{u}} \quad \int_0^{t_\mathrm{f}} (\boldsymbol{\epsilon}^\kappa)^\mathsf{T} \boldsymbol{Q}_1 \boldsymbol{\epsilon}^\kappa + Q_2 \cdot \sigma^2 + \boldsymbol{u}^\mathsf{T} \boldsymbol{S} \boldsymbol{u} \, \mathrm{d}t \tag{4.37a}$$

$$\text{subject to} \quad (4.36), \tag{4.37b}$$

where $\boldsymbol{Q}_1 \succcurlyeq 0$ and $Q_2 \geq 0$ are weighting factors for the transient error of the local restoration signals $\boldsymbol{\epsilon}^{\kappa}(t)$ and the overall balance $\sigma(t)$, respectively, and the control weight matrix $\boldsymbol{S} \succ 0$ penalizes the inputs $\boldsymbol{u}(t)$.

In fact, (4.37) entails an *optimal control problem* with quadratic performance index. This problem class is extensively studied in the following Chapter 5 by deriving explicit feedback control laws for the optimal control input $\boldsymbol{u}^{\star}$. Furthermore, the proposed methodology will be applicable to general Lagrange-type performance indices (cf. Definition A.7 in Appendix A.2) subject to general (nonlinear) PHS dynamics.

4.5 Analysis of the Closed-Loop System

This section studies the closed-loop equilibrium resulting from the interconnection between the physical plant system (3.49) and frequency and voltage controller (3.110) from Chapter 3 along with the dynamic pricing scheme from Chapter 4. It concludes with a stability analysis of the individual subsystems with respect to this equilibrium.

Theorem 4.20. *Each equilibrium of the closed-loop system* (3.49), (3.110), (3.142), (4.29), (4.11), (4.15) *has the following properties:*

a) *For all $i \in \mathcal{V}_{\mathrm{g}} \cap \mathcal{V}_{\mathcal{Z},k}$ with given prices $\Lambda_k = \lambda_i^{\star}$, the active power generation $p_{\mathrm{g},i}^{\star}$ is a solution of* (3.83).

b) *All frequency deviations are zero, i.e. $\boldsymbol{\omega}^{\star} = \mathbb{0}$.*

c) *For all $i \in \mathcal{V}_{\mathrm{g}}$, it holds that $U_i^{\star} = U_i^{\mathrm{set}\star}$ if no voltage limits* (3.83c) *are active.*

d) *BoP containment* (4.3) *is fulfilled.*

e) *Congestion is prevented, i.e. it holds that $|\vec{\boldsymbol{S}}| \leq \boldsymbol{S}^{\max}$ whenever the integral action feedback* (4.15) *is used for congestion management.*

Proof. a) Since (3.110a)–(3.110i) is the projected saddle-point flow of the convex optimization problem (3.83), it follows from Lemma 3.29 that each equilibrium of (3.110a)–(3.110i) with $\Lambda_k = \lambda_i^{\star}$is a solution of (3.83).

b) This statement immediately follows from Lemma 3.35.

c) For each $i \in \mathcal{V}_{\mathcal{S}}$, the equilibrium condition of (3.110g) equals

$$0 = \langle U_{\mathrm{f},i}^{\star} - \overline{U}_{\mathrm{f},i} \rangle_{\mu_{\mathcal{S}+,i}}^{+} = \begin{cases} U_{\mathrm{f},i}^{\star} - \overline{U}_{\mathrm{f},i}, & (U_{\mathrm{f},i}^{\star} - \overline{U}_{\mathrm{f},i} > 0) \text{ and } (\mu_{\mathcal{S}+,i}^{\star} > 0), \\ 0, & \text{otherwise.} \end{cases} \tag{4.38}$$

Since we assume that no voltage limits are active, it holds that $U_{\mathrm{f},i} < \overline{U}_{\mathrm{f},i}$. Thus, (4.38) states that $\mu_{\mathcal{S}+,i}^{\star} = 0$. The same reasoning can be applied to the steady-state conditions (3.110f), (3.110h), and (3.110i) to obtain $\boldsymbol{\mu}_{\mathcal{S}-} = \mathbb{0}$, $\boldsymbol{\mu}_{\mathcal{I}-} = \mathbb{0}$, and $\boldsymbol{\mu}_{\mathcal{I}+} = \mathbb{0}$. Using these results in conjunction with the equilibrium conditions of (3.142), we get $U_{\mathrm{g},i}^{\star} = (U_i^{\mathrm{set}})^{\star}$ for all $i \in \mathcal{V}_{\mathrm{g}}$, which is equivalent to c).

d) Since $\hat{\boldsymbol{D}}_{\mathrm{p}}$ is the incidence matrix of a weakly connected graph, each equilibrium of (4.29b) fulfills $a_1^\star = \ldots = a_N^\star =: a^\star$. Left-multiplying (4.29a) with $\mathbb{1}^\mathsf{T}$ yields $0 = -a^\star \left((\boldsymbol{\kappa}'')^\star\right)^\mathsf{T} \hat{\boldsymbol{p}}_{\mathrm{g}} + \mathbb{1}^\mathsf{T}\hat{\boldsymbol{p}}_\ell$, which is equal to the BoP containment condition (see (4.3)).

e) For any $m \in \hat{\mathcal{E}}_{\mathrm{p}}$, let $|S_m^\star| > S_m^{\max}$ hold such that e) is violated. Accordingly, $|\mathcal{C}_m| > 1$, and (4.15) implies that $\dot{\mho}_m = \frac{1}{\tau_{\mho_m}^{+}}\Re\{\mathcal{C}_m\} \neq 0$. Hence it follows by contradiction that each equilibrium of (3.49), (3.110), (3.142), (4.29), (4.11), (4.15) fulfills $|\vec{\boldsymbol{S}}| \leq \boldsymbol{S}^{\max}$. □

Remark 4.21. *Requirement c) in Theorem 4.20 can always be fulfilled by choosing $\overline{U}^\Delta$ sufficiently small, which prevents the limits on excitation and inverter output voltage from being reached.* ◇

Theorem 4.22. *If $\boldsymbol{\kappa}'' > \mathbb{0}$, $\hat{\boldsymbol{p}}_{\mathrm{g}} > \mathbb{0}$ and $\hat{\boldsymbol{p}}_\ell$ are constant, then the steady-state $(\boldsymbol{a}^\star, \boldsymbol{\nu}_{\mathrm{a}}^\star)$ of the balancing controller (4.29) is globally asymptotically stable.*

Proof. Let $\boldsymbol{\xi}_{\mathrm{M}} := (\boldsymbol{a}, \boldsymbol{\nu}_a)$ be the state of (4.29), and consider the Lyapunov function candidate

$$V(\boldsymbol{\xi}_{\mathrm{M}}) = \frac{1}{2}\Big((\boldsymbol{a} - \boldsymbol{a}^\star)^\mathsf{T}\boldsymbol{\tau}_a(\boldsymbol{a} - \boldsymbol{a}^\star) + (\boldsymbol{\nu}_a - \boldsymbol{\nu}_{\mathrm{a}}^\star)^\mathsf{T}\boldsymbol{\tau}_{\nu_a}(\boldsymbol{\nu}_a - \boldsymbol{\nu}_{\mathrm{a}}^\star)\Big).$$

Since $\boldsymbol{\tau}_{(\cdot)} \succ 0$, it follows that $V(\boldsymbol{\xi}_{\mathrm{M}}) > 0$ for all $\boldsymbol{\xi}_{\mathrm{M}} \neq \boldsymbol{\xi}_{\mathrm{M}}^\star$. The time derivative of $V(\boldsymbol{\xi}_{\mathrm{M}})$ along the trajectories of (4.29) equals

$$\begin{aligned}\dot{V}(\boldsymbol{\xi}_{\mathrm{M}}) &= -(\boldsymbol{a} - \boldsymbol{a}^\star)^\mathsf{T}(\boldsymbol{\kappa}'' \circ \hat{\boldsymbol{p}}_{\mathrm{g}} \circ (\boldsymbol{a} - \boldsymbol{a}^\star)) \\ &\quad + \underbrace{(\boldsymbol{a} - \boldsymbol{a}^\star)^\mathsf{T}\hat{\boldsymbol{D}}_{\mathrm{p}}(\boldsymbol{\nu}_a - \boldsymbol{\nu}_{\mathrm{a}}^\star) - (\boldsymbol{\nu}_a - \boldsymbol{\nu}_{\mathrm{a}}^\star)^\mathsf{T}\hat{\boldsymbol{D}}_{\mathrm{p}}^\mathsf{T}(\boldsymbol{a} - \boldsymbol{a}^\star)}_{=0} \qquad (4.39) \\ &= (\boldsymbol{a} - \boldsymbol{a}^\star)^\mathsf{T} \underbrace{-\mathrm{diag}(\boldsymbol{\kappa}'' \circ \hat{\boldsymbol{p}}_{\mathrm{g}})}_{\boldsymbol{T}}(\boldsymbol{a} - \boldsymbol{a}^\star). \qquad (4.40)\end{aligned}$$

As $\boldsymbol{\kappa}'' > \mathbb{0}$ and $\hat{\boldsymbol{p}}_{\mathrm{g}} > \mathbb{0}$ hold per definition, the matrix $\boldsymbol{T}$ in (4.40) is negative definite and thus $\dot{V}(\boldsymbol{\xi}_{\mathrm{M}}) \leq 0$. We define the set $E = \{\boldsymbol{\xi}_{\mathrm{M}} \in W \ : \ \dot{V}(\boldsymbol{\xi}_{\mathrm{M}}) = 0\} = \{(\boldsymbol{a}, \boldsymbol{\nu}_a) \ : \ \boldsymbol{a} = \boldsymbol{a}^\star\}$. The largest invariant set contained in E is $\{\boldsymbol{\xi}_{\mathrm{M}}^\star\}$. As (4.29) is derived using distributed consensus-based control and the centralized steady-state variable $a^\star$ is unique (i.e. the unique solution of (4.28)), it follows by Lemma B.1 that $\boldsymbol{\xi}_{\mathrm{M}}^\star$ is unique as well. Since the largest invariant set contained in E is a singleton and $V(\boldsymbol{\xi}_{\mathrm{M}})$ is radially unbounded, it thus follows from LaSalle's invariance principle that $\boldsymbol{\xi}_{\mathrm{M}}^\star$ is globally asymptotically stable [Kha02, Corollary 4.2]. □

Theorem 4.23. *If $\boldsymbol{\mho}$ and $\boldsymbol{\mathcal{A}}$ are constant, then the equilibrium $\boldsymbol{\phi}^\star$ of the zonal pricing controller (4.11) is globally asymptotically stable.*

Proof. Define $\boldsymbol{\phi}'(t) = \boldsymbol{\tau}_\phi \boldsymbol{\phi}(t)$ and note that $\boldsymbol{\tau}_\phi \succ 0$. From (4.11), it follows that $\mathbb{1}^\top \boldsymbol{\tau}_\phi \dot{\boldsymbol{\phi}} = \mathbb{1}^\top \dot{\boldsymbol{\phi}}' = 0$ for all $t > 0$ since $\boldsymbol{\mathcal{A}} = \hat{\boldsymbol{D}}_\mathrm{p} \mathrm{diag}\{|\vec{\boldsymbol{S}}|\} \hat{\boldsymbol{D}}_\mathrm{p}^\top$ and $\mathbb{1}^\top \hat{\boldsymbol{D}}_\mathrm{p} = \mathbb{0}^\top$. This means $\mathbb{1}^\top \boldsymbol{\phi}'(t) = \mathbb{1}^\top \boldsymbol{\phi}'(0)$ for all $t > 0$. Therefore the unique equilibrium of $\boldsymbol{\phi}'(t)$ is $\boldsymbol{\phi}'^\star = \bar{\boldsymbol{\phi}}'^\star + \bar{\phi}' \cdot \mathbb{1}$ with $\mathbb{1}^\top \bar{\boldsymbol{\phi}}'^\star = 0$ and $\bar{\phi}' = \mathbb{1}^\top \boldsymbol{\phi}(0)/N$. Using the Lyapunov function candidate $V(\boldsymbol{\phi}') = \frac{\boldsymbol{\tau}_\phi}{2}(\boldsymbol{\phi}' - \boldsymbol{\phi}'^\star)^\top(\boldsymbol{\phi}' - \boldsymbol{\phi}'^\star)$, we get $\dot{V}(\boldsymbol{\phi}') = -(\boldsymbol{\phi}' - \boldsymbol{\phi}'^\star)^\top \boldsymbol{\mathcal{A}} (\boldsymbol{\phi}' - \boldsymbol{\phi}'^\star)$ and $\dot{V}(\boldsymbol{\phi}') < 0$ for all $\boldsymbol{\phi}' \neq \boldsymbol{\phi}'^\star$ due to positive semidefiniteness of $\boldsymbol{\mathcal{A}}$ and $(\boldsymbol{\phi}' - \boldsymbol{\phi}'^\star) \notin \ker(\boldsymbol{\mathcal{A}})$. As $V(\boldsymbol{\phi}')$ is radially unbounded, it follows that $\boldsymbol{\phi}'^\star$ (and thus $\boldsymbol{\phi}^\star$) is globally asymptotically stable [Kha02, Theorem 4.2]. □

Remark 4.24. *For lossless power networks with* $U_i = \mathrm{const.}$ *and* $\vartheta_{ij} \approx 0$*, the Laplacian matrix can be assumed to be constant (cf.* (4.13) *in Remark 4.8). Thus, the zonal pricing controller* (4.11) *turns into a* distributed formation controller *[OSFM07], where the resulting state trajectory* $\boldsymbol{\phi}(t)$ *is a weighted average of the (time-varying) reference signal given by the inputs* $\boldsymbol{\psi}(t)$*, cf. [FYL06, Ren07, KvC+ 19].* ◊

Under the premise of timescale separation, the stability statements for the underlying frequency and voltage controller can be upheld. This feature is formalized in the next theorem.

Theorem 4.25. *Let* $\boldsymbol{x} = \mathrm{col}\{\boldsymbol{x}_\mathrm{p}, \boldsymbol{p}_\mathrm{g}, \boldsymbol{\mu}_{\mathrm{g}-}, \boldsymbol{\mu}_{\mathrm{g}+}, \boldsymbol{U}_\mathrm{f}, \boldsymbol{\mu}_{\mathcal{S}-}, \boldsymbol{\mu}_{\mathcal{S}+}, \boldsymbol{U}_\mathcal{I}, \boldsymbol{\mu}_{\mathcal{I}-}, \boldsymbol{\mu}_{\mathcal{I}+}\}$ *denote the state of* (3.49), (3.110). *Furthermore, let* $\boldsymbol{\kappa}$ *and* $\hat{\boldsymbol{U}}^{\mathrm{set}}$ *be fixed, the Hessian of* $H_\mathrm{p}(\boldsymbol{x}_\mathrm{p})$ *be positive definite in a neighborhood around* $\boldsymbol{x}_\mathrm{p}^\star$ *and let the conditions of Lemmas 3.39 and 3.41 hold in a neighborhood around* $\boldsymbol{x}^\star$*. Then the closed-loop equilibrium* $\boldsymbol{x}^\star$ *of the interconnection of the physical plant system* (3.49) *with the frequency controller* (3.110a), (3.110d)–(3.110e), (3.110j)–(3.110k) *and voltage controller* (3.110b)–(3.110c), (3.110f)–(3.110i) *is stable in a neighborhood around* $\boldsymbol{x}^\star$*.*

Proof. (see Appendix C.1) □

Corollary 4.26. *Let* $\boldsymbol{x} = \mathrm{col}\{\boldsymbol{x}_\mathrm{p}, \boldsymbol{p}_\mathrm{g}, \boldsymbol{\mu}_{\mathrm{g}-}, \boldsymbol{\mu}_{\mathrm{g}+}, \boldsymbol{U}_\mathrm{f}, \boldsymbol{\mu}_{\mathcal{S}-}, \boldsymbol{\mu}_{\mathcal{S}+}, \boldsymbol{U}_\mathcal{I}, \boldsymbol{\mu}_{\mathcal{I}-}, \boldsymbol{\mu}_{\mathcal{I}+}\}$ *denote the state of* (3.49), (3.110) *where* (3.49) *is* lossless, *i.e.* $\boldsymbol{G} = \boldsymbol{0}$*. Furthermore, let* $\boldsymbol{\kappa}$ *and* $\hat{\boldsymbol{U}}^{\mathrm{set}}$ *be fixed and the Hessian of* $H_\mathrm{p}(\boldsymbol{x}_\mathrm{p})$ *be positive definite in a neighborhood around* $\boldsymbol{x}_\mathrm{p}^\star$*. Then the closed-loop equilibrium* $\boldsymbol{x}^\star$ *of the interconnection of the physical plant system* (3.49) *with the frequency controller* (3.110a), (3.110d)–(3.110e), (3.110j)–(3.110k) *and voltage controller* (3.110b)–(3.110c), (3.110f)–(3.110i) *is stable in a neighborhood around* $\boldsymbol{x}^\star$*.*

Proof. With $\boldsymbol{G} = \boldsymbol{0}$, condition $W_2(\boldsymbol{x}) \leq 0$ in the proof of Theorem 4.25 is automatically fulfilled, since for lossless grids it holds that $\boldsymbol{\rho} = \mathbb{0}$ and $\boldsymbol{\mathcal{R}}_\mathrm{p}(\boldsymbol{x}_\mathrm{p}) = \boldsymbol{R}_\mathrm{p} \succcurlyeq 0$. □

The requirements for constant parameters stated in Theorem 4.22, Theorem 4.23 and Theorem 4.25 clearly indicate the ubiquitous need for *sufficient timescale separation* between the individual subsystems (see Subsection 3.4.1 and references therein). Even if there is no quantitative certificate, sufficient timescale separation can be managed conveniently by the controller parameters $\boldsymbol{\tau}_{(\cdot)}$. In particular, to meet the requirements of Theorem 4.25 that $\boldsymbol{\kappa}$ and $\hat{\boldsymbol{U}}^{\text{set}}$ are constant, $\boldsymbol{\tau}_{\phi}$ and $\boldsymbol{\tau}_{U^{\text{set}}}$ must be chosen sufficiently large compared to the controller parameters in (3.110). In order to meet the requirements of Theorem 4.23, $\boldsymbol{\tau}_{\breve{U}}^{-}$ and $\boldsymbol{\tau}_{\breve{U}}^{+}$ must be larger than $\boldsymbol{\tau}_{\phi}$. In order to cope with Theorem 4.22, $\boldsymbol{\tau}_{a}$ and $\boldsymbol{\tau}_{\nu_a}$ must be larger than $\boldsymbol{\tau}_{\phi}$ and the other controller parameters in (3.110). Thus, the individual controllers operate on three different timescales[63] as depicted in Fig. 4.2. We can clearly see the correlation between the three components in Fig. 4.2 to Contributions 1, 2, and 4 outlined in Fig. 2.3. In Fig. 4.7, the individual dependencies and signal flows between these three components are plotted in more detail. A comparison with today's separation of timescales shown in Fig. 2.2 indicates that by unifying all tasks into three sub-tasks, there is a sufficiently large "gap" between the individual controller parameters possible—specifically, a factor of at least 100 can be chosen between each $\boldsymbol{\tau}_{(\cdot)}$. For further classification and conclusions, please refer to Chapter 7.

4.6 Discussion

This chapter has presented a distributed control scheme for the automatic settlement of temporally and spatially differentiated prices. The developed continuous-time pricing scheme which is solely based upon neighbor-to-neighbor communication between CCs comprises of two conceptually distinct steps:

In the first step, a dynamic locational pricing scheme is provided by means of an automatic control of cell-specific participation factors $\kappa_k''(t)$. By modulating the uniform price $\Lambda^0(t)$ locally, the resulting pricing strategies are able to account for the current *scarcity of (marginal) production* in a specific location at a specific point in time. In Section 4.3, we have introduced different approaches with and without integral action for the practical choice of $\kappa''(t)$. In this context, special emphasis has been put on the development of a new real-time approach for distributed, flow-based congestion control. Through its "holistic" consideration of both active and reactive power flows along with integral action, the congestion controller enables the possibility to accurately incorporate (and thus exploit) the thermal load capacity of each separate inter-cell transmission line.

In the second step, *regularized* zonal prices $\kappa_k'(t)$ are generated out of $\kappa_k''(t)$ such that each resulting closed-loop equilibrium satisfies BoP containment. A salient feature of this regularization mechanism is that the allocation of WPs' power injections obtained in the first step are left unaltered, while at the same time all RPs are provided with a retail price $\Lambda_{\text{R}} = \lambda^\star$ that is independent of the individual cell and by which all transmission losses are recompensed instantaneously. The combination with the optimization-based frequency and voltage controller

[63] Note that these three timescales do not correspond to the classical hierarchy of primary, secondary and tertiary control. Instead, the controller (3.110) presented in Sections 3.4 and 3.5 covers all aspects of the classical frequency and voltage control, while the above control layers shown in Fig. 4.2 address additional (higher-level) objectives as suggested in Contributions 2 and 4.

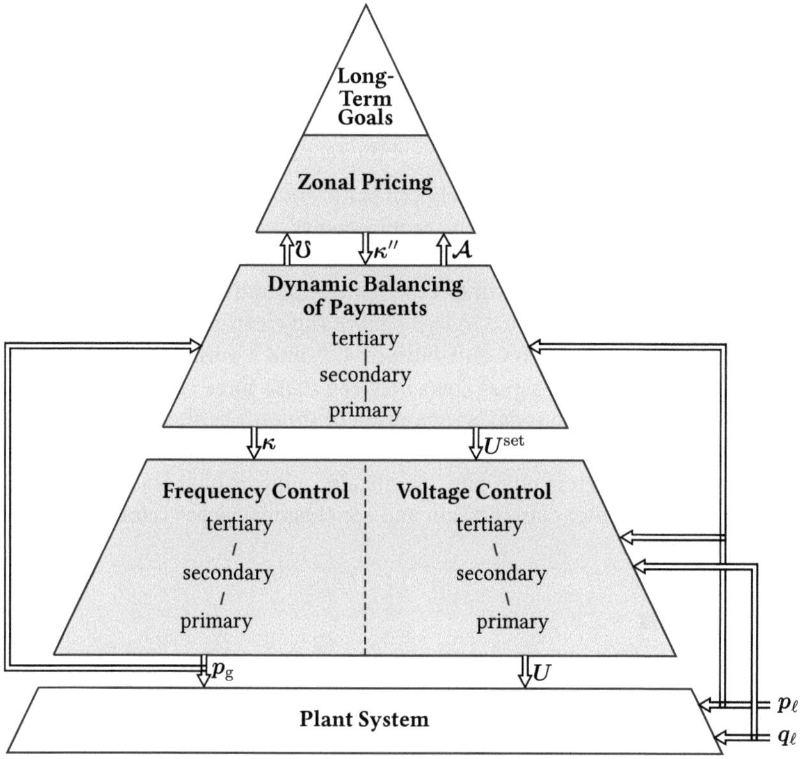

Figure 4.7: Controller structure for real-time zonal pricing with three different timescales. Subsystems that are depicted at the top act at slower timescales, those depicted at the bottom act at faster timescales.

(3.110) deployed in Chapter 3 provides economically efficient equilibria where the adherence of network constraints is fulfilled at market clearing without the need of a "global" (superordinate) authority, after-market settlements, or any other type of side payments. Moreover, due to the automatic regulation of wholesale and retail prices, no additional BRPs are necessary any longer.

So far, we have not developed an explicit procedure for determining the restoration signal $\epsilon^{\kappa}(t)$ mentioned in Section 4.1 and Subsection 4.4.2, which allows the resulting participation factors $\kappa_k = \kappa_k' + \epsilon_k^{\kappa}$ to even fulfill (4.2), so that any residual imbalances of costs and revenues that may have occured during transient processes can be settled in an optimal way with regard to the optimal control problem (4.37). The methodological tools necessary for this task are derived in the next chapter.

5 Optimal Control of Port-Hamiltonian Systems

This chapter is dedicated to the class of optimal control problems for input-state-output PHSs (A.13) with Lagrange-type objective functions (A.8) which is a generalization of the BoP restoration problem (4.37). Section 5.1 gives a brief introduction to main research branches in the field of PHS control. Section 5.2 presents a learning-based methodology for general optimal control problems with input-state-output PHSs. Section 5.3 extends the results from Section 5.2 to the multi-player case. The main results of Sections 5.2 and 5.3 give rise to Contribution 3. Section 5.4 first revisits the application of the proposed methodology to BoP restoration (4.37) in order to deliver the remaining steps of Contribution 2, and then outlines worthwhile avenues for further research[64].

5.1 Introduction

PHS modeling has had a major impact with respect to energy-based control design (see. e.g. [OvMM01, Kug01, OvME02, DBMS09, vJ14]. It opens a different perspective on viewing dynamical systems closer to physics and the fundamental concept of energy, which allowed for valuable advancements in science and engineering practice. Due to their specific energy-based structure with power-conjugated port variables $(\boldsymbol{u}, \boldsymbol{y})$ and a Hamiltonian $H(\boldsymbol{x})$ representing the total stored energy in the system, the concept of energy plays a central role in controller design for PHSs. In particular, as the PHS representation has a strong link to passivity properties (which is exemplarily seen in Subsection 3.4.3), the vast majority of control methods for PHSs rely on *passivity-based control* (PBC).

Besides feedback interconnection of passive systems [SJK97, p. 107ff.], a common theme in standard PBC is the passivation and asymptotic stabilization (under an additional detectability condition) via static state feedback $\boldsymbol{u}(\boldsymbol{x}) = -\boldsymbol{k}(\boldsymbol{x})$ [SJK97, Chapter 2.4; van17, Chapters 5,7; OGC04, OvCA08]. Both approaches are comparable in the sense that the controller parameters are not unique, but must be found implicitly by solving *matching PDEs*, which entails a high number of degrees of freedom and, more severely, a high degree of computational complexity. In order to reduce this complexity, the degrees of freedom are usually confined at certain points by setting a subset of the free parameters to a fixed value, thus leading to sub-categories such as algebraic or (non-)parameterized IDA-PBC, see [van17, Chapter 7] for a comprehensive survey on PBC methods for PHSs. For the passivity-based state feedback controller design, an *inverse optimality* property can be characterized in both continuous [SJK97, p. 107ff.; van17, Theorem 3.5.1] and discrete time [MNC15]. For input-state-output PHSs, it is known

64 Preliminary versions of the results of this chapter have been published in the journal papers [KJSSH21, KJSMH22].

that a static feedback controller is optimally stabilizing with respect to the specific performance index

$$\mathscr{J}(\boldsymbol{x}(t), \boldsymbol{u}(t)) = \frac{1}{2} \int_{0}^{t_\mathrm{f}} \ell(\boldsymbol{x}(t)) + \boldsymbol{u}^\mathsf{T}(t)\boldsymbol{u}(t) \, \mathrm{d}t \tag{5.1}$$

with

$$\ell(\boldsymbol{x}(t)) = -\left(\frac{\partial V(\boldsymbol{x}(t))}{\partial \boldsymbol{x}}\right)^\mathsf{T} \boldsymbol{f}(\boldsymbol{x}(t)) + \frac{1}{2}\boldsymbol{k}^\mathsf{T}(\boldsymbol{x}(t))\boldsymbol{k}(\boldsymbol{x}(t)), \tag{5.2}$$

if and only if the open-loop system is output feedback passive with a positive definite storage function $V(\boldsymbol{x}(t))$ and provided that an additional detectability condition is fulfilled, see [van17, p. 54ff.] and [SJK97, Theorem 3.30] for a detailed discussion. However, classical, PBC methods are in general not designed for optimal control problems in a more practical setup, where the "forward" solution to a given, arbitrary optimization problem with a general performance index (A.8) is sought. E.g. in (4.37), an optimal restoration signal $(\boldsymbol{\epsilon}^{\kappa})^{\star}(t)$ has to be determined that minimizes the sum of the state error (weighted by $\boldsymbol{Q}_1$ and Q_2) and control input error (weighted by $\boldsymbol{S}$).

In the following section, we consider the general case to (4.37) with an arbitrary Lagrangian performance index. This problem class is equal to (A.8) with $\mathbb{P} = \{1\}$, and thus refered to as the *single-player case* throughout this chapter.

5.2 Single-Player Case

This section presents a new continuous-time adaptive feedback controller for the optimal control of input-state-output PHSs (A.13) with respect to general Lagrangian performance indices (A.8) with $\mathbb{P} = \{1\}$. The proposed control law implements an online learning procedure which uses the Hamiltonian $H(\boldsymbol{x})$ of the system as an initial value function candidate. Subsection 5.2.1 formalizes the problem class and briefly discusses related work. Subsection 5.2.2 presents an analytical solution to a *modified optimal control* problem. Subsection 5.2.3 provides necessary and sufficient conditions for $H(\boldsymbol{x})$ to serve as initial value function candidate. Subsection 5.2.4 develops an adaptation strategy by which the modified optimal controller is able to learn the (original) optimal control law. For this adaptive optimal controller, Subsection 5.2.5 proves the asymptotic stability of the closed-loop equilibrium and gives an illustrative example. Subsection 5.2.6 summarizes the main result.

5.2.1 Problem Definition and Related Work

Definition 5.1 (Optimal Control Problem for PHSs). *Find the feedback control law $\boldsymbol{u}(\boldsymbol{x})$ as the solution of*

$$\min_{\boldsymbol{u}(t)} \quad \frac{1}{2}\int_0^{t_f} \left(\ell(\boldsymbol{x}(t)) + (\boldsymbol{u}(t))^{\mathsf{T}}\boldsymbol{S}(\boldsymbol{x})\boldsymbol{u}(t)\right)\mathrm{d}t \tag{5.3a}$$

$$\text{subject to} \quad \dot{\boldsymbol{x}}(t) = (\boldsymbol{J}(\boldsymbol{x}(t)) - \boldsymbol{R}(\boldsymbol{x}(t)))\frac{\partial H(\boldsymbol{x}(t))}{\partial \boldsymbol{x}(t)} + \boldsymbol{G}(\boldsymbol{x}(t))\boldsymbol{u}(t) \tag{5.3b}$$

with state vector $\boldsymbol{x} \in \mathbb{R}^n$, input vector $\boldsymbol{u} \in \mathbb{R}^m$, skew-symmetric interconnection matrix $\boldsymbol{J} \in \mathbb{R}^{n\times n}$, positive semidefinite dissipation matrix $\boldsymbol{R} \in \mathbb{R}^{n\times n}$, input matrix $\boldsymbol{G} \in \mathbb{R}^{n\times p}$, positive definite Hamiltonian $H(\boldsymbol{x}) : \mathbb{R}^n \to \mathbb{R}$, positive definite $\ell(\boldsymbol{x}) : \mathbb{R}^n \to \mathbb{R}$ and positive definite $\boldsymbol{S}(\boldsymbol{x}) \in \mathbb{R}^{m\times m}$.

In the case of *linear* input-state-output PHSs, the Hamiltonian $H(\boldsymbol{x})$ is quadratic, which allows to calculate an optimal controller by using state-dependent Riccati equations [PZCL11]. In [MJMF+09, GCS09, MFMC+08], the necessary conditions which follow from Pontryagin's Maximum Principle are used to derive an explicit expression for the optimal feedback controller, provided the Hamiltonian of the system is quadratic. The authors in [WHLM18] provide full- and reduced-order LQR controllers for linear input-state-output PHSs. Further extensions of LQ-optimal control for PHSs are given for stochastic or infinite-dimensional spaces [LW18] and boundary control systems [LLL20].

While there is a rich theory available for linear input-state-output PHSs, the general solution of optimal control problems for *nonlinear* input-state-output PHSs remains challenging due to the necessity of explicitly solving the HJB equation which is a nonlinear PDE and thus hard to solve. Recent publications use *adaptive dynamic programming* (ADP) methods to find an iterative solution of the HJB equation. In [ASF+20], a combination of neuro-fuzzy and backstepping control is proposed for adaptive control of input-state-output PHSs. A profound overview on recent adaptive and learning-based control methods for PHSs can be found in [NLJB16]. If the performance index of the optimal control problem has a specific structure and the system dynamics are given by a Hamiltonian system with controlled Hamiltonian $H(\boldsymbol{x}, \boldsymbol{u})$, iterative learning control [Fuj03, FHS03] and iterative feedback tuning methods [FK08b] have been proposed. For fully actuated mechanical PHSs, the authors in [OFM+20] propose an adaptive path-following controller from a training trajectory using Bayesian estimation and the authors in [ZDC20] propose a deep learning-based algorithm for optimal estimation of the system dynamics. [NLJB14] and [SBNL15] use actor-critic reinforcement learning schemes to minimize the error between the resulting closed-loop system and a given desired closed-loop system without the need of explicitly solving the matching PDE of the employed passivity-based controller. However, these approaches suffer from the dissipation obstacle[65], such that the Hamiltonian of the desired closed-loop system cannot be chosen freely.

[65] The dissipation obstacle [OvMM01] states that the Hamiltonian $H(\boldsymbol{x})$ can only be shaped for coordinates that are not affected by physical damping. This limits the applicability of these controller classes to certain physical domains, see e.g. the discussion in [ZOJS15].

For the more general class of *continuous-time input-affine nonlinear systems*, ADP methods [BJ16] are proposed where the optimal value function $V^\star(\boldsymbol{x})$ can be found iteratively using a weighted sum of basis functions [VL10, JJ13]. However, a proper set of initial weights leading to a stabilizing controller has to be found by educated guessing.

In summary, existing optimal control approaches of input-state-output PHSs hinge on very specific sub-classes of performance index, system dynamics, or both. Likewise, ADP methods for input-state-output PHSs as well as for the more general class of continuous-time, input-affine nonlinear systems are non-constructive in the sense that they require the intransparent guessing of initial weights for a stabilizing value function candidate. Hence, to the best of the author's knowledge, there exist no explicit control schemes for dynamic optimal control problems with generalized Lagrangian performance index and a general input-state-output PHS.

5.2.2 Modified Optimal Control for Port-Hamiltonian Systems

The initial step of our controller design is based on a trick originally outlined in [SK00]. By multiplying the dynamic constraints (5.3b) with the gradient of a *control-Lyapunov function* (CLF) $V(\boldsymbol{x})$, we obtain a MOC problem that allows for an analytical solution of an asymptotically stabilizing $\boldsymbol{u}(\boldsymbol{x}(t)) = -\boldsymbol{k}(\boldsymbol{x}(t))$.

Definition 5.2 (Control-Lyapunov Function (CLF) [FK08a, p. 46]). *Let* $\dot{\boldsymbol{x}} = \boldsymbol{f}(\boldsymbol{x}, \boldsymbol{u})$, $\boldsymbol{x} \in \mathbb{R}^n$, $\boldsymbol{u} \in \mathbb{R}^m$ *with* $\boldsymbol{f}(\mathbb{0}_n, \mathbb{0}_m) = \mathbb{0}_n$. *Then a radially unbounded, positive definite function* $V : \mathbb{R}^n \to \mathbb{R}$, *fulfilling*

$$\forall \boldsymbol{x} \neq \mathbb{0}_n : \qquad \inf_{\boldsymbol{u}} \left\{ \left(\frac{\partial V}{\partial \boldsymbol{x}}\right)^{\mathsf{T}} \boldsymbol{f}(\boldsymbol{x}, \boldsymbol{u}) \right\} < 0 \tag{5.4}$$

is said to be a CLF.

Remark 5.3. *For input-affine nonlinear systems*

$$\dot{\boldsymbol{x}}(t) = \boldsymbol{f}(\boldsymbol{x}(t)) + \boldsymbol{g}(\boldsymbol{x}(t))\boldsymbol{u}(t), \tag{5.5}$$

condition (5.4) *is equivalent to [KA01, p. 641]*

$$\left(\frac{\partial V}{\partial \boldsymbol{x}}\right)^{\mathsf{T}} \boldsymbol{G}(\boldsymbol{x}) = \mathbb{0}_m \Longrightarrow \left(\frac{\partial V}{\partial \boldsymbol{x}}\right)^{\mathsf{T}} \boldsymbol{f}(\boldsymbol{x}) \begin{cases} < 0, & \boldsymbol{x} \neq \mathbb{0}_n, \\ = 0, & \boldsymbol{x} = \mathbb{0}_n. \end{cases} \tag{5.6}$$

◇

It is well known that a CLF ensures the existence of an input $\boldsymbol{u} = -\boldsymbol{k}(\boldsymbol{x})$ such that the closed-loop system is asymptotically stable at the origin[66].

[66] The formal statement that the existence of a CLF is necessary and sufficient for stabilizability of a nonlinear system is given by Artstein's Theorem [Art83]. A first explicit control law that applies this finding in practice was provided by *Sontag's universal formula for stabilization* [Son89].

In particular, provided that a suitable CLF is given, it was shown in [SK00] that for general input-affine nonlinear systems (5.5), the following MOC problem can be solved explicitly:

Definition 5.4 (Modified Optimal Control (MOC) Problem for PHSs).

$$\min_{\boldsymbol{u}} \quad \frac{1}{2}\int_0^{t_\mathrm{f}} \left(\ell(\boldsymbol{x}) + \boldsymbol{u}^\mathsf{T}\boldsymbol{S}(\boldsymbol{x})\boldsymbol{u}\right)\mathrm{d}t \tag{5.7a}$$

$$\text{subject to} \quad \dot{V}(\boldsymbol{x}) = \left(\frac{\partial V(\boldsymbol{x})}{\partial \boldsymbol{x}}\right)^\mathsf{T}\left((\boldsymbol{J}(\boldsymbol{x}) - \boldsymbol{R}(\boldsymbol{x}))\frac{\partial H(\boldsymbol{x})}{\partial \boldsymbol{x}} + \boldsymbol{G}(\boldsymbol{x})\boldsymbol{u}\right), \tag{5.7b}$$

Lemma 5.5 (Modified Optimal Control Law). *Let* $V(\boldsymbol{x})$ *be a CLF of* (5.5). *Then an exact solution of the MOC problem* (5.7) *is given by the continuous control law*

$$\boldsymbol{u}^\star = \begin{cases} -\boldsymbol{S}^{-1}(\boldsymbol{x})\boldsymbol{G}^\mathsf{T}(\boldsymbol{x})\dfrac{\partial V}{\partial \boldsymbol{x}}\Upsilon(\boldsymbol{x}), & S_\Upsilon \neq 0, \\ \mathbb{0}, & S_\Upsilon = 0, \end{cases} \tag{5.8}$$

where

$$\Upsilon(\boldsymbol{x}) := \frac{f_\Upsilon + \sqrt{(f_\Upsilon)^2 + Q_\Upsilon \cdot S_\Upsilon}}{S_\Upsilon}, \tag{5.9}$$

$$f_\Upsilon(\boldsymbol{x}) := \left(\frac{\partial V}{\partial \boldsymbol{x}}\right)^\mathsf{T} \boldsymbol{f}(\boldsymbol{x}), \tag{5.10}$$

$$S_\Upsilon(\boldsymbol{x}) := \left(\frac{\partial V}{\partial \boldsymbol{x}}\right)^\mathsf{T} \boldsymbol{G}(\boldsymbol{x})\boldsymbol{S}^{-1}(\boldsymbol{x})\boldsymbol{G}^\mathsf{T}(\boldsymbol{x})\frac{\partial V}{\partial \boldsymbol{x}}, \tag{5.11}$$

$$Q_\Upsilon(\boldsymbol{x}) := \ell(\boldsymbol{x}). \tag{5.12}$$

Proof. The proof follows the lines of [SK00, Theorem 4.4]. For optimization problem (5.7), we get the Hamiltonian

$$\mathscr{H}(\boldsymbol{x}, \boldsymbol{u}, \Upsilon) = \frac{1}{2}\ell(\boldsymbol{x}) + \frac{1}{2}\boldsymbol{u}^\mathsf{T}\boldsymbol{S}(\boldsymbol{x})\boldsymbol{u} + \Upsilon \cdot \left(\frac{\partial V}{\partial x}\right)^\mathsf{T} \left(\boldsymbol{f}(\boldsymbol{x}) + \boldsymbol{G}(\boldsymbol{x})\boldsymbol{u}\right) \tag{5.13}$$

with scalar Lagrange multiplier Υ. Application of the control equation $\frac{\partial \mathscr{H}}{\partial \boldsymbol{u}} \stackrel{!}{=} \mathbb{0}$ leads to (5.8). As stated in [SK00, Theorem 4.3], a free final time and fixed final state in (5.7a) implies that for each optimal trajectory $(\boldsymbol{x}^\star, \Upsilon^\star)$ it has to hold that $\mathscr{H}(\boldsymbol{x}^\star, \boldsymbol{u}^\star, \Upsilon^\star) = 0$, which is equivalent to

$$\begin{aligned} &\frac{1}{2}(\Upsilon^\star)^2 \left(\frac{\partial V}{\partial \boldsymbol{x}}\right)^\mathsf{T} (\boldsymbol{x}^\star)\boldsymbol{G}(\boldsymbol{x}^\star)\boldsymbol{S}^{-1}(\boldsymbol{x}^\star)\boldsymbol{G}^\mathsf{T}(\boldsymbol{x}^\star)\frac{\partial V}{\partial \boldsymbol{x}}(\boldsymbol{x}^\star) \\ &- \Upsilon^\star \left(\frac{\partial V}{\partial \boldsymbol{x}}\right)^\mathsf{T} (\boldsymbol{x}^\star)\boldsymbol{f}(\boldsymbol{x}^\star) - \frac{1}{2}r(\boldsymbol{x}^\star) = 0. \end{aligned} \tag{5.14}$$

Since (5.14) is a quadratic function in Υ, it has the explicit solution

$$\Upsilon(\boldsymbol{x}) = \begin{cases} \dfrac{f_\Upsilon \pm \sqrt{(f_\Upsilon)^2 + Q_\Upsilon \cdot S_\Upsilon}}{S_\Upsilon}, & \left(\dfrac{\partial V}{\partial \boldsymbol{x}}\right)^\mathsf{T} \boldsymbol{G}(\boldsymbol{x}) \neq \mathbb{0}, \quad (5.15a) \\ -\dfrac{Q_\Upsilon}{2 f_\Upsilon}, & \left(\dfrac{\partial V}{\partial \boldsymbol{x}}\right)^\mathsf{T} \boldsymbol{G}(\boldsymbol{x}) = \mathbb{0} \quad (5.15b) \end{cases}$$

with f_Υ, Q_Υ, S_Υ as in (5.10)–(5.12). Note that the "+" solution in (5.15a) implies $\Upsilon(\boldsymbol{x}) > 0$, thus the "−" solution in (5.15a) is discarded (cf. (5.9)) since it implies $\Upsilon(\boldsymbol{x}) < 0$, which always leads to an unstable solution. Moreover, note that $\left(\frac{\partial V}{\partial \boldsymbol{x}}\right)^\mathsf{T} \boldsymbol{G}(\boldsymbol{x}) = \mathbb{0}$ implies $f_\Upsilon < 0$, since $V(\boldsymbol{x})$ is a CLF. It can be shown by de L'Hospital's rule [Sac01, pp. 88, 186] that the Lagrange multiplier $\Upsilon(\boldsymbol{x})$ in (5.15) is continuous even for the case $\left(\frac{\partial V}{\partial \boldsymbol{x}}\right)^\mathsf{T} \boldsymbol{G}(\boldsymbol{x}) = \mathbb{0}$. Hence, it can be fully described with the compact notation (5.9). □

5.2.3 Control-Lyapunov Functions for Port-Hamiltonian Systems

In the following, we investigate under which conditions the Hamiltonian $H(\boldsymbol{x})$ is a CLF.

Lemma 5.6. *Consider an input-state-output PHS as in* (A.13). *Then $H(\boldsymbol{x})$ is a CLF for* (A.13), *if and only if*

$$\forall \boldsymbol{x} \in \mathcal{X}_{\boldsymbol{G}}: \quad \left(\frac{\partial H}{\partial \boldsymbol{x}}\right)^\mathsf{T} \boldsymbol{R}(\boldsymbol{x}) \frac{\partial H}{\partial \boldsymbol{x}} > 0, \tag{5.16}$$

where $\mathcal{X}_{\boldsymbol{G}} := \{\boldsymbol{x} \in \mathbb{R}^n : \boldsymbol{G}^\mathsf{T}(\boldsymbol{x}) \frac{\partial H}{\partial \boldsymbol{x}} = \boldsymbol{0},\ \boldsymbol{x} \neq \mathbb{0}\}$.

Proof. Since $\boldsymbol{J}(\boldsymbol{x})$ is skew-symmetric and $H(\boldsymbol{x})$ is a CLF, it holds by definition for all $\boldsymbol{x} \neq \mathbb{0}$ (cf. (5.4)) that

$$\begin{aligned} &\inf_{\boldsymbol{u}} \{\dot{H}(\boldsymbol{x})\} \\ &= \inf_{\boldsymbol{u}} \left\{ \left(\frac{\partial H}{\partial \boldsymbol{x}}\right)^\mathsf{T} \left((\boldsymbol{J}(\boldsymbol{x}) - \boldsymbol{R}(\boldsymbol{x})) \frac{\partial H}{\partial \boldsymbol{x}} + \boldsymbol{G}(\boldsymbol{x}) \boldsymbol{u} \right) \right\} \\ &= \inf_{\boldsymbol{u}} \left\{ -\left(\frac{\partial H}{\partial \boldsymbol{x}}\right)^\mathsf{T} \boldsymbol{R}(\boldsymbol{x}) \frac{\partial H}{\partial \boldsymbol{x}} + \left(\frac{\partial H}{\partial \boldsymbol{x}}\right)^\mathsf{T} \boldsymbol{G}(\boldsymbol{x}) \boldsymbol{u} \right\} \\ &< 0. \end{aligned} \tag{5.17}$$

If $\left(\frac{\partial H}{\partial \boldsymbol{x}}\right)^\mathsf{T} \boldsymbol{G}(\boldsymbol{x}) \neq \mathbb{0}$, then there always exists an input $\boldsymbol{u}'$ such that $\left(\frac{\partial H}{\partial \boldsymbol{x}}\right)^\mathsf{T} \boldsymbol{G}(\boldsymbol{x}) \boldsymbol{u}' < \mathbb{0}$ is fulfilled. If $\left(\frac{\partial H}{\partial \boldsymbol{x}}\right)^\mathsf{T} \boldsymbol{G}(\boldsymbol{x}) = \mathbb{0}$, then the first term inside the brackets in (5.17) needs to be negative whenever $\boldsymbol{x} \neq \mathbb{0}$, i.e.

$$\left(\frac{\partial H}{\partial \boldsymbol{x}}\right)^\mathsf{T} \boldsymbol{G}(\boldsymbol{x}) = \mathbb{0} \quad \Rightarrow \quad -\left(\frac{\partial H}{\partial \boldsymbol{x}}\right)^\mathsf{T} \boldsymbol{R}(\boldsymbol{x}) \frac{\partial H}{\partial \boldsymbol{x}} < 0, \tag{5.18}$$

which is equivalent to (5.16). Conversely, if (5.16) is satisfied, then the definition of a CLF is automatically fulfilled since $\mathcal{X}_{\boldsymbol{G}}$ is the space where $\left(\frac{\partial V}{\partial \boldsymbol{x}}\right)^{\top} \boldsymbol{G}(\boldsymbol{x}) = \mathbb{0}$ in Definition 5.26 and hence (5.16) is equivalent to (5.6). □

Corollary 5.7. *Consider an input-state-output PHS as in* (A.13). *Then $H(\boldsymbol{x})$ is a CLF for* (A.13), *if and only if* (A.13) *is zero-state detectable.*

Proof. Since the input-state-output PHS (5.3b) is equipped with the passive output (A.13b), it holds that $\mathcal{X}_G = \{\boldsymbol{x} \in \mathbb{R}^n : \boldsymbol{y}(\boldsymbol{x}) = 0\}$ and thus condition (5.16) is identical with the definition of a zero-state detectable input-affine nonlinear system (5.5) given in [van17, p. 47]. □

Corollary 5.8. *Consider an input-state-output PHS as in* (A.13). *Then $H(\boldsymbol{x})$ is a CLF, if* $\operatorname{rank}(\boldsymbol{R}(\boldsymbol{x})) = n$ *holds for all $\boldsymbol{x} \in \mathbb{R}^n$.*

Proof. Trivially, if $\boldsymbol{R}(\boldsymbol{x})$ has full rank, then $\left(\frac{\partial H}{\partial \boldsymbol{x}}\right)^{\top} \boldsymbol{R}(\boldsymbol{x}) \frac{\partial H}{\partial \boldsymbol{x}}$ is positive and thus (5.16) is fulfilled for all $\boldsymbol{x} \in \mathbb{R}^n$, which also implies for all $\boldsymbol{x} \in \mathcal{X}_{\boldsymbol{G}}$. □

Next, we introduce a necessary and sufficient condition under which the Hamiltonian of a *linear* input-state-output PHS is a CLF:

Lemma 5.9. *Consider the linear input-state-output PHS dynamics*

$$\dot{\boldsymbol{x}} = (\boldsymbol{J} - \boldsymbol{R}) \frac{\partial H}{\partial \boldsymbol{x}} + \boldsymbol{G}(\boldsymbol{x})\boldsymbol{u} \tag{5.19}$$

with $\boldsymbol{J} = -\boldsymbol{J}^{\top}$, $\boldsymbol{R} \succcurlyeq 0$, $H(\boldsymbol{x}) = \frac{1}{2}\boldsymbol{x}^{\top}\boldsymbol{Q}\boldsymbol{x}$ and $\boldsymbol{Q} \succ 0$. Then $H(\boldsymbol{x})$ is a CLF, if and only if

$$\ker\{\boldsymbol{G}^{\top}\boldsymbol{Q}\} \cap \ker\{\boldsymbol{Q}^{\top}\boldsymbol{R}\boldsymbol{Q}\} = \emptyset. \tag{5.20}$$

Proof. For linear input-state-output PHSs, it holds that $\frac{\partial H}{\partial \boldsymbol{x}} = \boldsymbol{Q}\boldsymbol{x}$ and thus $\mathcal{X}_{\boldsymbol{G}} = \{\boldsymbol{x} \in \mathbb{R}^n : \boldsymbol{G}^{\top}\frac{\partial H}{\partial \boldsymbol{x}} = \boldsymbol{0}\}$ is equal to $\ker\{\boldsymbol{G}^{\top}\boldsymbol{Q}\}$, since

$$\forall\, \boldsymbol{x} \in \ker\{\boldsymbol{G}^{\top}\boldsymbol{Q}\} : \qquad \boldsymbol{G}^{\top}\frac{\partial H}{\partial \boldsymbol{x}} = \boldsymbol{G}^{\top}\boldsymbol{Q}\boldsymbol{x} = \mathbb{0}. \tag{5.21}$$

Consequently, (5.16) in Lemma 5.6 reads as follows:

$$\forall \boldsymbol{x} \in \ker\{\boldsymbol{G}^{\top}\boldsymbol{Q}\} : \qquad -\left(\frac{\partial H}{\partial \boldsymbol{x}}\right)^{\top} \boldsymbol{R}\frac{\partial H}{\partial \boldsymbol{x}} = -(\boldsymbol{Q}\boldsymbol{x})^{\top}\boldsymbol{R}\boldsymbol{Q}\boldsymbol{x} < 0. \tag{5.22}$$

First, it is important to note that $-(\boldsymbol{Q}\boldsymbol{x})^{\top}\boldsymbol{R}\boldsymbol{Q}\boldsymbol{x}$ in (5.22) is always ≤ 0 for $\boldsymbol{x} \in \mathbb{R}^n$, since the product $\boldsymbol{Q}^{\top}\boldsymbol{R}\boldsymbol{Q}$ of positive semidefinite matrices is again positive semidefinite [HJ12,

p. 431]. Secondly, since $\boldsymbol{x}^\top \boldsymbol{M}\boldsymbol{x} = 0 \iff \boldsymbol{x} \in \ker\{\boldsymbol{M}\}$ for each $\boldsymbol{M} \succcurlyeq 0$ (cf. [HJ12, Observation 7.1.6]), it follows that the equality $\boldsymbol{x}^\top \boldsymbol{Q}^\top \boldsymbol{R}\boldsymbol{Q}\boldsymbol{x} = 0$ holds if and only if $\boldsymbol{x} \in \ker\{\boldsymbol{Q}^\top \boldsymbol{R}\boldsymbol{Q}\}$. Consequently, (5.22) holds if and only if $\ker\{\boldsymbol{G}^\top \boldsymbol{Q}\}$ and $\ker\{\boldsymbol{Q}^\top \boldsymbol{R}\boldsymbol{Q}\}$ are disjoint. □

Taking Lemma 5.6 into account, MOC can be applied to input-state-output PHSs in a straightforward manner[67]. However, as a consequence of the modification, the MOC law is optimal to an unintentionally modified objective function. The next lemma further evaluates to what extent the controller (5.8)–(5.12) is also optimal with respect to the original optimal control problem (5.3).

Lemma 5.10. *Consider an input-state-output PHS as in* (A.13). *The (modified optimal) controller* (5.8)–(5.12) *is optimal with respect to* (5.3), *if*

$$\forall \boldsymbol{x} \in \mathbb{R}^n : \qquad \Upsilon(\boldsymbol{x}) = 1. \tag{5.23}$$

Proof. If $S_\Upsilon \neq 0$, then it follows from (5.11) that $\boldsymbol{G}^\top \frac{\partial V}{\partial \boldsymbol{x}} \neq \mathbb{0}$. Thus (5.23) is equivalent to $S_\Upsilon - 2f_\Upsilon = Q_\Upsilon$, and substitution of (5.10)–(5.12) leads to

$$\frac{\ell(\boldsymbol{x})}{2} - \frac{1}{2}\left(\frac{\partial V}{\partial \boldsymbol{x}}\right)^\top \boldsymbol{G}(\boldsymbol{x})\boldsymbol{S}^{-1}(\boldsymbol{x})\boldsymbol{G}^\top(\boldsymbol{x})\frac{\partial V}{\partial \boldsymbol{x}} + \left(\frac{\partial V}{\partial \boldsymbol{x}}\right)^\top \boldsymbol{f}(\boldsymbol{x}) = 0, \tag{5.24}$$

which is equivalent to the HJB equation for time-invariant systems. The function $V(\boldsymbol{x})$ solving (5.24) is the value function $V^\star(\boldsymbol{x})$. This means, if a CLF is found for which $\Upsilon(\boldsymbol{x}) = 1$ holds for all $\boldsymbol{x} \in \mathbb{R}^n$, then this CLF is equal to the value function $V^\star$ and thus (5.8) is an optimal control input for (5.3). If $S_\Upsilon = 0$, then $\boldsymbol{G}^\top(\boldsymbol{x})\frac{\partial V}{\partial \boldsymbol{x}} = \mathbb{0}$. Thus, (5.23) is equivalent to $\Upsilon(\boldsymbol{x}) = -\frac{Q_\Upsilon}{2f_\Upsilon} \overset{!}{=} 1$, which also leads to $S_\Upsilon - 2f_\Upsilon = Q_\Upsilon$. □

A less restrictive requirement can be derived by allowing $\Upsilon(\boldsymbol{x})$ to have an arbitrary but fixed positive value:

Corollary 5.11. *The (modified optimal) controller* (5.8)–(5.12) *is optimal with respect to* (5.3), *if*

$$\forall \boldsymbol{x} \in \mathbb{R}^n : \qquad \Upsilon(\boldsymbol{x}) = c, \qquad c \in \mathbb{R}_{>0}. \tag{5.25}$$

In this case, the value function $V^\star(\boldsymbol{x})$ is a c-multiple of the CLF $V(\boldsymbol{x})$, i.e. $V^\star(\boldsymbol{x}) = c \cdot V(\boldsymbol{x})$.

Proof. (trivial) □

[67] For a special subclass of input-state-output PHSs, it can even be shown that apart from $H(\boldsymbol{x})$, there exist *no other* viable CLF candidates (see Lemma D.1 in Appendix D.1).

Remark 5.12. *Lemma 5.10 states that if $\Upsilon(\boldsymbol{x}) \neq 1$, then the chosen CLF $V(\boldsymbol{x})$ does not solve the HJB equation* (5.24) *and is thus not equivalent to the value function of* (5.3). *Consequently, the resulting controller* (5.8)–(5.12) *is not optimal with respect to* (5.3) *unless $\Upsilon(x) = 1$.* ◇

Remark 5.13. *From Corollary 5.11 we can conclude that unless $\Upsilon(\boldsymbol{x})$ converges to a constant value and remains constant even after a disturbance, the chosen CLF cannot be equivalent to the value function. Consequently the resulting controller* (5.8)–(5.12) *is not optimal with respect to* (5.3). *The stationarity of $\Upsilon(t)$ over time can therefore be interpreted as an* indicator of fitness *with respect to the original problem* (5.3). ◇

Example 6 (DC Motor)

Consider the linear model of a DC motor from [vJ14, p. 29].

$$\dot{\boldsymbol{x}} = \begin{bmatrix} -R & -K \\ K & -b \end{bmatrix} \frac{\partial H(\boldsymbol{x})}{\partial \boldsymbol{x}} + \begin{bmatrix} 1 \\ 0 \end{bmatrix} u, \tag{5.26a}$$

$$y = \begin{bmatrix} 1 & 0 \end{bmatrix} \frac{\partial H(\boldsymbol{x})}{\partial \boldsymbol{x}}, \tag{5.26b}$$

where $H(\boldsymbol{x}) = 1/2(x_1^2 + x_2^2)$, with u being the input voltage, y being the output current, $b > 0$ being the mechanical damping constant, $R > 0$ being the resistance, and $K > 0$ being the gyrator constant. Now choose $b = R = K = 1$ and consider the MOC problem (5.7) *with $\ell(\boldsymbol{x}) = x_1^2 + x_2^2$ and $S(\boldsymbol{x}) = 1$ subject to* (5.26). *By numerically solving the algebraic Riccati equation, we can calculate the value function as $V^\star(\boldsymbol{x}) = 1/2\boldsymbol{x}^\top \boldsymbol{P}\boldsymbol{x} \approx 0.21x_1^2 + 0.02x_1x_2 + 0.24x_2^2$.*

Fig. 5.1 shows the resulting trajectories of $\Upsilon(t)$ depending on different chosen CLFs $V(\boldsymbol{x})$ in (5.7):

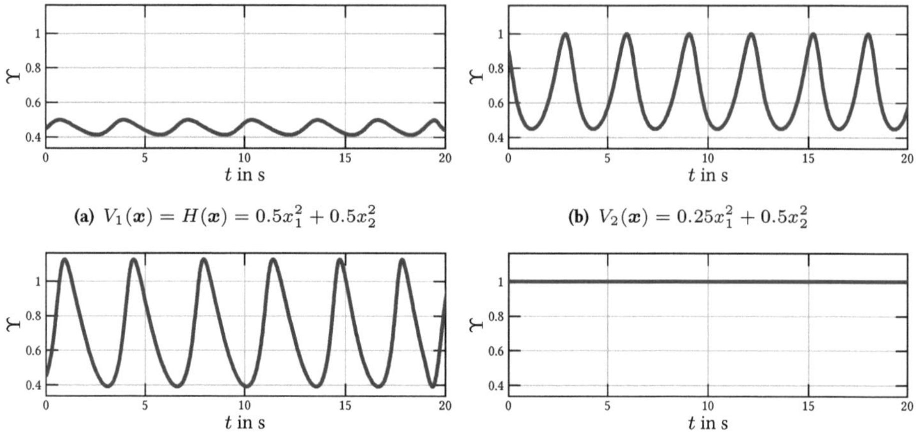

(a) $V_1(\boldsymbol{x}) = H(\boldsymbol{x}) = 0.5x_1^2 + 0.5x_2^2$

(b) $V_2(\boldsymbol{x}) = 0.25x_1^2 + 0.5x_2^2$

(c) $V_3(\boldsymbol{x}) = 0.5x_1^2 + 0.25x_2^2$

(d) $V_4(\boldsymbol{x}) = V^\star(\boldsymbol{x}) = 0.21x_1^2 + 0.02x_1x_2 + 0.24x_2^2$

Figure 5.1: Trajectory of Υ for different CLFs $V(\boldsymbol{x})$.

It can be seen that as $V(\boldsymbol{x})$ is closer to the value function $V^\star(\boldsymbol{x})$, Υ is closer to 1. Since $V_4(\boldsymbol{x})$ is equal to $V^\star(\boldsymbol{x})$, Υ in Fig. 5.1d is equal to 1.

5.2.4 Adaptive Optimal Control for Port-Hamiltonian Systems

Even if $H(\boldsymbol{x})$ qualifies as a CLF as described in Subsection 5.2.3, choosing $V(\boldsymbol{x}) = H(\boldsymbol{x})$ will generally result in $\Upsilon(\boldsymbol{x}) \neq 1$. With Lemma 5.10, this implies that the HJB equation (5.24) is not fulfilled and thus the resulting controller (5.8)–(5.12) is not optimal with respect to the original problem (5.3). For this purpose, in this subsection, we propose an *extended* CLF $V(\boldsymbol{x}, \boldsymbol{w})$ as a linear combination of $H(\boldsymbol{x})$ and a weighted sum of basis functions. Later in this subsection, we devise a gradient-based adaptation strategy of the weighting factors $\boldsymbol{w}$ such that $V(\boldsymbol{x}, \boldsymbol{w})$ fulfills the condition of Lemma 5.10, and present an adaptive optimal controller for (5.3) based on $V(\boldsymbol{x}, \boldsymbol{w})$. In the following Subsection 5.2.5, we then show that the equilibrium of the closed-loop system is (asymptotically) stable.

Extended CLF

Definition 5.14 (Extended CLF). *Let $\Xi : \mathbb{R}^n \to \mathbb{R}^r$ be a vector of basis functions. Then $V : \mathbb{R}^n \times \mathbb{R}^r \to \mathbb{R}$ with*

$$V(\boldsymbol{x}, \boldsymbol{w}) = H(\boldsymbol{x}) + \boldsymbol{w}^\mathsf{T}\Xi(\boldsymbol{x}), \qquad w \in \mathbb{R}^r \tag{5.27}$$

is said to be an extended CLF.

Assumption 5.15. *The basis functions* $\Xi(\boldsymbol{x})$ *in* (5.27) *are chosen such that there exists a vector* $\boldsymbol{w}^\star$ *of* optimal weights *that allows the actual value function to be parameterized as* $V^\star(\boldsymbol{x}) = H(\boldsymbol{x}) + (\boldsymbol{w}^\star)^\mathsf{T}\Xi(\boldsymbol{x})$.

Remark 5.16. *With the Weierstraß approximation theorem, Assumption 5.15 is admissible if the number of basis functions is large (see e.g. [VL10, p. 881; LLW14, p. 1020f.]). This fact will allow us to characterize the deviation from the optimal solution by the distance between* $\boldsymbol{w}(t)$ *and* $\boldsymbol{w}^\star$. ◇

With $V(\boldsymbol{x}, \boldsymbol{w})$ as in (5.27), we obtain

$$\frac{\partial V(\boldsymbol{x}, \boldsymbol{w})}{\partial \boldsymbol{x}} = \frac{\partial H}{\partial \boldsymbol{x}} + \left(\frac{\partial \Xi}{\partial \boldsymbol{x}}\right)^\mathsf{T} \boldsymbol{w}, \tag{5.28}$$

and accordingly the MOC law (5.8) takes the form

$$\boldsymbol{u}^\star = \begin{cases} -\boldsymbol{S}^{-1}(\boldsymbol{x})\boldsymbol{G}^\mathsf{T}(\boldsymbol{x})\dfrac{\partial V}{\partial \boldsymbol{x}}\Upsilon(\boldsymbol{x}, \boldsymbol{w}), & S'_\Upsilon \neq 0, \\ \mathbb{0}, & S'_\Upsilon = 0, \end{cases} \tag{5.29}$$

where

$$\Upsilon(\boldsymbol{x}, \boldsymbol{w}) := \frac{f'_\Upsilon + \sqrt{(f'_\Upsilon)^2 + Q'_\Upsilon \cdot S'_\Upsilon}}{S'_\Upsilon}, \tag{5.30}$$

$$f'_\Upsilon(\boldsymbol{x}) := \left(\frac{\partial V}{\partial \boldsymbol{x}}\right)^\mathsf{T} (\boldsymbol{J}(\boldsymbol{x}) - \boldsymbol{R}(\boldsymbol{x}))\frac{\partial H}{\partial \boldsymbol{x}} - \left(\frac{\partial V}{\partial \boldsymbol{w}}\right)^\mathsf{T} \dot{\boldsymbol{w}}, \tag{5.31}$$

$$S'_\Upsilon(\boldsymbol{x}) := \left(\frac{\partial V}{\partial \boldsymbol{x}}\right)^\mathsf{T} \boldsymbol{K}(\boldsymbol{x})\frac{\partial V}{\partial \boldsymbol{x}}, \tag{5.32}$$

$$Q'_\Upsilon(\boldsymbol{x}) := \ell(\boldsymbol{x}), \tag{5.33}$$

$$\boldsymbol{K}(\boldsymbol{x}) := \boldsymbol{G}(\boldsymbol{x})\boldsymbol{S}^{-1}(\boldsymbol{x})\boldsymbol{G}^\mathsf{T}(\boldsymbol{x}). \tag{5.34}$$

Employing the same reasoning as in Lemma 5.10, we will study how to check whether a given CLF (5.27) is equivalent to the value function $V^\star$ in more detail.

Lemma 5.17. *Let* $V(\boldsymbol{x}) = H(\boldsymbol{x}) + (\boldsymbol{w}^\diamond)^\mathsf{T}\Xi(\boldsymbol{x})$ *be a given CLF with* $\boldsymbol{w}^\diamond \in \mathbb{R}^r$. *Then* $V(\boldsymbol{x})$ *is equivalent to the value function* $V^\star(\boldsymbol{x})$ *of* (5.3), *if and only if*

$$\forall \boldsymbol{x} \in \mathbb{R}^n : \qquad (\boldsymbol{x}, \boldsymbol{w}^\diamond) \in \mathcal{Q}(\boldsymbol{x}, \boldsymbol{w}), \tag{5.35}$$

where

$$\mathcal{Q}(\boldsymbol{x}.\boldsymbol{w}) = \left\{(\boldsymbol{x},\boldsymbol{w}) \in \mathbb{R}^n \times \mathbb{R}^r : \boldsymbol{w}^\mathsf{T}\boldsymbol{A}(\boldsymbol{x})\boldsymbol{w} + \boldsymbol{a}^\mathsf{T}(\boldsymbol{x})\boldsymbol{w} + a(\boldsymbol{x}) = 0\right\}, \tag{5.36}$$

$$\boldsymbol{A}(\boldsymbol{x}) = \frac{\partial \Xi}{\partial \boldsymbol{x}}\boldsymbol{K}(\boldsymbol{x})\left(\frac{\partial \Xi}{\partial \boldsymbol{x}}\right)^\mathsf{T}, \tag{5.37}$$

$$\boldsymbol{a}(\boldsymbol{x}) = 2\Bigg(\left(\frac{\partial H}{\partial \boldsymbol{x}}\right)^\mathsf{T}\boldsymbol{K}(\boldsymbol{x})\left(\frac{\partial \Xi}{\partial \boldsymbol{x}}\right)^\mathsf{T} - \left(\frac{\partial H}{\partial \boldsymbol{x}}\right)^\mathsf{T}(\boldsymbol{J}(\boldsymbol{x}) - \boldsymbol{R}(\boldsymbol{x}))\left(\frac{\partial \Xi}{\partial \boldsymbol{x}}\right)^\mathsf{T}\Bigg), \tag{5.38}$$

$$a(\boldsymbol{x}) = \left(\frac{\partial H}{\partial \boldsymbol{x}}\right)^\mathsf{T}\boldsymbol{K}(\boldsymbol{x})\frac{\partial H}{\partial \boldsymbol{x}} - 2\left(\frac{\partial H}{\partial \boldsymbol{x}}\right)^\mathsf{T}(\boldsymbol{J}(\boldsymbol{x}) - \boldsymbol{R}(\boldsymbol{x}))\frac{\partial H}{\partial \boldsymbol{x}} - \ell(\boldsymbol{x}). \tag{5.39}$$

Proof. Let $\boldsymbol{w}^\star \in \mathbb{R}^r$ be the optimal weighting vector. According to Lemma 5.10, it thus holds that $\Upsilon(\boldsymbol{x},\boldsymbol{w}^\star) = 1$ for all $\boldsymbol{x} \in \mathbb{R}^n$. As shown in the proof of Lemma 5.10, this condition is equivalent to $S'_\Upsilon - 2f'_\Upsilon - Q'_\Upsilon \stackrel{!}{=} 0$. Substitution of (5.30)–(5.34) yields

$$\begin{aligned} 0 = {} & (\boldsymbol{w}^\star)^\mathsf{T}\frac{\partial \Xi}{\partial \boldsymbol{x}}\boldsymbol{K}(\boldsymbol{x})\left(\frac{\partial \Xi}{\partial \boldsymbol{x}}\right)^\mathsf{T}\boldsymbol{w}^\star + 2\Bigg(\left(\frac{\partial H}{\partial \boldsymbol{x}}\right)^\mathsf{T}\boldsymbol{K}(\boldsymbol{x})\left(\frac{\partial \Xi}{\partial \boldsymbol{x}}\right)^\mathsf{T} \\ & - \left(\frac{\partial H}{\partial \boldsymbol{x}}\right)^\mathsf{T}(\boldsymbol{J}(\boldsymbol{x}) - \boldsymbol{R}(\boldsymbol{x}))\left(\frac{\partial \Xi}{\partial \boldsymbol{x}}\right)^\mathsf{T}\Bigg)\boldsymbol{w}^\star \\ & + \left(\frac{\partial H}{\partial \boldsymbol{x}}\right)^\mathsf{T}\boldsymbol{K}(\boldsymbol{x})\frac{\partial H}{\partial \boldsymbol{x}} - 2\left(\frac{\partial H}{\partial \boldsymbol{x}}\right)^\mathsf{T}(\boldsymbol{J}(\boldsymbol{x}) - \boldsymbol{R}(\boldsymbol{x}))\frac{\partial H}{\partial \boldsymbol{x}} - \ell(\boldsymbol{x}). \end{aligned} \tag{5.40}$$

With $\mathcal{Q}(\boldsymbol{x},\boldsymbol{w})$ as in (5.36), condition (5.40) can be written as

$$\forall \boldsymbol{x} \in \mathbb{R}^n : \qquad (\boldsymbol{x},\boldsymbol{w}^\star) \in \mathcal{Q}(\boldsymbol{x},\boldsymbol{w}). \tag{5.41}$$

Since the given CLF $V(\boldsymbol{x})$ is equal to $V^\star(\boldsymbol{x})$ if and only if $\boldsymbol{w}^\diamond = \boldsymbol{w}^\star$, this is equivalent to (5.35). □

Adaptation of the Extended Control-Lyapunov Function

For each fixed $\boldsymbol{x} \in \mathbb{R}^n$, the set $\mathcal{Q}(\boldsymbol{x},\boldsymbol{w})$ in (5.36) is a quadric. The basic idea of the following approach relies on the fact that, according to Lemma 5.17, the optimal weighting vector $\boldsymbol{w}^\star$ is contained in each quadric, i.e. $(\boldsymbol{w}^\star)^\mathsf{T}\boldsymbol{A}(\boldsymbol{x})\boldsymbol{w}^\star + \boldsymbol{a}^\mathsf{T}(\boldsymbol{x})\boldsymbol{w}^\star + a(\boldsymbol{x}) = 0$ holds for all $\boldsymbol{x} \in \mathbb{R}^n$. Thus for each arbitrary but fixed $\boldsymbol{x} \in \mathbb{R}^n$, $\boldsymbol{w}^\star$ can be characterized as the minimizer of an objective function $\mathcal{J}^0_{\mathrm{w}}(\boldsymbol{x},\boldsymbol{w}) := (\boldsymbol{w}^\mathsf{T}\boldsymbol{A}(\boldsymbol{x})\boldsymbol{w} + \boldsymbol{a}^\mathsf{T}(\boldsymbol{x})\boldsymbol{w} + a(\boldsymbol{x}))^2$. Moreover, with $\boldsymbol{w}^\star = \arg\min_{\boldsymbol{w}}\left\{\mathcal{J}^0_{\mathrm{w}}(\boldsymbol{x},\boldsymbol{w})\right\}$ and due to the fact that $\boldsymbol{w}^\star$ is contained in each quadric $\mathcal{Q}(\boldsymbol{x},\boldsymbol{w})$, it follows that

$$\{\boldsymbol{w}^\star\} \subseteq \bigcap_{\boldsymbol{x}\in\mathbb{R}^n} \arg\min_{\boldsymbol{w}}\left\{\mathcal{J}^0_{\mathrm{w}}(\boldsymbol{x},\boldsymbol{w})\right\}. \tag{5.42}$$

If the basis functions are linearly independent, then the intersection of all quadrics is a singleton and thus

$$\{\boldsymbol{w}^*\} \equiv \bigcap_{\boldsymbol{x}\in\mathbb{R}^n} \arg\min_{\boldsymbol{w}} \left\{\mathcal{J}_{\mathrm{w}}^0(\boldsymbol{x},\boldsymbol{w})\right\}. \tag{5.43}$$

Since $\mathcal{J}_{\mathrm{w}}^0$ is continuously differentiable, this allows for an adaptation of $\boldsymbol{w}(t)$ by a simple gradient flow

$$\dot{\boldsymbol{w}} = -\alpha \cdot \frac{\partial \mathcal{J}_{\mathrm{w}}^0(\boldsymbol{x},\boldsymbol{w})}{\partial \boldsymbol{w}}, \tag{5.44}$$

where $\alpha > 0$ defines the *learning rate*. However, note that $\mathcal{J}_{\mathrm{w}}^0$ is in general *not* strictly convex around $\boldsymbol{w}^\star$, which hampers convergence of (5.44) towards $\boldsymbol{w}^\star$ and thus necessitates additional conditions for a sufficient exploration of the state space. To circumvent these requirements, which are often very hard to evaluate, we formulate an extended objective function $\mathcal{J}_{\mathrm{w}}(\boldsymbol{x},\boldsymbol{w})$ providing strict convexity with respect to $\boldsymbol{w}$ in a neighborhood of $\boldsymbol{w}^\star$ as follows:

Definition 5.18 (Extended Objective Function). *Let*

$$\mathsf{Q}(\boldsymbol{x},\boldsymbol{w}) = \boldsymbol{w}^\top \boldsymbol{A}(\boldsymbol{x})\boldsymbol{w} + \boldsymbol{a}^\top(\boldsymbol{x})\boldsymbol{w} + a(\boldsymbol{x}) \tag{5.45}$$

with $\boldsymbol{A}(\boldsymbol{x})$, $\boldsymbol{a}(\boldsymbol{x})$, $a(\boldsymbol{x})$ as in (5.37)–(5.39) *be the corresponding quadratic function to the quadric $\mathcal{Q}(\boldsymbol{x},\boldsymbol{w})$ in* (5.36). *Then with $\boldsymbol{c}_1,\ldots,\boldsymbol{c}_r \in \mathbb{R}^n$, the function*

$$\begin{aligned}\mathcal{J}_{\mathrm{w}}(\boldsymbol{x},\boldsymbol{w}) &= \mathcal{J}_{\mathrm{w}}^0(\boldsymbol{x}+\boldsymbol{c}_1,\boldsymbol{w}) + \cdots + \mathcal{J}_{\mathrm{w}}^0(\boldsymbol{x}+\boldsymbol{c}_r,\boldsymbol{w}) \\ &= (\mathsf{Q}(\boldsymbol{x}+\boldsymbol{c}_1,\boldsymbol{w}))^2 + \cdots + (\mathsf{Q}(\boldsymbol{x}+\boldsymbol{c}_r,\boldsymbol{w}))^2\end{aligned} \tag{5.46}$$

composed by a linear combination of shifted objective functions $\mathcal{J}_{\mathrm{w}}^0(\boldsymbol{x},\boldsymbol{w})$ is said to be an extended objective function.

Lemma 5.19. *Let $\mathcal{J}_{\mathrm{w}}(\boldsymbol{x},\boldsymbol{w})$ be an extended objective function as in* (5.46). *Then $\mathcal{J}_{\mathrm{w}}(\boldsymbol{x},\boldsymbol{w})$ is (locally) strictly convex in an open neighborhood $\mathcal{M}$ of the optimal weights $\boldsymbol{w}^\star$ for all $\boldsymbol{x} \in \mathbb{R}^n$, if and only if the vectors*

$$\boldsymbol{v}_l = 2\boldsymbol{A}(\boldsymbol{x}+\boldsymbol{c}_l)\boldsymbol{w}^\star + \boldsymbol{a}(\boldsymbol{x}+\boldsymbol{c}_l) \tag{5.47}$$

with $l = 1,\ldots,r$ are linearly independent.

Proof. Without loss of generality, we choose $\boldsymbol{c}_1 = \mathbb{0}$. As strict convexity with respect to $\boldsymbol{w}$ needs to be shown, the Hessian of $\mathcal{J}_{\mathrm{w}}^0$ and $\mathcal{J}_{\mathrm{w}}$ (see(5.46)) are studied. The Hessian of $\mathcal{J}_{\mathrm{w}}^0$ can be written as

$$\begin{aligned}\frac{\partial^2 \mathcal{J}_{\mathrm{w}}^0(\boldsymbol{x},\boldsymbol{w})}{\partial \boldsymbol{w}^2} &= 2\big(2\boldsymbol{A}(\boldsymbol{x})\boldsymbol{w} + \boldsymbol{a}(\boldsymbol{x})\big)\big(2\boldsymbol{A}(\boldsymbol{x})\boldsymbol{w} + \boldsymbol{a}(\boldsymbol{x})\big)^\top \\ &\quad + 4\boldsymbol{A}(\boldsymbol{x})\big(\boldsymbol{w}^\top\boldsymbol{A}(\boldsymbol{x})\boldsymbol{w} + \boldsymbol{a}(\boldsymbol{x})\boldsymbol{w} + a(\boldsymbol{x})\big).\end{aligned} \tag{5.48}$$

Since the optimal weights $\boldsymbol{w}^\star$ need to be part of each quadric $\mathcal{Q}(\boldsymbol{x}, \boldsymbol{w})$ regardless of the state $\boldsymbol{x}$ (see Lemma 5.17), the second summand in (5.48) is equal to zero for $\boldsymbol{w} = \boldsymbol{w}^\star$ and accordingly

$$\left.\frac{\partial^2 \mathcal{J}_{\mathrm{w}}^0(\boldsymbol{x}, \boldsymbol{w})}{\partial \boldsymbol{w}^2}\right|_{\boldsymbol{w}=\boldsymbol{w}^\star} = 2\boldsymbol{v}_1\boldsymbol{v}_1^\mathsf{T} \tag{5.49}$$

with $\boldsymbol{v}_1 = 2\boldsymbol{A}(\boldsymbol{x})\boldsymbol{w}^\star + \boldsymbol{a}(\boldsymbol{x})$. Since the Hessian (5.49) is only composed by the multiplication of two vectors, which yields a matrix with identical but scaled row vectors, it is positive semidefinite and has rank one.

For the Hessian of the shifted objective function $\mathcal{J}_{\mathrm{w}}^0(\boldsymbol{x}+\boldsymbol{c}_2, \boldsymbol{w})$, we obtain in a similar manner

$$\left.\frac{\partial^2 \mathcal{J}_{\mathrm{w}}^0(\boldsymbol{x}+\boldsymbol{c}_2, \boldsymbol{w})}{\partial \boldsymbol{w}^2}\right|_{\boldsymbol{w}=\boldsymbol{w}^\star} = 2\boldsymbol{v}_2\boldsymbol{v}_2^\mathsf{T} \tag{5.50}$$

with $\boldsymbol{v}_2 = 2\boldsymbol{A}(\boldsymbol{x}+\boldsymbol{c}_2)\boldsymbol{w}^\star + \boldsymbol{a}(\boldsymbol{x}+\boldsymbol{c}_2)$. Note that the rank of the matrix (5.50) is one regardless of the shifting $\boldsymbol{c}_2$. The linear combination $\mathcal{J}_{\mathrm{w}}^1(\boldsymbol{x}, \boldsymbol{w}) := \mathcal{J}_{\mathrm{w}}^0(\boldsymbol{x}, \boldsymbol{w}) + \mathcal{J}_{\mathrm{w}}^0(\boldsymbol{x}+\boldsymbol{c}_2, \boldsymbol{w})$ leads to the Hessian

$$\left.\frac{\partial^2 \mathcal{J}_{\mathrm{w}}^1}{\partial \boldsymbol{w}^2}\right|_{\boldsymbol{w}=\boldsymbol{w}^\star} = 2\boldsymbol{v}_1\boldsymbol{v}_1^\mathsf{T} + 2\boldsymbol{v}_2\boldsymbol{v}_2^\mathsf{T}. \tag{5.51}$$

The same applies to linear combinations with more than two summands due to the linearity property of differentiation. It can be seen that (5.51) has a maximum rank of two. As strict convexity of $\mathcal{J}_{\mathrm{w}}$ is required, full rank r needs to be satisfied for the Hessian of $\mathcal{J}_{\mathrm{w}}(\boldsymbol{x}, \boldsymbol{w})$ at $\boldsymbol{w} = \boldsymbol{w}^\star$. Thus the question arises, in which case the increase of summands implies a rank increase of the Hessian. Each matrix in (5.49) or (5.50) describes a linear map $\mathbb{R}^r \to \mathbb{R}^r$ of rank one and its image is a subspace of $\mathbb{R}^r$ of dimension one. If the image of $\frac{\partial^2 \mathcal{J}_{\mathrm{w}}^1}{\partial \boldsymbol{w}^2}$ in (5.51) is of dimension two, the rank automatically increases, since $\dim(\mathrm{im}(\boldsymbol{M})) = \mathrm{rank}(\boldsymbol{M})$ for an arbitrary matrix $\boldsymbol{M}$. With the dimension formula for the sum of subspaces [Bos14, p. 47], it follows that $\dim(\mathcal{U}_1 + \mathcal{U}_2) = \dim(\mathcal{U}_1) + \dim(\mathcal{U}_2) - \dim(\mathcal{U}_1 \cap \mathcal{U}_2)$ for two arbitrary subspaces $\mathcal{U}_1$ and $\mathcal{U}_2$. By setting $\mathcal{U}_1 = \mathrm{im}(\boldsymbol{v}_1{\boldsymbol{v}_1}^\mathsf{T})$ and $\mathcal{U}_2 = \mathrm{im}(\boldsymbol{v}_2{\boldsymbol{v}_2}^\mathsf{T})$ it follows that only if $\dim(\mathcal{U}_1 \cap \mathcal{U}_2) = 0$, the summation of the two matrices $\boldsymbol{v}_1{\boldsymbol{v}_1}^\mathsf{T}$ and $\boldsymbol{v}_2{\boldsymbol{v}_2}^\mathsf{T}$ leads to a rank increase. For each $l \in \{1, \dots, r\}$, the rank-one matrix $\boldsymbol{v}_l{\boldsymbol{v}_l}^\mathsf{T} = (v_{l_1} \cdot \boldsymbol{v}_l, \cdots, v_{l_r} \cdot \boldsymbol{v}_l)^\mathsf{T}$ is formed by weighted rows of ${\boldsymbol{v}_l}^\mathsf{T}$ with the respective components v_{l_i}, $i = 1, \dots, r$ of $\boldsymbol{v}_l$,

$$\boldsymbol{v}_l{\boldsymbol{v}_l}^\mathsf{T} = \begin{bmatrix} v_{l_1} \cdot {\boldsymbol{v}_l}^\mathsf{T} \\ \vdots \\ v_{l_r} \cdot {\boldsymbol{v}_l}^\mathsf{T} \end{bmatrix}, \tag{5.52}$$

and hence its image spans the subspace $\mathcal{U}_l = \mathrm{im}(\boldsymbol{v}_l{\boldsymbol{v}_l}^\mathsf{T}) = \{\mu \cdot \boldsymbol{v}_l \mid \mu \in \mathbb{R}\}$. With regard to $\mathcal{J}_{\mathrm{w}}^1$ in (5.51) it can be seen that in order to let both sets $\mathcal{U}_1$ and $\mathcal{U}_2$ be disjunct, the linear independence of both vectors $\boldsymbol{v}_1$ and $\boldsymbol{v}_2$ is necessary. Graphically, the subspace $\mathrm{im}(\boldsymbol{v}_l{\boldsymbol{v}_l}^\mathsf{T})$ is a straight line in $\mathbb{R}^r$, and linear independence leads to non-coinciding straight lines such that $\dim(\mathcal{U}_1 \cap \mathcal{U}_2) = 0$. Applied to $\mathcal{J}_{\mathrm{w}}$ it becomes clear that $\mathcal{J}_{\mathrm{w}}$ has a Hessian with full rank, if the vectors $\boldsymbol{v}_l$ in (5.47) with $l = 1, \dots, r$ are linearly independent. Hence, each vector $\boldsymbol{v}_l$ induces a matrix, implying an increase of one for the rank of the Hessian $\frac{\partial^2 \mathcal{J}_{\mathrm{w}}}{\partial \boldsymbol{w}^2}$, which leads to full rank and thus positive definiteness of $\frac{\partial^2 \mathcal{J}_{\mathrm{w}}}{\partial \boldsymbol{w}^2}$. Moreover, since $\mathcal{J}_{\mathrm{w}}$ is a C^2 function, positive definiteness of the Hessian is preserved for all $\boldsymbol{w} \in \mathcal{M}$ around $\boldsymbol{w}^\star$. □

Remark 5.20. *Lemma 5.19 states that the minimum number of summands in $\mathcal{J}_{\mathrm{w}}$ to achieve strict convexity is r. From a practical point of view, however, it is preferable to compose $\mathcal{J}_{\mathrm{w}}$ from more than r summands in order to enhance convergence properties.* ♢

Due to the fact that $\mathtt{Q}(\boldsymbol{x}, \boldsymbol{w}^\star) = 0$ is fulfilled for all $\boldsymbol{x} \in \mathbb{R}^n$, the optimal weighing factor $\boldsymbol{w}^\star$ is characterized by the strictly convex optimization problem $\boldsymbol{w}^\star = \arg\min_{\boldsymbol{w}} \{\mathcal{J}_{\mathrm{w}}(\boldsymbol{x}, \boldsymbol{w})\}$. Thus, a given weighting vector $\boldsymbol{w}$ can be adapted by the gradient flow

$$\dot{\boldsymbol{w}} = -\alpha \cdot \frac{\partial \mathcal{J}_{\mathrm{w}}(\boldsymbol{x}, \boldsymbol{w})}{\partial \boldsymbol{w}}, \qquad \alpha > 0. \tag{5.53}$$

Example 6 (DC Motor) – *cont'd*

Consider again the DC motor model (5.26) *and choose the extended CLF*

$$V(\boldsymbol{x}, \boldsymbol{w}) = H(\boldsymbol{x}) + \begin{bmatrix} w_1 & w_2 & w_3 \end{bmatrix} \begin{bmatrix} x_1^2 \\ x_1 x_2 \\ x_2^2 \end{bmatrix}. \tag{5.54}$$

Fig. 5.2 shows the contour plot of the resulting objective function $\mathcal{J}_{\mathrm{w}}^0(\boldsymbol{x}, \boldsymbol{w})$ for fixed $w_3 = w_3^\star$ and some states:

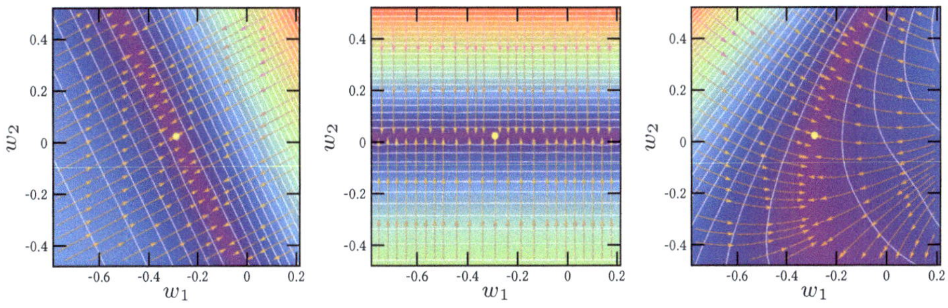

(a) $x_1 = -1, x_2 = 1$ **(b)** $x_1 = 0, x_2 = 1$ **(c)** $x_1 = 1.5, x_2 = 0$

Figure 5.2: Contour plot of $\mathcal{J}_{\mathrm{w}}^0(\boldsymbol{x}, \boldsymbol{w})$ with $w_3 = w_3^\star$.

The yellow dot marks the optimal point $(w_1^\star, w_2^\star)^\mathsf{T}$ in each case. It can be seen that for each of the states in Figs. 5.2a–c, the optimal point is a minimizer of $\mathcal{J}_{\mathrm{w}}^0(\boldsymbol{x}, \boldsymbol{w})$. However, $\mathcal{J}_{\mathrm{w}}^0(\boldsymbol{x}, \boldsymbol{w})$ has no strict minimizer for any of the states in Fig. 5.2a–c, which indicates that $\mathcal{J}_{\mathrm{w}}^0(\boldsymbol{x}, \boldsymbol{w})$ is not strictly convex around $\boldsymbol{w}^\star$.

Now consider the extended objective function $\mathcal{J}_{\mathrm{w}}(\boldsymbol{x}, \boldsymbol{w})$ *with shifts* $\boldsymbol{c}_0 = (0,0)^\mathsf{T}$, $\boldsymbol{c}_1 = (-0.5,-0.5)^\mathsf{T}$, $\boldsymbol{c}_2 = (0,1.5)^\mathsf{T}$, *and* $\boldsymbol{c}_3 = (1.5,0)^\mathsf{T}$. *Fig. 5.3 shows the contour plot of* $\mathcal{J}_{\mathrm{w}}(\boldsymbol{x}, \boldsymbol{w})$ *for fixed* $w_3 = w_3^\star$ *and some states:*

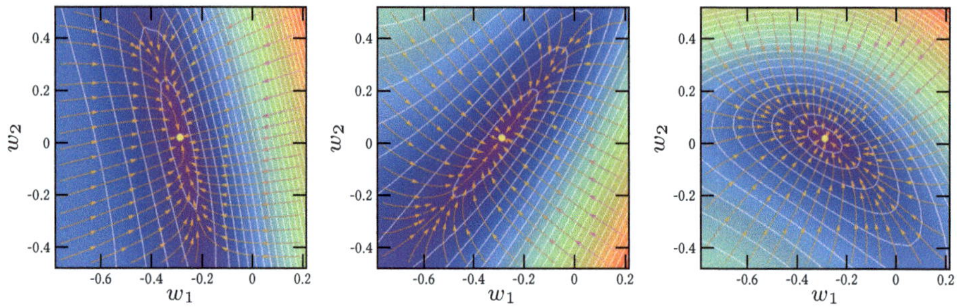

(a) $x_1 = -1, x_2 = 1$ (b) $x_1 = 0, x_2 = 1$ (c) $x_1 = 1.5, x_2 = 0$

Figure 5.3: Contour plot of $\mathcal{J}_{\mathrm{w}}(\boldsymbol{x}, \boldsymbol{w})$ with $w_3 = w_3^\star$.

In Figs. 5.3a–c, the contour lines around $(w_1^\star, w_2^\star)$ *are closed, indicating that* $\mathcal{J}_{\mathrm{w}}(\boldsymbol{x}, \boldsymbol{w})$ *is strictly convex around* $\boldsymbol{w}^\star$.

Now the adaptation procedure (5.53) *is applied with both* $\mathcal{J}_{\mathrm{w}}^0(\boldsymbol{x}, \boldsymbol{w})$ *and* $\mathcal{J}_{\mathrm{w}}(\boldsymbol{x}, \boldsymbol{w})$. *Fig. 5.4 shows the resulting trajectories of* $\boldsymbol{w}$ *and* Υ.

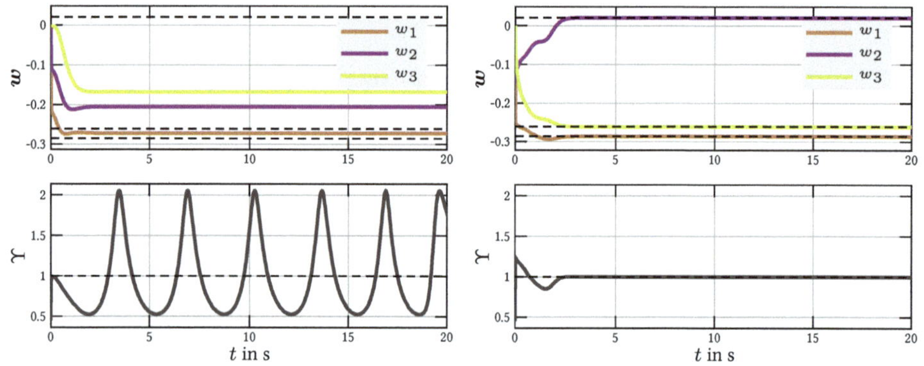

(a) Adaptation using $\mathcal{J}_{\mathrm{w}}^0(\boldsymbol{x}, \boldsymbol{w})$ (b) Adaptation using $\mathcal{J}_{\mathrm{w}}(\boldsymbol{x}, \boldsymbol{w})$

Figure 5.4: Adaptation results of the weights $\boldsymbol{w}$ depending on the chosen objective function.

While in Fig. 5.4a, $\boldsymbol{w}(t)$ *does not converge to the optimal weights, in Fig. 5.4b* $\boldsymbol{w}^\star$ *is reached after a learning phase of* 3 s. *Correspondingly, the Lagrange multiplier* $\Upsilon(t)$ *in Fig. 5.4a is instationary, while in Fig. 5.4b it converges to* 1 *after having completed the learning phase.*

Fig. 5.5 shows the trajectories of $\boldsymbol{w}(t)$ and $\Upsilon(t)$ for the original objective function $\mathcal{J}^0_\mathrm{w}(\boldsymbol{x}, \boldsymbol{w})$, if an additive voltage source $u_d(t) = \sin(t/2)$ serves as an excitation signal of the DC motor (5.26).

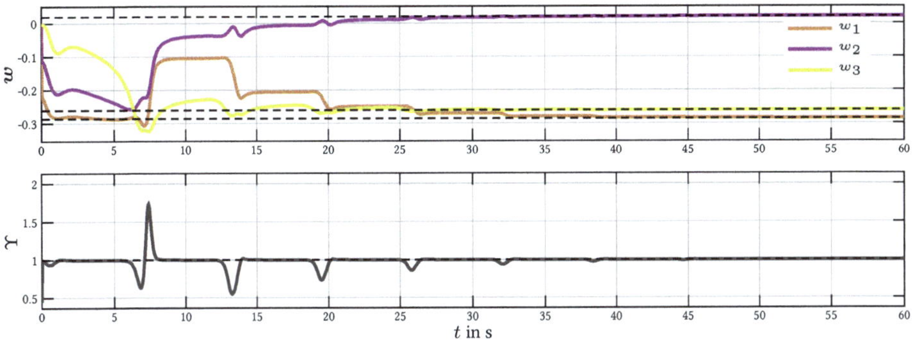

Figure 5.5: Adaptation results of the weights $\boldsymbol{w}$ with $\mathcal{J}^0_\mathrm{w}(\boldsymbol{x}, \boldsymbol{w})$ and additive excitation input.

In this case, convergence to $\boldsymbol{w} = \boldsymbol{w}^\star$ and $\Upsilon = 1$ is achieved after 45 s. *This demonstrates that convergence and thus optimality with respect to the original problem* (5.3) *can obtained even with a locally non-strictly convex objective function $\mathcal{J}^0_\mathrm{w}(\boldsymbol{x}, \boldsymbol{w})$ in* (5.53), *provided that there is an additive input signal which ensures a persistent excitation of $\boldsymbol{x}(t)$ and thus a persistent movement of the value of $\mathcal{J}^0_\mathrm{w}(\boldsymbol{x}, \boldsymbol{w})$ such that $\boldsymbol{w}^\star$ becomes the unique minimizer for all t.*

5.2.5 Stability of the Closed-Loop System

The combination of the open-loop input-state-output PHS (5.3b), the extended CLF (5.27), the MOC law (5.29), the adaptation procedure (5.53), and the shorthand notation (5.30)–(5.34) results in the following closed-loop system:

$$\dot{\boldsymbol{x}} = (\boldsymbol{J}(\boldsymbol{x}) - \boldsymbol{R}(\boldsymbol{x}))\frac{\partial H(\boldsymbol{x})}{\partial \boldsymbol{x}} + \boldsymbol{G}(\boldsymbol{x})\boldsymbol{u}^\star, \tag{5.55a}$$

$$\boldsymbol{u}^\star = \begin{cases} \boldsymbol{S}^{-1}(\boldsymbol{x})\boldsymbol{G}^\mathsf{T}(\boldsymbol{x})\dfrac{\partial V}{\partial \boldsymbol{x}} \cdot \Upsilon(\boldsymbol{x}, \boldsymbol{w}), & S'_\Upsilon \neq 0, \\ \mathbb{0}, & S'_\Upsilon = 0, \end{cases} \tag{5.55b}$$

$$\dot{\boldsymbol{w}} = -\alpha \cdot \frac{\partial \mathcal{J}_\mathrm{w}(\boldsymbol{x}, \boldsymbol{w})}{\partial \boldsymbol{w}}, \tag{5.55c}$$

$$\boldsymbol{x}_0 = \boldsymbol{x}_0, \tag{5.55d}$$

$$\boldsymbol{w}_0 = \mathbb{0}_r. \tag{5.55e}$$

To perform a stability analysis of the equilibrium $(\mathbb{0}, \boldsymbol{w}^\star)$ of (5.55), $V(\boldsymbol{x}, \boldsymbol{w})$ is used as a Lyapunov function candidate to prove that

$$\forall\ (\boldsymbol{x}, \boldsymbol{w}) \neq (\mathbb{0}, \boldsymbol{w}^\star): \qquad V(\boldsymbol{x}, \boldsymbol{w}) > 0, \tag{5.56}$$

$$\forall\ (\boldsymbol{x}, \boldsymbol{w}) \in \mathbb{R}^n \times \mathbb{R}^r: \qquad \dot{V}(\boldsymbol{x}, \boldsymbol{w}) \leq 0. \tag{5.57}$$

While the proof of (5.57) can be handled by the following Lemma 5.21, statement (5.56) (detailed later in Lemma 5.23) requires some additional preparatory work.

Lemma 5.21. *Consider the closed-loop system* (5.55) *starting at* $(\boldsymbol{x}_0, \boldsymbol{w}_0) \in \mathbb{R}^n \times \mathbb{R}^r$. *Then*

$$\forall\, (\boldsymbol{x}, \boldsymbol{w}) \in \mathbb{R}^n \times \mathbb{R}^r: \qquad \dot{V}(\boldsymbol{x}, \boldsymbol{w}) \leq 0, \tag{5.58}$$

i.e. $V(\boldsymbol{x}, \boldsymbol{w})$ *decreases monotonically over time.*

Proof. Applying the chain rule to (5.27) and inserting (5.55a), (5.55b), (5.55c), we get

$$\begin{aligned}
\dot{V}(\boldsymbol{x}, \boldsymbol{w}) &= \left(\frac{\partial V(\boldsymbol{x}, \boldsymbol{w})}{\partial \boldsymbol{x}}\right)^{\mathsf{T}} \dot{\boldsymbol{x}} + \left(\frac{\partial V(\boldsymbol{x}, \boldsymbol{w})}{\partial \boldsymbol{w}}\right)^{\mathsf{T}} \dot{\boldsymbol{w}} \\
&= \left(\frac{\partial V(\boldsymbol{x}, \boldsymbol{w})}{\partial \boldsymbol{x}}\right)^{\mathsf{T}} (\boldsymbol{J}(\boldsymbol{x}) - \boldsymbol{R}(\boldsymbol{x})) \frac{\partial H}{\partial \boldsymbol{x}} - \left(\frac{\partial V(\boldsymbol{x}, \boldsymbol{w})}{\partial \boldsymbol{w}}\right)^{\mathsf{T}} \alpha \frac{\partial \mathcal{J}_{\mathrm{w}}(\boldsymbol{x}, \boldsymbol{w})}{\partial \boldsymbol{w}} && (5.59) \\
&\quad - \left(\frac{\partial V(\boldsymbol{x}, \boldsymbol{w})}{\partial \boldsymbol{x}}\right)^{\mathsf{T}} \boldsymbol{G}(\boldsymbol{x}) \boldsymbol{S}^{-1}(\boldsymbol{x}) \boldsymbol{G}^{\mathsf{T}}(\boldsymbol{x}) \frac{\partial V(\boldsymbol{x}, \boldsymbol{w})}{\partial \boldsymbol{x}} \Upsilon && (5.60) \\
&= \left(\frac{\partial V(\boldsymbol{x}, \boldsymbol{w})}{\partial \boldsymbol{x}}\right)^{\mathsf{T}} (\boldsymbol{J}(\boldsymbol{x}) - \boldsymbol{R}(\boldsymbol{x})) \frac{\partial H}{\partial \boldsymbol{x}} - \left(\frac{\partial V(\boldsymbol{x}, \boldsymbol{w})}{\partial \boldsymbol{w}}\right)^{\mathsf{T}} \alpha \frac{\partial \mathcal{J}_{\mathrm{w}}(\boldsymbol{x}, \boldsymbol{w})}{\partial \boldsymbol{w}} \\
&\quad - S'_\Upsilon \cdot \frac{f'_\Upsilon + \sqrt{(f'_\Upsilon)^2 + Q'_\Upsilon \cdot S'_\Upsilon}}{S'_\Upsilon} \\
&= f'_\Upsilon - f'_\Upsilon - \sqrt{(f'_\Upsilon)^2 + Q'_\Upsilon \cdot S'_\Upsilon} \\
&= -\sqrt{(f'_\Upsilon)^2 + Q'_\Upsilon \cdot S'_\Upsilon}. && (5.61)
\end{aligned}$$

Since $S'_\Upsilon \geq 0$ and $Q'_\Upsilon \geq 0$ holds (see (5.32) and (5.33)), $\dot{V}(\boldsymbol{x}, \boldsymbol{w})$ is nonpositive for all $(\boldsymbol{x}, \boldsymbol{w}) \in \mathbb{R}^n \times \mathbb{R}^r$. □

Next, statement (5.56) (positive definiteness of $V(\boldsymbol{x}, \boldsymbol{w})$) has to be evaluated. Despite the fact that $H(\boldsymbol{x})$ is positive definite by definition, $V(\boldsymbol{x}, \boldsymbol{w}) = H(\boldsymbol{x}) + \boldsymbol{w}^{\mathsf{T}} \Xi(\boldsymbol{x})$ may be nonpositive, if "$+\boldsymbol{w}^{\mathsf{T}} \Xi(\boldsymbol{x})$" is negative for some $(\boldsymbol{x}, \boldsymbol{w}) \in \mathbb{R}^n \times \mathbb{R}^r$. In particular, we thus have to prove that $V(\boldsymbol{x}(t), \boldsymbol{w}(t))$ is still positive for all $t \geq t_0$ with $\boldsymbol{x} \neq \mathbb{0}$. This is stated in Lemma 5.23 with the help of the following Lemma 5.22.

Lemma 5.22. *Let* $V^\star(\boldsymbol{x}) = H(\boldsymbol{x}) + (\boldsymbol{w}^\star)^{\mathsf{T}} \Xi(\boldsymbol{x})$ *be the value function of optimization problem* (5.3) *and let the conditions of Lemma 5.19 hold with* $\boldsymbol{w}_0 \in \mathcal{M}$. *Then, all weighting factors* $\boldsymbol{w}(t)$ *asymptotically converge to the optimal ones, i.e.* $\boldsymbol{w}(t)$ *fulfills*

$$\lim_{t \to \infty} \|\boldsymbol{w}(t) - \boldsymbol{w}^\star\|_2 = 0. \tag{5.62}$$

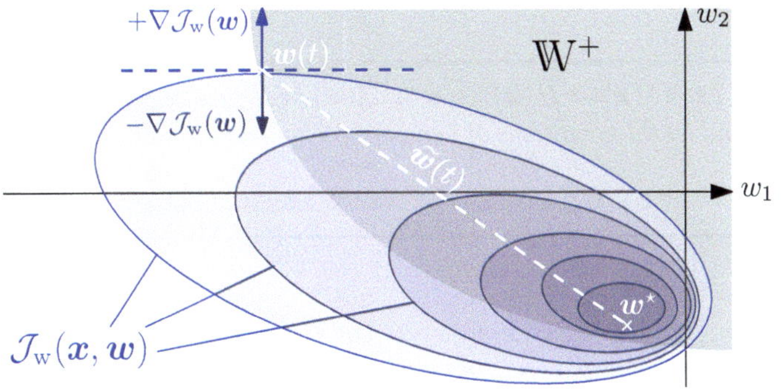

Figure 5.6: Contour plot of $\mathcal{J}_{\mathrm{w}}(\boldsymbol{x}, \boldsymbol{w})$ for fixed $\boldsymbol{x}$.

Proof. If the conditions of Lemma 5.19 hold with $\boldsymbol{w}_0 \in \mathcal{M}$, then $\mathcal{J}_{\mathrm{w}}(\boldsymbol{x}, \boldsymbol{w})$ is strictly convex with respect to $\boldsymbol{w}$ in an open neighborhood $\mathcal{M}$ of the optimizer $\boldsymbol{w}^\star$ for each arbitrary but fixed $\boldsymbol{x}$, i.e. for all $t \geq 0$ we have

$$(\boldsymbol{w}_1(t) - \boldsymbol{w}_2(t))^{\mathsf{T}} \left(\left. \frac{\partial \mathcal{J}_{\mathrm{w}}(\boldsymbol{x}(t), \boldsymbol{w})}{\partial \boldsymbol{w}} \right|_{\boldsymbol{w}_1(t)} - \left. \frac{\partial \mathcal{J}_{\mathrm{w}}(\boldsymbol{x}(t), \boldsymbol{w})}{\partial \boldsymbol{w}} \right|_{\boldsymbol{w}_2(t)} \right) > 0. \tag{5.63}$$

With $\boldsymbol{w}_1(t) = \boldsymbol{w}(t)$ and $\boldsymbol{w}_2(t) = \boldsymbol{w}^\star$, (5.63) reads as

$$(\boldsymbol{w}(t) - \boldsymbol{w}^\star)^{\mathsf{T}} \left(\left. \frac{\partial \mathcal{J}_{\mathrm{w}}(\boldsymbol{x}(t), \boldsymbol{w})}{\partial \boldsymbol{w}} \right|_{\boldsymbol{w}(t)} \right) > 0. \tag{5.64}$$

Insertion of (5.55c) in (5.64) yields $(\boldsymbol{w}(t) - \boldsymbol{w}^\star)^{\mathsf{T}} \dot{\boldsymbol{w}}(t) < 0$. With $\widetilde{\boldsymbol{w}}(t) := \boldsymbol{w}(t) - \boldsymbol{w}^\star$, this is equivalent to

$$(\widetilde{\boldsymbol{w}}(t))^{\mathsf{T}} \dot{\widetilde{\boldsymbol{w}}}(t) < 0. \tag{5.65}$$

By using the chain rule, (5.65) can be transformed to

$$\frac{1}{2} \cdot \frac{\mathrm{d}}{\mathrm{d}t} \left\{ (\widetilde{\boldsymbol{w}}(t))^{\mathsf{T}} \widetilde{\boldsymbol{w}}(t) \right\} < 0, \tag{5.66}$$

where $(\widetilde{\boldsymbol{w}}(t))^{\mathsf{T}} \widetilde{\boldsymbol{w}}(t) = \|\widetilde{\boldsymbol{w}}(t)\|_2^2$. Multiplying (5.66) by two and applying the square root on both sides, we finally obtain $\frac{\mathrm{d}}{\mathrm{d}t} \|\widetilde{\boldsymbol{w}}(t)\|_2 < 0$, i.e. the distance $\|\boldsymbol{w}(t) - \boldsymbol{w}^\star\|_2$ decreases strictly monotonically with time for all $t \geq 0$. This results in (5.62). □

With the help of the above Lemma 5.22, we can prove that $V(\boldsymbol{x}(t), \boldsymbol{w}(t))$ is indeed a positive-definite function:

Lemma 5.23. *Let $V^\star(\boldsymbol{x}) = H(\boldsymbol{x}) + (\boldsymbol{w}^\star)^\mathsf{T}\boldsymbol{\Xi}(\boldsymbol{x})$ be the value function of optimization problem* (5.3) *and let the conditions of Lemma 5.19 hold with $\boldsymbol{w}_0 \in \mathcal{M}$. Then*

$$\forall\, t \geq 0: \qquad V(\boldsymbol{x}(t), \boldsymbol{w}(t)) \succ 0. \tag{5.67}$$

Proof. Let $\mathbb{W}^+ = \{\boldsymbol{w} \in \mathbb{R}^r : V(\boldsymbol{x}, \boldsymbol{w}) > 0\ \forall \boldsymbol{x} \neq \mathbb{0}\}$ denote the set of $\boldsymbol{w} \in \mathbb{R}^r$ where $V(\boldsymbol{x}, \boldsymbol{w}) > 0$ is fulfilled for *all* $\boldsymbol{x} \neq \mathbb{0}$. As it trivially holds that $\boldsymbol{w}^\star \in \mathbb{W}^+$ and $\mathbb{0} \in \mathbb{W}^+$, and as $\mathbb{W}^+$ is an open set (see Lemma D.2 in Appendix D.1), it can be concluded that $\mathbb{0} \in \mathrm{int}(\mathbb{W}^+)$ and $\boldsymbol{w}^\star \in \mathrm{int}(\mathbb{W}^+)$, which is illustrated in Fig. 5.6[68]. Hence, there exists an $\varepsilon > 0$ such that the ball $\mathcal{B}(\boldsymbol{w}^\star, \varepsilon) = \{\boldsymbol{w} \in \mathbb{R}^r : \|\boldsymbol{w} - \boldsymbol{w}^\star\|_2 \leq \varepsilon\}$ lies completely within $\mathbb{W}^+$. According to (5.66), $\|\boldsymbol{w}(t) - \boldsymbol{w}^\star\|_2$ is strictly decreasing with time. Consequently, there exists a $T \geq 0$ such that $\|\boldsymbol{w}(T) - \boldsymbol{w}^\star\|_2 = \varepsilon$, i.e. $\boldsymbol{x}$ intersects the surface of the ball. Since $\|\boldsymbol{w}(t) - \boldsymbol{w}^\star\|_2$ is strictly decreasing, $\boldsymbol{w}(t)$ will remain within the ball for all $t > T$, thus

$$\forall t > T: \qquad \boldsymbol{w}(T) \in \mathbb{W}^+. \tag{5.68}$$

Since $\mathbb{W}^+$ is the set of parameters $\boldsymbol{w}$ where $V(\boldsymbol{x}, \boldsymbol{w})$ is positive definite for *all* $\boldsymbol{x} \in \mathbb{R}^n$, (5.68) implies that $V(\boldsymbol{x}(t), \boldsymbol{w}(t)) \succ 0$ holds for all $t > T$. With $V(\boldsymbol{x}(0), \boldsymbol{w}(0)) = V(\boldsymbol{x}_0, \mathbb{0}_r) = H(\boldsymbol{x}_0) \succ 0$ and due to the fact that $V(\boldsymbol{x}(t), \boldsymbol{w}(t))$ is continuous and $\dot{V}(\boldsymbol{x}(t), \boldsymbol{w}(t))$ is monotonically decreasing according to Lemma 5.21, it holds that $V(\boldsymbol{x}(t), \boldsymbol{w}(t)) \succ 0$ for all $t \geq 0$. □

As a consequence of Lemma 5.21 and Lemma 5.23, $V(\boldsymbol{x}(t), \boldsymbol{w}(t))$ is a suitable Lyapunov function. With this finding, the main statement of this section regarding stability and asymptotic stability of the closed-loop equilibrium can be formulated.

Theorem 5.24. *Let $V^\star(\boldsymbol{x}) = H(\boldsymbol{x}) + (\boldsymbol{w}^\star)^\mathsf{T}\boldsymbol{\Xi}(\boldsymbol{x})$ be the value function of optimization problem* (5.3) *and let the conditions of Lemma 5.19 hold with $\boldsymbol{w}_0 \in \mathcal{M}$. Then $\boldsymbol{x} = \mathbb{0}, \boldsymbol{w} = \boldsymbol{w}^\star$ is a stable equilibrium of* (5.55).

If additionally one of the following conditions holds:

(1) The autonomous system $\dot{\boldsymbol{x}} = (\boldsymbol{J}(\boldsymbol{x}) - \boldsymbol{R}(\boldsymbol{x}))\frac{\partial H}{\partial \boldsymbol{x}}$ is asymptotically stable with respect to the origin $\boldsymbol{x} = \mathbb{0}$,

(2) $\boldsymbol{G}(\boldsymbol{x})$ has full rank,

then $\boldsymbol{x} = \mathbb{0}, \boldsymbol{w} = \boldsymbol{w}^\star$ is an asymptotically stable equilibrium of (5.55).

[68] Note that $\mathbb{0} \in \mathrm{int}(\mathbb{W}^+)$ and $\boldsymbol{w}^\star \in \mathrm{int}(\mathbb{W}^+)$ does not necessarily imply that $\boldsymbol{w}(t) \in \mathbb{W}^+$ holds for *all* t. Fig. 5.6 shows an illustrative contour plot of $\mathcal{J}_\mathrm{w}(\boldsymbol{x}, \boldsymbol{w})$ for a fixed $\boldsymbol{x}$. Of course, $\boldsymbol{w}^\star = \arg\min \mathcal{J}_\mathrm{w}(\boldsymbol{x}, \boldsymbol{w})$. However, depending on the shape of $\mathcal{J}_\mathrm{w}$, it may be possible that the descent direction $-\nabla\mathcal{J}_\mathrm{w}$ is pointing out of $\mathbb{W}^+$, which may yield $\boldsymbol{w}(t') \notin \mathbb{W}^+$ for some $t' > t$. Despite the fact that $\boldsymbol{w}(t)$ may be temporarily outside of $\mathbb{W}^+$, it follows from (5.68) that for a sufficiently large but finite $T \geq 0$, $\boldsymbol{w}(t)$ always lies within $\mathbb{W}^+$.

Proof. According to Lemma 5.23, $V(\boldsymbol{x}, \boldsymbol{w})$ is positive definite and according to Lemma 5.21, $\dot{V}(\boldsymbol{x}, \boldsymbol{w})$ is negative semidefinite. As such, $V(\boldsymbol{x}, \boldsymbol{w})$ is a Lyapunov function for the equilibrium $(\mathbb{0}, \boldsymbol{w}^\star)$ of (5.55), which is consequently a stable equilibrium [Kha02, Theorem 4.1].

To prove asymptotic stability of $(\mathbb{0}, \boldsymbol{w}^\star)$, recall Lemma 5.22, which states that $\boldsymbol{w}(t)$ converges strictly monotonically to $\boldsymbol{w}^\star$. Now let $E = \{\boldsymbol{x} \in \mathbb{R}^n : \dot{V}(\boldsymbol{x}(t), \boldsymbol{w}^\star) = 0\}$ be the set of states where $V(\boldsymbol{x}, \boldsymbol{w})$ is constant and $\boldsymbol{w} = \boldsymbol{w}^\star$. With regard to the individual summands in (5.61), we get $E = \{\boldsymbol{x} \in \mathbb{R}^n : (f'_\Upsilon = 0) \wedge ((Q'_\Upsilon = 0) \vee (S'_\Upsilon = 0)), \boldsymbol{w} = \boldsymbol{w}^\star\}$. With $Q'_\Upsilon = \ell(\boldsymbol{x}) \succ 0$, statement $Q'_\Upsilon = 0$ is equivalent to $\boldsymbol{x} = \mathbb{0}$, which implies $f'_\Upsilon = 0$. Accordingly, E can be simplified to $E = \{\boldsymbol{x} \in \mathbb{R}^n : (\boldsymbol{x} = \mathbb{0}) \vee ((f'_\Upsilon = 0) \wedge (S'_\Upsilon = 0)), \boldsymbol{w} = \boldsymbol{w}^\star\}$. Based on this definition, conditions (1) and (2) of Theorem 5.24 are then obtained as follows:

(1) According to LaSalle's invariance principle, all trajectories $\boldsymbol{x}(t)$ with $\dot{V}(\boldsymbol{x}(t), \boldsymbol{w}(t)) = 0$ converge to the largest invariant set contained in E [Kha02, Theorem 4.4]. With (5.55b), $S'_\Upsilon = 0$ implies $\boldsymbol{u} = \mathbb{0}$. Due to the assumption that the autonomous system is asymptotically stable with respect to $\boldsymbol{x} = \mathbb{0}$, the largest invariant set in E is a singleton. Thus $\boldsymbol{x} = \mathbb{0}, \boldsymbol{w} = \boldsymbol{w}^\star$ is an asymptotically stable equilibrium of (5.55) [Kha02, Corollary 4.1].

(2) if $\boldsymbol{G}(\boldsymbol{x})$ has full rank, then $\boldsymbol{G}(\boldsymbol{x})\boldsymbol{S}^{-1}(\boldsymbol{x})\boldsymbol{G}^\mathsf{T}(\boldsymbol{x}) \succ 0$ and hence $S'_\Upsilon = 0$ only holds for $\boldsymbol{x} = \mathbb{0}$. Thus E is equal to $\{\mathbb{0}\}$, which implies that $\boldsymbol{x} = \mathbb{0}, \boldsymbol{w} = \boldsymbol{w}^\star$ is an asymptotically stable equilibrium of (5.55) [Kha02, Corollary 4.1]. □

Example 7 (Synchronous Motor)

We apply the presented method to the nonlinear model of a non-salient permanent magnet synchronous motor (PMSM) in rotating reference (dq) frame from [OPA⁺11]:

$$L_\mathrm{s}\dot{I}_\mathrm{d} = -R_\mathrm{s}I_\mathrm{d} - \Omega L_\mathrm{s}I_\mathrm{q} + U_\mathrm{d}, \tag{5.69a}$$

$$L_\mathrm{s}\dot{I}_\mathrm{q} = -R_\mathrm{s}I_\mathrm{q} + \Omega L_\mathrm{s}I_\mathrm{d} - \Theta\Omega + U_\mathrm{q}, \tag{5.69b}$$

$$J\dot{\Omega} = n_\mathrm{pp}\Theta I_\mathrm{q} - f\Omega - M_\ell, \tag{5.69c}$$

where I_d and I_q are the stator currents, U_d and U_q are the stator voltages, Ω is the angular velocity, M_ℓ is the load torque, L_s is the stator inductance, R_s is the stator winding resistance, Θ is the magnetic flux, J is the moment of inertia, f is the mechanic friction constant, and n_pp is the number of pole pairs.

The objective of the control system to be designed is to follow a desired angular velocity Ω^des with zero flux current $I_\mathrm{d}^\mathrm{des} = 0$ for a given load torque M_ℓ^des. Denote by $(U_\mathrm{d}^\mathrm{des}, U_\mathrm{q}^\mathrm{des}, I_\mathrm{q}^\mathrm{des})$ the resulting open-loop equilibrium values and define $\boldsymbol{x} := \boldsymbol{F}(I_\mathrm{d} - i_\mathrm{d}^\mathrm{des}, I_\mathrm{q} - i_\mathrm{q}^\mathrm{des}, \Omega - \Omega^\mathrm{des})^\mathsf{T}$ with $\boldsymbol{F} := \mathrm{diag}\{L_\mathrm{s}, L_\mathrm{s}, J/n_\mathrm{pp}\}$, and $\boldsymbol{u} := (U_\mathrm{d}, U_\mathrm{q}, M_\ell)^\mathsf{T}$. Then, the error dynamics of (5.69) *can be modeled by the input-state-output PHS* (5.3b), *where $\boldsymbol{G} = \mathrm{diag}\{1, 1, -1/n_\mathrm{pp}\}$,*

$H(\boldsymbol{x}) = 1/2(\boldsymbol{x}^\mathsf{T}\boldsymbol{F}^{-1}\boldsymbol{x})$,

$$\boldsymbol{J}(\boldsymbol{x}) = \begin{bmatrix} 0 & \frac{Ln_{\mathrm{pp}}x_3}{J} + L_{\mathrm{s}}\Omega^{\mathrm{des}} & \frac{L_{\mathrm{s}}I_{\mathrm{q}}^{\mathrm{des}}}{2} \\ -\frac{L_{\mathrm{s}}n_{\mathrm{pp}}x_3}{J} - L_{\mathrm{s}}\Omega^{\mathrm{des}} & 0 & -\Theta \\ -\frac{L_{\mathrm{s}}I_{\mathrm{q}}^{\mathrm{des}}}{2} & \Theta & 0 \end{bmatrix},$$

$$\boldsymbol{R} = \begin{bmatrix} R_{\mathrm{s}} & 0 & -\frac{L_{\mathrm{s}}I_{\mathrm{q}}^{\mathrm{des}}}{2} \\ 0 & R_{\mathrm{s}} & 0 \\ -\frac{L_{\mathrm{s}}I_{\mathrm{q}}^{\mathrm{des}}}{2} & 0 & \frac{f}{n_{\mathrm{pp}}} \end{bmatrix}.$$

We choose $L_{\mathrm{s}} = 0.5$, $R_{\mathrm{s}} = 0.5$, $\Theta = 1$, $J = 2$, $f = 0.1$, $n_{\mathrm{pp}} = 2$, *and consider the optimal control problem* (5.3), *where*

$$\boldsymbol{Q} = \begin{bmatrix} 2.25 & 0 & -\frac{I_{\mathrm{q}}^{\mathrm{des}}}{2} \\ 0 & 2.1 & 0 \\ -\frac{I_{\mathrm{q}}^{\mathrm{des}}}{2} & 0 & 3.175 \end{bmatrix}, \qquad \boldsymbol{S} = \begin{bmatrix} 4 & 0 & 0 \\ 0 & 10 & 0 \\ 0 & 0 & 0.02 \end{bmatrix}.$$

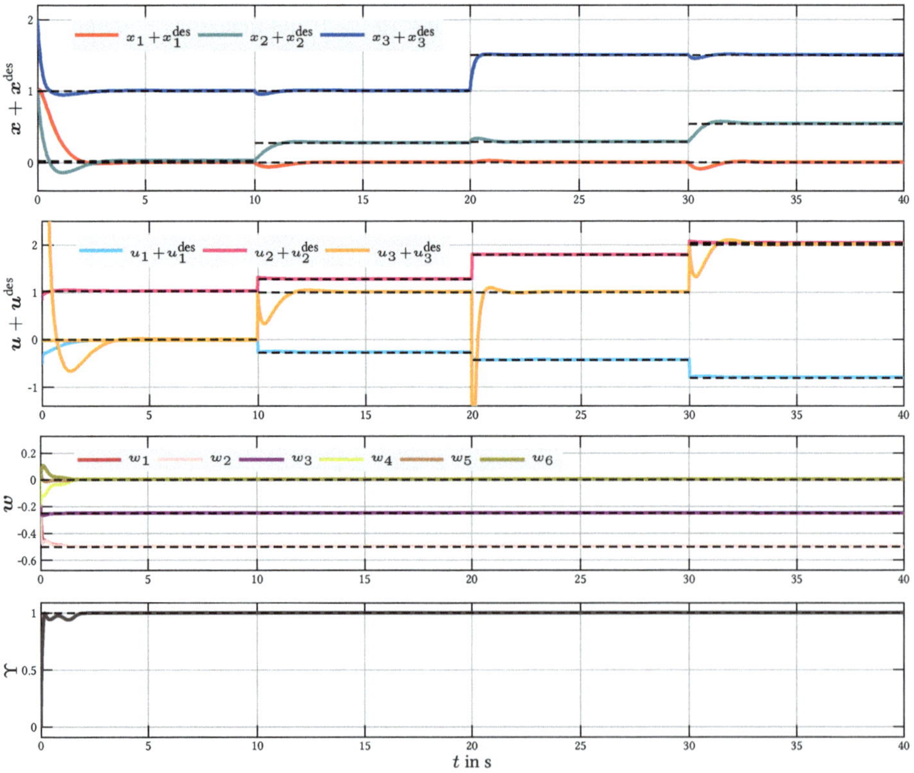

Figure 5.7: Controller performance for PMSM example.

Since the condition of Corollary 5.8 is fulfilled, the Hamiltonian $H(\boldsymbol{x})$ is a CLF. The basis functions for the extended CLF are chosen to $\Xi(\boldsymbol{x}) = (x_1^2, x_2^2, x_3^2, x_1x_2, x_2x_3, x_1x_3)^\top$*. For the above optimization problem, it can be shown that the exact solution of the HJB equation* (5.24) *is given by* $V^\star(\boldsymbol{x}) = 0.5(x_1^2+x_2^2+2x_3^2)$*, which implies* $\boldsymbol{w}^\star = (-0.5, -0.5, -0.25, 0, 0, 0)^\top$*. Now the control system* (5.55) *is initialized at* $\boldsymbol{x}_0 = \mathbb{1}$*,* $\boldsymbol{w}_0 = \mathbb{0}$*. The shifts in* $\mathcal{J}_\mathrm{w}$ *are set to* $c_1 = \mathbb{0}$*,* $c_2 = (1,0,0)^\top$*,* $c_3 = (0,1,0)^\top$*,* $c_4 = (0,0,1)^\top$*,* $c_5 = (-1,-1,0)^\top$*,* $c_6 = (0,-1,-1)^\top$*,* $c_7 = (-1,0,-1)^\top$*, and the learning rate is set to* $\alpha = 0.5$*. At* $t = 10$ *and* $t = 30$*, the load torque* M_ℓ^des *increases by* 1*, and at* $t = 20$*, the desired angular voltage* Ω^des *increases by* 0.5*.*

Fig. 5.7 shows the trajectories of $\boldsymbol{x}(t)$*,* $\boldsymbol{u}(t)$*,* $\boldsymbol{w}(t)$*, and* Υ*. It can be seen that after* $2\,\mathrm{s}$*, the learning process is completed and both* $\boldsymbol{w}$ *and* Υ *have reached their optimal values. Correspondingly, the trajectories of* $\boldsymbol{x}$ *and* $\boldsymbol{u}$ *converge to the solution associated to the optimal controller, once the learning process is completed. This demonstrates that the proposed control strategy is capable of adapting the optimal parameters after a single learning phase.*

To investigate the effects of an incorrect choice of basis functions, we repeat the simulation with $\Xi'(\boldsymbol{x}) = (x_1^3, x_2^3, x_3^3, x_1x_2, x_2x_3, x_1x_3)^\top$*, i.e.* $H(\boldsymbol{x}) + \boldsymbol{w}^\top\Xi'(\boldsymbol{x})$ *does not fit the structure of* $V^\star(\boldsymbol{x})$ *and hence Assumption 5.15 is violated. The results are shown in Fig. 5.8.*

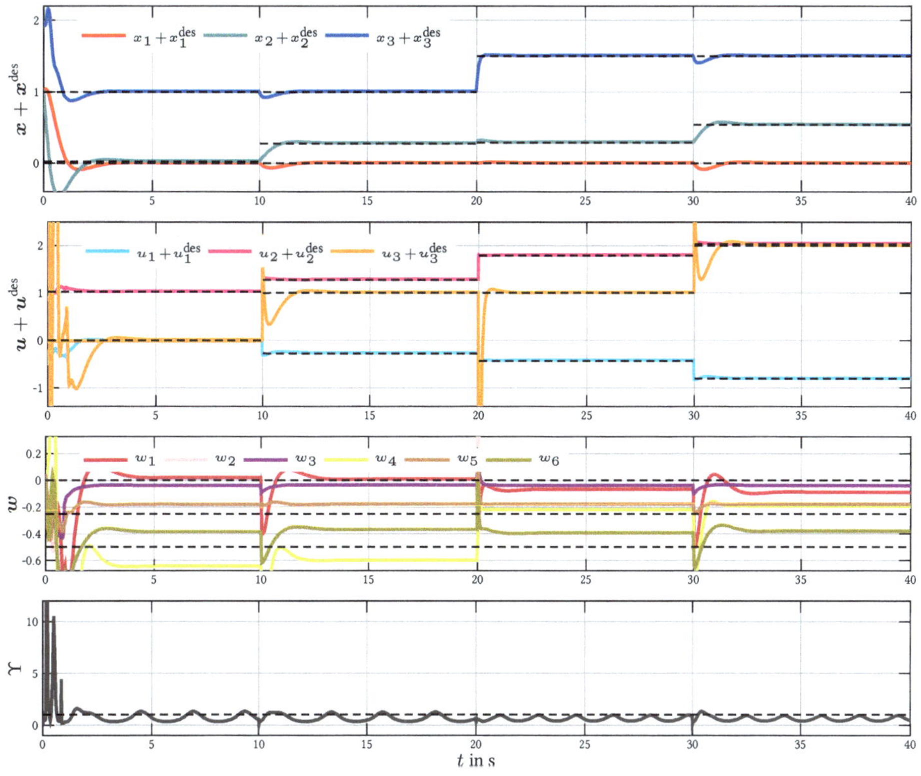

Figure 5.8: Controller performance if Assumption 5.15 is violated.

The trajectory of Υ reveals a remarkable and distinct oscillatory behavior. However, after 5 s, *the weighting factors $\boldsymbol{w}$ converge to a certain value $\boldsymbol{w}^\diamond$. After the step increases of M_ℓ^{des} and Ω^{des}, the weighting factors do not remain at their previous values. Both the oscillation of $\Upsilon(t)$ and the fluctuation of $\boldsymbol{w}(t)$ after the disturbances imply that $H(\boldsymbol{x}) + (\boldsymbol{w}^\diamond)^\mathsf{T}\boldsymbol{\Xi}'(\boldsymbol{x})$ is not equal to the value function for this specific set of basis functions. However, the results show that even if $\boldsymbol{\Xi}'(\boldsymbol{x})$ is not accurate, the proposed controller is able to learn suitable weighting factors for a "suboptimal" control, with the oscillation of $\Upsilon(t)$ being interpreted as an indicator of lack of fitness for the chosen basis functions in $\boldsymbol{\Xi}'(\boldsymbol{x})$ (see the discussion in Remark 5.13).*

5.2.6 Summary

In this section, we have introduced a continuous-time adaptive feedback control scheme for dynamic optimization problems with generalized Lagrangian performance indices (A.8) and general input-state-output PHSs (A.13). Employing the initial value function guess $V(\boldsymbol{x}) = H(\boldsymbol{x})$ generates an admissible controller which is already stabilizing. While applying the CLF-based MOC approach to PHSs, this section shows that the port-Hamiltonian modeling gives access to a *natural CLF candidate*, which can serve as a helpful tool for constructing *universal formulae for stabilization* [Son89, CB04][69].

Subsequently, a gradient adaptation of the critic weights $\boldsymbol{w}$ based on the Lagrange multiplier $\Upsilon(t)$ enables a learning of the (unknown) value function $V^\star(\boldsymbol{x})$ and thus the optimal control strategy $\boldsymbol{u}^\star$. Asymptotic stability of the closed-loop equilibrium is shown by the main Theorem 5.24. Although a reasonable choice of basis functions is nontrivial, simulations show that even if $\boldsymbol{\Xi}(\boldsymbol{x})$ is inaccurate, the proposed controller is able to stabilize the system and provide near-optimal solution trajectories. Furthermore, the optimality of the adapted control law can be assessed for the case of a bad choice of basis functions via $\Upsilon(t)$. However, a rigorous perturbation analysis for systems where parameterization $\boldsymbol{\Xi}(\boldsymbol{x})$ does not fit the structure of $V^\star(\boldsymbol{x})$ remains an open research question.

5.3 Multi-Player Case

If a dynamical system is controlled by two or more actuators with differing interests and goals, a scenario emerges which can be modeled as a noncooperative differential game (see Definition A.2 in Appendix A.2). This section presents an adaptive controller for such noncooperative differential games with dynamics modeled as general nonlinear input-state-output PHSs. The proposed controller is able to learn feedback Nash strategies (cf. Definition A.6, see also Lemma A.8) for general Lagrange-type performance indices (A.8).

Subsection 5.3.1 formalizes the problem class by extending the single-player case from Section 5.2 to N players, and briefly discusses related work. Subsection 5.3.2 provides an analytical solution of the so-called *modified differential game*. Subsection 5.3.3 analyzes the asymptotic

[69] Sontag's universal formula for stabilization itself is a special case of the modified optimal controller (5.8). It can be obtained by choosing $\ell(\boldsymbol{x}) = \left(\frac{\partial V}{\partial \boldsymbol{x}}\right)^\mathsf{T} \boldsymbol{G}(\boldsymbol{x})(\boldsymbol{S}(\boldsymbol{x}))^{-1}\boldsymbol{G}^\mathsf{T}(\boldsymbol{x})\frac{\partial V}{\partial \boldsymbol{x}}$ and $\boldsymbol{S}(\boldsymbol{x}) = \boldsymbol{I}$.

stability of the closed-loop equilibrium resulting from the modified differential game. Subsection 5.3.4 presents necessary and sufficient conditions for $H(\boldsymbol{x})$ to serve as admissible CLF. Subsection 5.3.5 develops an adaptation strategy which enables to learn the original (non-modified) differential game (A.8). Subsection 5.3.6 proves the asymptotic stability of the resulting feedback Nash equilibrium and gives an illustrative example. Finally, Subsection 5.3.7 summarizes the main result of this section.

5.3.1 Problem Definition and Related Work

Definition 5.25 (Multi-Player Optimal Control Problem for PHSs). *For all $i \in \mathbb{P}$, find the feedback control laws $\boldsymbol{u}_i(\boldsymbol{x})$ as the solutions of*

$$\min_{\boldsymbol{u}_i(t)} \quad \frac{1}{2}\int_0^{t_f} \ell_i(\boldsymbol{x}(t)) + (\boldsymbol{u}_i(t))^\mathsf{T}\boldsymbol{S}_i(\boldsymbol{x}(t))\boldsymbol{u}_i(t)\,\mathrm{d}t \tag{5.70a}$$

$$\text{subject to} \quad \dot{\boldsymbol{x}}(t) = (\boldsymbol{J}(\boldsymbol{x}(t)) - \boldsymbol{R}(\boldsymbol{x}(t)))\frac{\partial H(\boldsymbol{x}(t))}{\partial \boldsymbol{x}(t)} + \sum_{i=1}^{N} \boldsymbol{G}_i(\boldsymbol{x}(t))\boldsymbol{u}_i(t) \tag{5.70b}$$

with set of players $\mathbb{P} = \{1,\dots,N\}$, state vector $\boldsymbol{x} \in \mathbb{R}^n$, input vectors $\boldsymbol{u}_i \in \mathbb{R}^{p_i}$, skew-symmetric interconnection matrix $\boldsymbol{J}(\boldsymbol{x}) \in \mathbb{R}^{n\times n}$, positive semidefinite dissipation matrix $\boldsymbol{R}(\boldsymbol{x}) \in \mathbb{R}^{n\times n}$, input matrices $\boldsymbol{G}_i(\boldsymbol{x}) \in \mathbb{R}^{n\times p_i}$, positive definite and radially unbounded Hamiltonian $H : \mathbb{R}^n \to \mathbb{R}_{\geq 0}$; smooth, positive definite $\ell_i : \mathbb{R}^n \to \mathbb{R}_{\geq 0}$ and positive definite $\boldsymbol{S}_i(\boldsymbol{x}) \in \mathbb{R}^{p_i\times p_i}$. Without loss of generality, we assume that $H(\boldsymbol{x})$ has its minimum at $\boldsymbol{x} = \mathbb{0}$ with $H(\mathbb{0}) = 0$.

To find a feedback Nash strategy for (5.70), however, a system of HJB equations [Bre11] has to be solved in order to get a set of coupled value functions $V_i^\star(\boldsymbol{x})$ (one for each specific player), cf. Lemma A.8. Since an analytical solution of this problem is currently not known, it is crucial to exploit the structure of the optimization problems and/or the underlying plant system. For example, in linear-affine systems with quadratic objective functions (LQ games), the value functions are quadratic forms and thus the set of HJB equations translates into a tractable set of (algebraic) matrix-Riccati ODEs [PMC79, WSE99, AL13, WHLM18], for which extensive research has been undertaken in recent decades, see [Eng06] and [ES13] for a profound introduction to solution methods for LQ games.

For the more general case of input-affine nonlinear systems, methods derived from ADP provide viable means for solving the system of coupled HJB equations. In this case, approximators (critics) for the players' value functions are applied, whereby the weights of the critics are continuously adjusted based on measurements of the current state vector and the current inputs of the other players. Depending on the selected ADP approach, however, initial stabilizing weights [VL11, LLW14, MLS14, VMH16] or a suitable Lyapunov function [ZCL13, MNSS18] are required to guarantee convergence of the weights and stability of the closed-loop system. This makes the deduction of a general methodology difficult, especially for large-scale systems.

In the following, we present an adaptive feedback Nash strategy for the class of nonlinear input-state-output PHSs. In the previous Section 5.2, it was shown for the single-player case that $H(\boldsymbol{x})$ inherently qualifies as a value function candidate. Again inspired by [SK00], in the remainder of this section particular attention is paid to determine the deviation of the current suboptimal solution of player i from the optimal (Nash) strategy using a certain Lagrange multiplier Υ_i. Based on an evaluation of Υ_i, the respective critic weights are updated simultaneously using a continuous-time gradient flow. Finally, the main theorem of this section (Theorem 5.53) will show convergence of the critic weights towards the Nash solution and asymptotic stability of the closed-loop system.

5.3.2 Modified Differential Game

As with the single-player case, an analytical solution of the original differential game (5.70) is generally intractable [Bre11, p. 390ff.], [LVS12, p. 278]. However, inspired by the basic idea of [SK00], projection of (5.70b) using a suitable CLF allows to formulate a set of modified optimization problems that makes obtaining an analytical solution considerably easier. For this purpose, we first introduce the new definition of a CLF in the context of differential games and then present an analytical solution of the set of modified optimization problems.

Definition 5.26 (CLFs for Multi-Player Scenarios). *A radially unbounded, smooth, positive definite function $V_i : \mathbb{R}^n \to \mathbb{R}$ is said to be a* control-Lyapunov function associated to input $i \in \mathbb{P}$ *for the system*

$$\dot{\boldsymbol{x}} = \boldsymbol{f}(\boldsymbol{x}, \boldsymbol{u}_1, \ldots, \boldsymbol{u}_N), \qquad \boldsymbol{x} \in \mathbb{R}^n, \quad \boldsymbol{u}_i \in \mathbb{R}^{p_i}, \quad i \in \mathbb{P}, \tag{5.71}$$

if $\boldsymbol{f}(\mathbb{0}_n, \mathbb{0}_{p_1}, \ldots, \mathbb{0}_{p_N}) = \mathbb{0}_n$ and

$$\forall \boldsymbol{x} \neq \mathbb{0}_n : \qquad \inf_{\boldsymbol{u}_i} \left\{ \left(\frac{\partial V_i}{\partial \boldsymbol{x}} \right)^{\mathsf{T}} \boldsymbol{f}(\boldsymbol{x}, \boldsymbol{u}_1, \ldots, \boldsymbol{u}_N) \right\} < 0. \tag{5.72}$$

Remark 5.27. *For input-affine nonlinear systems*

$$\dot{\boldsymbol{x}} = \boldsymbol{f}(\boldsymbol{x}) + \sum_{i=1}^{N} \boldsymbol{G}_i(\boldsymbol{x}) \boldsymbol{u}_i, \qquad \boldsymbol{x} \in \mathbb{R}^n, \quad \boldsymbol{u}_i \in \mathbb{R}^{p_i}, \tag{5.73}$$

condition (5.72) *is equivalent to*

$$\left(\frac{\partial V_i}{\partial \boldsymbol{x}} \right)^{\mathsf{T}} \boldsymbol{G}_i(\boldsymbol{x}) = \mathbb{0}_{p_i} \implies \left(\frac{\partial V_i}{\partial \boldsymbol{x}} \right)^{\mathsf{T}} \left(\boldsymbol{f}(\boldsymbol{x}) + \sum_{\substack{j=1 \\ j \neq i}}^{N} \boldsymbol{G}_j(\boldsymbol{x}) \boldsymbol{u}_j \right) \begin{cases} < 0, & \boldsymbol{x} \neq \mathbb{0}_n, \\ = 0, & \boldsymbol{x} = \mathbb{0}_n. \end{cases} \tag{5.74}$$

◇

Now the original problem (5.70) is modified by left-multiplying (5.70b) with $\left(\frac{\partial V_i}{\partial \boldsymbol{x}}\right)^\mathsf{T}$:

Lemma 5.28. *For all $i \in \mathbb{P}$, let $V_i(\boldsymbol{x})$ be a CLF associated to input $\boldsymbol{u}_i$ of* (5.70)*. Then an exact solution of the modified differential game*

$$\min_{\boldsymbol{u}_i} \quad \left\{\frac{1}{2}\int_0^{t_\mathrm{f}} \ell_i(\boldsymbol{x}) + \boldsymbol{u}_i^\mathsf{T}\boldsymbol{S}_i(\boldsymbol{x})\boldsymbol{u}_i \mathrm{d}t\right\} \tag{5.75a}$$

$$\text{subject to} \quad \dot{V}_i(\boldsymbol{x}) = \left(\frac{\partial V_i}{\partial \boldsymbol{x}}\right)^\mathsf{T}\left((\boldsymbol{J}-\boldsymbol{R})\frac{\partial H}{\partial \boldsymbol{x}} + \sum_{i=1}^{N}\boldsymbol{G}_i(\boldsymbol{x})\boldsymbol{u}_i\right) \tag{5.75b}$$

with $\ell_i(\boldsymbol{x}) \succcurlyeq 0$, $\boldsymbol{S}_i(\boldsymbol{x}) \succ 0$ is given by

$$\boldsymbol{u}_i^\star(\boldsymbol{x}) = -(\boldsymbol{S}_i(\boldsymbol{x}))^{-1}\boldsymbol{G}_i^\mathsf{T}(\boldsymbol{x})\frac{\partial V_i(\boldsymbol{x})}{\partial \boldsymbol{x}} \cdot \Upsilon_i(t), \qquad i \in \mathbb{P}, \tag{5.76}$$

$$0 = q_i^0(\boldsymbol{x}) + \Upsilon_i q_i^1(\boldsymbol{x}) + \Upsilon_i^2 q_i^2(\boldsymbol{x}) + \Upsilon_i \sum_{\substack{j=1\\ j\neq i}}^{N} q_{ij}^2(\boldsymbol{x})\Upsilon_j, \qquad i \in \mathbb{P}, \tag{5.77}$$

where

$$q_i^0(\boldsymbol{x}) := \frac{1}{2}\ell_i(\boldsymbol{x}), \tag{5.78}$$

$$q_i^1(\boldsymbol{x}) := \left(\frac{\partial V_i}{\partial \boldsymbol{x}}\right)^\mathsf{T}(\boldsymbol{J}-\boldsymbol{R})\frac{\partial H}{\partial \boldsymbol{x}}, \tag{5.79}$$

$$q_i^2(\boldsymbol{x}) := -\frac{1}{2}\left(\frac{\partial V_i}{\partial \boldsymbol{x}}\right)^\mathsf{T}\boldsymbol{G}_i(\boldsymbol{x})(\boldsymbol{S}_i(\boldsymbol{x}))^{-1}\boldsymbol{G}_i^\mathsf{T}(\boldsymbol{x})\frac{\partial V_i}{\partial \boldsymbol{x}}, \tag{5.80}$$

$$q_{ij}^2(\boldsymbol{x}) := -\left(\frac{\partial V_i}{\partial \boldsymbol{x}}\right)^\mathsf{T}\boldsymbol{G}_j(\boldsymbol{x})(\boldsymbol{S}_j(\boldsymbol{x}))^{-1}\boldsymbol{G}_j^\mathsf{T}(\boldsymbol{x})\frac{\partial V_j}{\partial \boldsymbol{x}}. \tag{5.81}$$

Proof. For each $i \in \mathbb{P}$, the Hamiltonian of optimization problem (5.75) is given by

$$\mathscr{H}_i(\boldsymbol{x},\boldsymbol{u},\boldsymbol{\Upsilon}) = \frac{1}{2}\ell_i(\boldsymbol{x}) + \frac{1}{2}\boldsymbol{u}_i^\mathsf{T}\boldsymbol{S}_i(\boldsymbol{x})\boldsymbol{u}_i + \Upsilon_i\dot{V}_i(\boldsymbol{x}) \tag{5.82}$$

$$= \frac{1}{2}\ell_i(\boldsymbol{x}) + \frac{1}{2}\boldsymbol{u}_i^\mathsf{T}\boldsymbol{S}_i(\boldsymbol{x})\boldsymbol{u}_i + \Upsilon_i\left(\frac{\partial V_i}{\partial \boldsymbol{x}}\right)^\mathsf{T}\left((\boldsymbol{J}-\boldsymbol{R})\frac{\partial H}{\partial \boldsymbol{x}} + \sum_{j=1}^{N}\boldsymbol{G}_j(\boldsymbol{x})\boldsymbol{u}_j\right), \tag{5.83}$$

where Υ_i is the (scalar) Lagrange multiplier corresponding to the constraint (5.75b) and $\boldsymbol{\Upsilon} := \mathrm{col}_i\{\Upsilon_i\}$. The optimal control $\boldsymbol{u}_i^\star$ can be computed by

$$\left.\frac{\partial \mathscr{H}_i(\boldsymbol{x},\boldsymbol{u},\boldsymbol{\Upsilon})}{\partial \boldsymbol{u}_i}\right|_{\boldsymbol{u}_i^\star} = 0, \tag{5.84}$$

which yields

$$\boldsymbol{u}_i^\star(\boldsymbol{x}) = -(\boldsymbol{S}_i(\boldsymbol{x}))^{-1}\boldsymbol{G}_i^\mathsf{T}(\boldsymbol{x})\frac{\partial V_i}{\partial \boldsymbol{x}}\Upsilon_i. \tag{5.85}$$

From the HJB equation for time-invariant systems

$$\mathscr{H}_i(\boldsymbol{x}, \boldsymbol{u}^\star, \boldsymbol{\Upsilon}^\star) = 0, \qquad i \in \mathbb{P}, \tag{5.86}$$

we get

$$\begin{aligned}\forall i \in \mathbb{P}: \quad 0 = \frac{1}{2}\ell_i(\boldsymbol{x}) + \Upsilon_i^\star \left(\frac{\partial V_i}{\partial \boldsymbol{x}}\right)^\mathsf{T} (\boldsymbol{J} - \boldsymbol{R})\frac{\partial H}{\partial \boldsymbol{x}} \\ -(\Upsilon_i^\star)^2 \frac{1}{2}\left(\frac{\partial V_i}{\partial \boldsymbol{x}}\right)^\mathsf{T} \boldsymbol{G}_i(\boldsymbol{x})(\boldsymbol{S}_i(\boldsymbol{x}))^{-1}\boldsymbol{G}_i^\mathsf{T}(\boldsymbol{x})\frac{\partial V_i}{\partial \boldsymbol{x}} \\ -\Upsilon_i^\star \left(\frac{\partial V_i}{\partial \boldsymbol{x}}\right)^\mathsf{T} \sum_{\substack{j=1\\ j\neq i}}^{N} \boldsymbol{G}_j(\boldsymbol{x})(\boldsymbol{S}_j(\boldsymbol{x}))^{-1}\boldsymbol{G}_j^\mathsf{T}(\boldsymbol{x})\frac{\partial V_j}{\partial \boldsymbol{x}}\Upsilon_j^\star. \end{aligned} \tag{5.87}$$

With the definitions (5.78)–(5.81), this is equivalent to (5.76). □

Remark 5.29. *The resulting closed-loop system*

$$\dot{\boldsymbol{x}} = (\boldsymbol{J} - \boldsymbol{R})\frac{\partial H}{\partial \boldsymbol{x}} - \boldsymbol{G}_1(\boldsymbol{S}_1(\boldsymbol{x}))^{-1}\boldsymbol{G}_1^\mathsf{T}\frac{\partial V_1}{\partial \boldsymbol{x}}\Upsilon_1 - \cdots - \boldsymbol{G}_N(\boldsymbol{S}_N(\boldsymbol{x}))^{-1}\boldsymbol{G}_N^\mathsf{T}\frac{\partial V_N}{\partial \boldsymbol{x}}\Upsilon_N, \tag{5.88}$$

$$0 = q_1^0(\boldsymbol{x}) + \Upsilon_1 q_1^1(\boldsymbol{x}) + \Upsilon_1^2 q_1^2(\boldsymbol{x}) + \Upsilon_1 \sum_{j=2}^{N} q_{1j}^2(\boldsymbol{x})\Upsilon_j, \tag{5.89}$$

$$0 = q_2^0(\boldsymbol{x}) + \Upsilon_2 q_2^1(\boldsymbol{x}) + \Upsilon_2^2 q_2^2(\boldsymbol{x}) + \Upsilon_2 \sum_{\substack{j=1\\ j\neq 2}}^{N} q_{2j}^2(\boldsymbol{x})\Upsilon_j, \tag{5.90}$$

$$\vdots$$

$$0 = q_N^0(\boldsymbol{x}) + \Upsilon_N q_N^1(\boldsymbol{x}) + \Upsilon_N^2 q_N^2(\boldsymbol{x}) + \Upsilon_N \sum_{\substack{j=1\\ j\neq N}}^{N} q_{Nj}^2(\boldsymbol{x})\Upsilon_j \tag{5.91}$$

constitutes a differential-algebraic system of equations of index 1. ◇

According to Definition 5.26, the CLF property of $V_i(\boldsymbol{x})$ is highly dependent on the other inputs $\boldsymbol{u}_{\neg i}$. Thus, (5.72) must apply to all possible inputs $\boldsymbol{u}_1, \ldots, \boldsymbol{u}_i, \ldots, \boldsymbol{u}_N$. This is generally a very strict requirement, if those inputs are assumed to be unconstrained in any way. However, since in the context of the modified differential game (5.75) all inputs are restricted according to (5.85), a much less restrictive requirement on the vector $(V_1, \ldots, V_N)^\mathsf{T}$ of CLFs can be imposed:

Definition 5.30 (Admissible Vector of CLFs). *A vector* $\boldsymbol{V}(\boldsymbol{x}) = (V_1(\boldsymbol{x}), \ldots, V_p(\boldsymbol{x}))^\mathsf{T}$ *is said to be an* admissible vector of CLFs *for* (5.73), *if* (5.72) *is fulfilled for all* $i \in \mathbb{P}$ *with* $\boldsymbol{u}_i(\boldsymbol{x}) = \boldsymbol{u}_i^\star(\boldsymbol{x})$ *according to* (5.85), *i.e. if*

$$\forall \boldsymbol{x} \neq \mathbb{0}_n : \qquad \inf_{\boldsymbol{u}_i^\star} \left\{ \left(\frac{\partial V_i}{\partial \boldsymbol{x}} \right)^\mathsf{T} \boldsymbol{f}(\boldsymbol{x}, \boldsymbol{u}_1^\star, \ldots, \boldsymbol{u}_N^\star) \right\} < 0 \tag{5.92}$$

holds for all $i \in \mathbb{P}$.

Remark 5.31. *For input-affine nonlinear systems, condition* (5.92) *is equivalent to*

$$\left(\frac{\partial V_i}{\partial \boldsymbol{x}} \right)^\mathsf{T} \boldsymbol{G}_i(\boldsymbol{x}) = \mathbb{0}_{p_i}$$

$$\Longrightarrow \left(\frac{\partial V_i}{\partial \boldsymbol{x}} \right)^\mathsf{T} \left(\boldsymbol{f}(\boldsymbol{x}) - \sum_{\substack{j=1 \\ j \neq i}}^{N} \boldsymbol{G}_j(\boldsymbol{x}) (\boldsymbol{S}_j(\boldsymbol{x}))^{-1} \boldsymbol{G}_j^\mathsf{T}(\boldsymbol{x}) \frac{\partial V_j}{\partial \boldsymbol{x}} \cdot \Upsilon_j \right) \begin{cases} < 0, & \boldsymbol{x} \neq \mathbb{0}_n, \\ = 0, & \boldsymbol{x} = \mathbb{0}_n \end{cases} \tag{5.93}$$

for all $i \in \mathbb{P}$, $\boldsymbol{x} \in \mathbb{R}^n$. ◇

Since the individual inputs $\boldsymbol{u}_i(\boldsymbol{x})$ are interdependent due to coupling by the scalars $\Upsilon_1, \ldots, \Upsilon_N$ via the control laws (5.76)–(5.77), it is sufficient to require that $\boldsymbol{V}(\boldsymbol{x})$ is an admissible vector of CLFs instead of the (stricter) requirement that each individual $V_i(\boldsymbol{x})$ is a CLF in accordance with Definition 5.26. Lemma 5.28 can thus be relaxed as follows:

Corollary 5.32. *Let* $\boldsymbol{V}(\boldsymbol{x})$ *be an admissible vector of CLFs for* (5.70). *Then an exact solution of the modified differential game* (5.75) *is given by* (5.76)–(5.81).

Proof. Trivially, if $\boldsymbol{u}_i(\boldsymbol{x})$ is given by (5.76)–(5.77), then (5.74) and (5.93) are equivalent. □

Suboptimality of the Modified Differential Game

With shorthand notations $\boldsymbol{u} = \mathrm{col}_i\{\boldsymbol{u}_i\}$ and $\boldsymbol{\Upsilon} = \mathrm{col}_i\{\Upsilon_i\}$, the solution $(\boldsymbol{u}^\star, \boldsymbol{\Upsilon}^\star)$ generated by the MOC law (5.76)–(5.81) can be characterized as the Nash equilibrium of a differential game with scaled objective functions $\mathscr{J}_{\Upsilon,i}(\boldsymbol{x})$:

Lemma 5.33. *Let* $\boldsymbol{V}(\boldsymbol{x})$ *be an admissible vector of CLFs for* (5.70). *Then each solution* $(\boldsymbol{u}^\star, \boldsymbol{\Upsilon})$ *fulfilling* (5.76)–(5.77) *constitutes a Nash equilibrium with regard to the scaled* *objective functions*

$$\mathscr{J}_{\Upsilon,i}(\boldsymbol{x}) = \int_0^{t_\mathrm{f}} \frac{1}{2\Upsilon_i} \cdot \left(\ell_i(\boldsymbol{x}(t)) + (\boldsymbol{u}_i(t))^\mathsf{T} \boldsymbol{S}_i(\boldsymbol{x}(t))\boldsymbol{u}_i(t)\right) \mathrm{d}t, \qquad i \in \mathbb{P} \tag{5.94}$$

subject to the dynamic constraint (5.70b).

Proof. For each $i \in \mathbb{P}$, the Hamiltonian of the scaled optimization problem (5.94) subject to (5.70b) equals

$$\mathscr{H}_i^{\text{scaled}}(\boldsymbol{x}, \boldsymbol{u}, \boldsymbol{\xi}_i) = \frac{1}{2\Upsilon_i}\left(\ell_i(\boldsymbol{x}) + \boldsymbol{u}_i^\mathsf{T} \boldsymbol{S}_i(\boldsymbol{x})\boldsymbol{u}_i\right) + \boldsymbol{\xi}_i^\mathsf{T}\left((\boldsymbol{J} - \boldsymbol{R})\frac{\partial H}{\partial \boldsymbol{x}} + \sum_{j=1}^{N} \boldsymbol{G}_j(\boldsymbol{x})\boldsymbol{u}_j\right), \tag{5.95}$$

where $\boldsymbol{\xi}_i \in \mathbb{R}^n$ is the Lagrange multiplier corresponding to the individual optimization problem of player $i \in \mathbb{P}$. For the optimal control $\boldsymbol{u}_i^{\text{scaled}\star}$, it has to hold that

$$\left.\frac{\partial \mathscr{H}_i^{\text{scaled}}(\boldsymbol{x}, \boldsymbol{u}, \boldsymbol{\xi}_i)}{\partial \boldsymbol{u}_i}\right|_{\boldsymbol{u}_i^{\text{scaled}\star}} = \mathbb{0}_{p_i}, \tag{5.96}$$

which yields

$$\boldsymbol{u}_i^{\text{scaled}\star}(\boldsymbol{x}) = -(\boldsymbol{S}_i(\boldsymbol{x}))^{-1}\boldsymbol{G}_i^\mathsf{T}(\boldsymbol{x})\boldsymbol{\xi}_i\Upsilon_i. \tag{5.97}$$

From the HJB equation for time-invariant systems

$$\mathscr{H}_i^{\text{scaled}}(\boldsymbol{x}, \boldsymbol{u}^\star, \boldsymbol{\xi}_i^\star) = 0, \tag{5.98}$$

we get

$$\begin{aligned}\forall i \in \mathbb{P}: \qquad 0 = {} & \frac{1}{2\Upsilon_i}\ell_i(\boldsymbol{x}) + (\boldsymbol{\xi}_i^\star)^\mathsf{T}(\boldsymbol{J} - \boldsymbol{R})\frac{\partial H}{\partial \boldsymbol{x}} - \frac{1}{2}\Upsilon_i(\boldsymbol{\xi}_i^\star)^\mathsf{T}\boldsymbol{G}_i(\boldsymbol{x})(\boldsymbol{S}_i(\boldsymbol{x}))^{-1}\boldsymbol{G}_i^\mathsf{T}(\boldsymbol{x})\boldsymbol{\xi}_i^\star \\ & -(\boldsymbol{\xi}_i^\star)^\mathsf{T}\sum_{\substack{j=1\\ j\neq i}}^{N} \boldsymbol{G}_j(\boldsymbol{x})(\boldsymbol{S}_j(\boldsymbol{x}))^{-1}\boldsymbol{G}_j^\mathsf{T}(\boldsymbol{x})\boldsymbol{\xi}_j^\star\Upsilon_j. \end{aligned} \tag{5.99}$$

Now recall the HJB equation for the modified differential game in (5.87). Define

$$\boldsymbol{\xi}_i^\star = \frac{\partial V_i}{\partial \boldsymbol{x}}, \qquad i \in \mathbb{P}, \tag{5.100}$$

which results in (5.87) and (5.99) being equivalent. Accordingly, each solution $\boldsymbol{\Upsilon}^\star$ of (5.87) corresponding to the modified differential game problem in (5.75) is a solution of (5.98), with (5.100) associated to the new, scaled problem (5.94) in Lemma 5.33, and vice versa. By inserting $\boldsymbol{\xi}_i^\star$ and $\boldsymbol{\Upsilon}^\star$ in (5.97), we can conclude that $\boldsymbol{u}_i^\star(\boldsymbol{x})$ in (5.76) is a Nash strategy of player $i \in \mathbb{P}$ for the scaled differential game (5.94) subject to (5.70b), i.e. $\boldsymbol{u}_i^\star(\boldsymbol{x}) = \boldsymbol{u}_i^{\text{scaled}\star}(\boldsymbol{x})$ holds for all $i \in \mathbb{P}$. □

Corollary 5.34. *Let $\boldsymbol{V}(\boldsymbol{x})$ be an admissible vector of CLFs for* (5.70). *Then each solution* $(\boldsymbol{u}^\star, \boldsymbol{\Upsilon}^\star)$ *fulfilling* (5.76)–(5.77) *constitutes a Nash equilibrium with regard to the original problem* (5.70), *if* $\Upsilon_i^\star(\boldsymbol{x}) = 1$.

Proof. Trivially, by comparison between the original differential game (5.70) and the scaled differential game (5.94) subject to (5.70b), which is enforced by the controller (5.76)–(5.77), it follows that all components in $\boldsymbol{\Upsilon}$ must be equal to one so that both differential games are equivalent. □

Remark 5.35. *From Corollary 5.34, it follows that, unless $\boldsymbol{\Upsilon}(\boldsymbol{x})$ converges to a constant value $\mathbb{1}$ and remains constant even after a disturbance, the chosen set of CLFs $V_i(\boldsymbol{x})$ cannot be equivalent to the players' value functions. In this case, the resulting controllers* (5.76)–(5.77) *do not constitute a Nash strategy with regard to the original problem* (5.70). *Conversely, the fluctuation of $\boldsymbol{\Upsilon}(t)$ over time can therefore again be interpreted as an indicator of lack of fitness of the control strategy $\boldsymbol{u}_i^\star(\boldsymbol{x})$ with respect to* (5.70) *as already reasoned in Remark 5.13 for the single-player case.* ♢

Example 8 (Nonlinear Differential Game)

Consider the two-player differential game (5.70) *with*

$$\mathscr{J}_1 = \frac{1}{2}\int_0^{t_f} \left(\boldsymbol{x}^\mathsf{T} \begin{bmatrix} 16 + 8x_2^2 & -2x_2^2 \\ -2x_2^2 & 5 \end{bmatrix} \boldsymbol{x} + 4 \cdot u_1^2 \right) \mathrm{d}t, \tag{5.101a}$$

$$\mathscr{J}_2 = \frac{1}{2}\int_0^{t_f} \left(\boldsymbol{x}^\mathsf{T} \begin{bmatrix} 5 + 2x_2^2 & 2.5 \\ 2.5 & 12 \end{bmatrix} \boldsymbol{x} + u_2^2 \right) \mathrm{d}t \tag{5.101b}$$

subject to the nonlinear input-state-output PHS

$$\dot{\boldsymbol{x}} = \Bigg(\underbrace{\begin{bmatrix} 0 & -\frac{1}{2} \\ \frac{1}{2} & 0 \end{bmatrix}}_{\boldsymbol{J}(\boldsymbol{x})} - \underbrace{\begin{bmatrix} 1 + x_2^2 & \frac{1}{2} \\ \frac{1}{2} & 2 \end{bmatrix}}_{\boldsymbol{R}(\boldsymbol{x})} \Bigg) \frac{\partial H}{\partial \boldsymbol{x}} + \underbrace{\begin{bmatrix} 1 \\ 0 \end{bmatrix}}_{\boldsymbol{G}_1(\boldsymbol{x})} u_1 + \underbrace{\begin{bmatrix} 1 \\ 1 \end{bmatrix}}_{\boldsymbol{G}_2(\boldsymbol{x})} u_2, \tag{5.102}$$

where $H(\boldsymbol{x}) = \frac{1}{2}x_1^2 + \frac{1}{2}x_2^2$.

The corresponding value functions of (5.101a) *and* (5.101b) *are*

$$V_1^\star(\boldsymbol{x}) = 2x_1^2 - 2x_1x_2 + x_2^2, \tag{5.103a}$$

$$V_2^\star(\boldsymbol{x}) = \frac{1}{2}x_1^2 + x_2^2, \tag{5.103b}$$

which can be calculated by using the converse HJB approach [NP96].

The system (5.102) is initialized at $\boldsymbol{x} = (0,0)^\mathsf{T}$. *Fig. 5.9 shows the resulting trajectories of* $\Upsilon_1(t)$ *and* $\Upsilon_2(t)$ *for different vectors of CLFs* $\boldsymbol{V}(\boldsymbol{x})$.

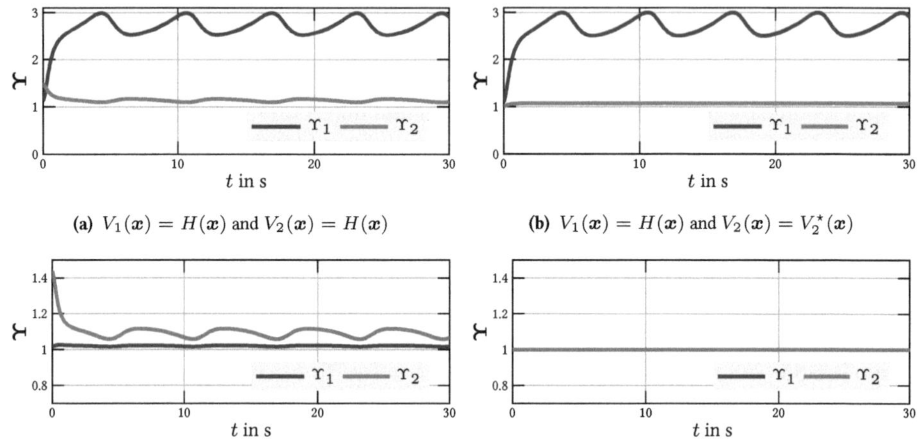

(a) $V_1(\boldsymbol{x}) = H(\boldsymbol{x})$ and $V_2(\boldsymbol{x}) = H(\boldsymbol{x})$

(b) $V_1(\boldsymbol{x}) = H(\boldsymbol{x})$ and $V_2(\boldsymbol{x}) = V_2^\star(\boldsymbol{x})$

(c) $V_1(\boldsymbol{x}) = V_1^\star(\boldsymbol{x})$ and $V_2(\boldsymbol{x}) = H(\boldsymbol{x})$

(d) $V_1(\boldsymbol{x}) = V_1^\star(\boldsymbol{x})$ and $V_2(\boldsymbol{x}) = V_2^\star(\boldsymbol{x})$

Figure 5.9: Trajectory of $\boldsymbol{\Upsilon}$ for different vectors of CLFs $\boldsymbol{V}(\boldsymbol{x})$.

In Figs. 5.9a–c, $\boldsymbol{\Upsilon}(t)$ *is varying with time, while in Fig. 5.9d it holds for all t that* $\boldsymbol{\Upsilon}(t) = \mathbb{1}$. *Moreover, it can be seen that the closer* $\boldsymbol{V}(\boldsymbol{x})$ *is to* $\boldsymbol{V}^\star(\boldsymbol{x})$, *the closer* $\boldsymbol{\Upsilon}$ *is to* $\mathbb{1}$.
However, $V_i(\boldsymbol{x}) = V_i^\star(\boldsymbol{x})$ *does not necessarily result in* $\Upsilon_i(t) = 1$, *which can be seen e.g. in Fig. 5.9b: Although* $V_2(\boldsymbol{x}) = V_2^\star(\boldsymbol{x})$ *holds,* Υ_2 *approaches a value* ≈ 1.1.
This effect is based on the dynamic coupling of both players and the nature of a Nash equilibrium: Nash strategies optimize the individual objective function of each player while assuming that the other players also act in an optimal way. However, since $V_1(\boldsymbol{x}) \neq V_1^\star(\boldsymbol{x})$ *in Fig. 5.9, the strategy of player 2 is not a Nash strategy which is indicated by* $\Upsilon_2(t) \neq \mathbb{1}$ *according to Corollary 5.34. The same holds for player 1 in Fig. 5.9c where* $V_1(\boldsymbol{x}) = V_1^\star(\boldsymbol{x})$ *and* $\Upsilon_1 \approx 1.017$.

5.3.3 Stability of the Modified Differential Game

In this section, we show the relation between the sign of Υ_i and asymptotic stability of the closed-loop system (5.70b), (5.76), (5.77). In particular, we show in the next lemma that an asymptotically stabilizing controller can always be constructed by choosing a positive solution for Υ_i. Subsequently, we study the existence of a positive solution in Lemma 5.37.

Lemma 5.36. *Let* $\boldsymbol{V}(\boldsymbol{x})$ *be an admissible vector of CLFs for* (5.70) *and* $\ell_i(\boldsymbol{x}) \succ 0$. *Then the equilibrium* $\boldsymbol{x} = \mathbb{0}$ *of the closed-loop system* (5.70b), (5.76)–(5.77) *is globally asymptotically stable, if* $\Upsilon_i > 0$.

Proof. Let $\Upsilon_i > 0$ holds. Inserting $\boldsymbol{u}_i(\boldsymbol{x}) = \boldsymbol{u}_i^\star(\boldsymbol{x})$ in (5.82) and application of (5.86) yields

$$\forall i \in \mathbb{P}: \qquad \dot{V}_i(\boldsymbol{x}) = -\frac{1}{2\Upsilon_i}\left(\ell_i(\boldsymbol{x}) + (\boldsymbol{u}_i^\star)^\top(\boldsymbol{x})\boldsymbol{S}(\boldsymbol{x})\boldsymbol{u}_i^\star(\boldsymbol{x})\right) \leq -\frac{1}{2\Upsilon_i}\ell_i(\boldsymbol{x}) \prec 0. \tag{5.104}$$

Thus for each $i \in \mathbb{P}$, the Lyapunov function candidate $V_i(\boldsymbol{x})$ fulfills $V_i(\boldsymbol{x}) \succ 0$ and $\dot{V}_i(\boldsymbol{x}) \prec 0$ and is radially unbounded, which means that $\boldsymbol{x} = \mathbb{0}$ is a globally asymptotically stable equilibrium of (5.70b), (5.76)–(5.77) [Kha02, Theorem 4.2]. □

Lemma 5.37. *If* (5.77) *has a real solution, then there exists exactly one solution* $\boldsymbol{\Upsilon}_+$ *with* $\boldsymbol{\Upsilon}_+ > \mathbb{0}$.

Proof. Let $\boldsymbol{\Upsilon} \in \mathbb{R}^N$ be a solution of (5.77). For each $i \in \mathbb{P}$, (5.77) takes the form

$$\underbrace{q_i^2(\boldsymbol{x})}_{a_i} \cdot \Upsilon_i^2 + \underbrace{\left(q_i^1(\boldsymbol{x}) + \sum_{\substack{j=1\\ j\neq i}}^{N} q_{ij}^2(\boldsymbol{x})\Upsilon_j\right)}_{b_i} \cdot \Upsilon_i + \underbrace{q_i^0(\boldsymbol{x})}_{c_i} = 0. \tag{5.105}$$

Thus, a solution of (5.105) is given by

$$\Upsilon_i = \begin{cases} \dfrac{-b_i \pm \sqrt{b_i^2 - 4a_i c_i}}{2a_i}, & a_i \neq 0, \\ -\dfrac{c_i}{b_i}, & a_i = 0. \end{cases} \tag{5.106}$$

Due to the fact that $q_i^0(\boldsymbol{x}) \succ 0$, the inequality $c_i > 0$ holds for all $\boldsymbol{x} \neq \mathbb{0}$. Moreover, $a_i \preccurlyeq 0$, which follows directly from (5.80). Thus if $a_i < 0$, then the "$-$" solution in the first row of (5.106) always leads to $\Upsilon_i > 0$. Otherwise, if $a_i = 0$, then with the CLF property (5.93) of $V_i(\boldsymbol{x})$ we get

$$b_i = \left(\frac{\partial V_i}{\partial \boldsymbol{x}}\right)^\top \left(\boldsymbol{f}(\boldsymbol{x}) + \sum_{\substack{j=1\\ j\neq i}}^{N} \boldsymbol{G}_j(\boldsymbol{x})\boldsymbol{u}_j\right) < 0, \tag{5.107}$$

which again results in $\Upsilon_i > 0$. □

In the following, we assume that the positive solution $\boldsymbol{\Upsilon}_+$ of (5.77) is chosen. We finish this subsection with two concluding remarks.

Remark 5.38. *The continuity of* $\Upsilon_i(\boldsymbol{x})$ *can be shown by applying de L'Hospital's rule to* (5.106). ♢

Remark 5.39. *By using the alternative Lyapunov function candidate $\sum_{i\in\mathbb{P}} V_i(\boldsymbol{x})$, it can be shown that asymptotic stability is still preserved, if $\ell_i(\boldsymbol{x}) \succ 0$, $\ell_{\neg i}(\boldsymbol{x}) \succcurlyeq 0$, and correspondingly, if $\Upsilon_i^+ > 0$, $\Upsilon_{\neg i}^+ \geq 0$ holds for* some $i \in \mathbb{P}$. *However, for the sake of brevity and to avoid case distinctions in the proofs, we will proceed with the stricter requirement $\ell_i(\boldsymbol{x}) \succ 0$ for all $i \in \mathbb{P}$.* ◇

5.3.4 Admissible Vectors of Control-Lyapunov Functions for Input-State-Output Port-Hamiltonian Systems

As mentioned above, finding an appropriate CLF for a given system is in general very difficult, even in the single-player case. Therefore, in this subsection, we will investigate under which condition the Hamiltonian $H(\boldsymbol{x})$ can be used as a CLF $V_i(\boldsymbol{x})$, in particular, under which conditions the vector $H(\boldsymbol{x}) \cdot \mathbb{1}$ is an admissible vector of CLFs.

Lemma 5.40. *The vector $\boldsymbol{V}_0(\boldsymbol{x}) = \mathrm{col}_i\{H(\boldsymbol{x})\} = H(\boldsymbol{x}) \cdot \mathbb{1}$ is an admissible vector of CLFs for the input-state-output PHS* (5.70b), *if and only if*

$$\forall \boldsymbol{x} \in \mathcal{X}_{\boldsymbol{G}_i}: \ \left(\frac{\partial H}{\partial \boldsymbol{x}}\right)^{\mathsf{T}} \boldsymbol{R}(\boldsymbol{x}) \frac{\partial H}{\partial \boldsymbol{x}} + \sum_{\substack{j=1 \\ j\neq i}}^{N} \left(\frac{\partial H}{\partial \boldsymbol{x}}\right)^{\mathsf{T}} \boldsymbol{G}_j(\boldsymbol{x}) (\boldsymbol{S}_j(\boldsymbol{x}))^{-1} \boldsymbol{G}_j^{\mathsf{T}}(\boldsymbol{x}) \frac{\partial H}{\partial \boldsymbol{x}} \cdot \Upsilon_j > 0, \tag{5.108}$$

where

$$\mathcal{X}_{\boldsymbol{G}_i} = \{\boldsymbol{x} \in \mathbb{R}^n \ : \ \boldsymbol{G}_i^{\mathsf{T}}(\boldsymbol{x}) \frac{\partial H}{\partial \boldsymbol{x}} = \mathbb{0}, \boldsymbol{x} \neq \mathbb{0}\}. \tag{5.109}$$

Proof. With $\mathcal{X}_{\boldsymbol{G}_i}$ as in (5.109), condition (5.93) is equivalent to

$$\forall \, i \in \mathbb{P}, \, \boldsymbol{x} \in \mathcal{X}_{\boldsymbol{G}_i} : \quad \left(\frac{\partial V_i}{\partial \boldsymbol{x}}\right)^{\mathsf{T}} \left(\boldsymbol{f}(\boldsymbol{x}) + \sum_{\substack{j=1 \\ j\neq i}}^{N} \boldsymbol{G}_j(\boldsymbol{x}) \boldsymbol{u}_j \right) \begin{cases} < 0, & \boldsymbol{x} \neq \mathbb{0}_n, \\ = 0, & \boldsymbol{x} = \mathbb{0}_n. \end{cases} \tag{5.110}$$

Now with

$$
\begin{aligned}
&\left(\frac{\partial V_i}{\partial \boldsymbol{x}}\right)^\mathsf{T} \left(\boldsymbol{f}(\boldsymbol{x}) + \sum_{\substack{j=1\\ j\neq i}}^{N} \boldsymbol{G}_j(\boldsymbol{x})\boldsymbol{u}_j \right) \\
&= \left(\frac{\partial H}{\partial \boldsymbol{x}}\right)^\mathsf{T} (\boldsymbol{J}(\boldsymbol{x}) - \boldsymbol{R}(\boldsymbol{x}))\frac{\partial H}{\partial \boldsymbol{x}} - \sum_{\substack{j=1\\ j\neq i}}^{N} \left(\frac{\partial H}{\partial \boldsymbol{x}}\right)^\mathsf{T} \boldsymbol{G}_j(\boldsymbol{x})(\boldsymbol{S}_j(\boldsymbol{x}))^{-1}\boldsymbol{G}_j^\mathsf{T}(\boldsymbol{x})\frac{\partial H}{\partial \boldsymbol{x}} \cdot \Upsilon_j \\
&= -\left(\frac{\partial H}{\partial \boldsymbol{x}}\right)^\mathsf{T} \boldsymbol{R}(\boldsymbol{x})\frac{\partial H}{\partial \boldsymbol{x}} - \sum_{\substack{j=1\\ j\neq i}}^{N} \left(\frac{\partial H}{\partial \boldsymbol{x}}\right)^\mathsf{T} \boldsymbol{G}_j(\boldsymbol{x})(\boldsymbol{S}_j(\boldsymbol{x}))^{-1}\boldsymbol{G}_j^\mathsf{T}(\boldsymbol{x})\frac{\partial H}{\partial \boldsymbol{x}} \cdot \Upsilon_j,
\end{aligned} \tag{5.111}
$$

the vector $\boldsymbol{V}_0(\boldsymbol{x}) = \mathrm{col}_i\{H(\boldsymbol{x})\} = H(\boldsymbol{x}) \cdot \mathbb{1}$ is an admissible vector of CLFs for the input-state-output PHS (5.70b), if and only if (5.108) is fulfilled. □

Corollary 5.41. *The vector* $\boldsymbol{V}_0(\boldsymbol{x}) = \mathrm{col}_i\{H(\boldsymbol{x})\} = H(\boldsymbol{x}) \cdot \mathbb{1}$ *is an admissible vector of CLFs for the input-state-output PHS* (5.70b), *if and only if* (5.70b) *with passive outputs*

$$
\boldsymbol{y}_i = \boldsymbol{G}_i^\mathsf{T}(\boldsymbol{x})\frac{\partial H(\boldsymbol{x})}{\partial \boldsymbol{x}} \tag{5.112}
$$

is zero-state detectable.

Proof. Let $\boldsymbol{V}_0(\boldsymbol{x}) = \mathrm{col}_i\{H(\boldsymbol{x})\} = H(\boldsymbol{x}) \cdot \mathbb{1}$ be an admissible vector of CLFs for the input-state-output PHS (5.70b). Then (5.108) is fulfilled (and vice versa). Due to the fact that $\mathcal{X}_{\boldsymbol{G}_i} = \{\boldsymbol{x} \in \mathbb{R}^n : \boldsymbol{G}_i^\mathsf{T}(\boldsymbol{x})\frac{\partial H}{\partial \boldsymbol{x}} = \mathbb{0}, \boldsymbol{x} \neq \mathbb{0}\}$, (5.108) is equivalent to

$$
\forall \boldsymbol{x} \in \{\boldsymbol{x} \in \mathbb{R}^n : \boldsymbol{y}_i(\boldsymbol{x}) = \mathbb{0}\} : \qquad \dot{H}(\boldsymbol{x}) < 0. \tag{5.113}
$$

With the Lyapunov function candidate $H(\boldsymbol{x}) \succ 0$, this means that (5.70b), (5.112) is zero-state detectable (cf. [van17, p. 47]). □

Although the lemmas presented so far are necessary and sufficient, they are generally difficult to verify. Therefore, the following lemma presents a more practical condition which can be directly evaluated based on the given system and input matrices in (5.70b):

Lemma 5.42. *The vector* $\boldsymbol{V}_0(\boldsymbol{x}) = \mathrm{col}_i\{H(\boldsymbol{x})\} = H(\boldsymbol{x}) \cdot \mathbb{1}$ *is an admissible vector of CLFs for the input-state-output PHS* (5.70b), *if* $H(\boldsymbol{x})$ *is pseudoconvex and at least one of the following conditions holds:*

1. *The dissipation matrix* $\boldsymbol{R}(\boldsymbol{x})$ *has full rank* n.
2. *At least two input matrices* $\boldsymbol{G}_{i_1}(\boldsymbol{x}), \boldsymbol{G}_{i_2}(\boldsymbol{x})$ *have full rank* n.

Proof. Since $H(\boldsymbol{x})$ is pseudoconvex, there exists no $\boldsymbol{x}' \neq \mathbb{0}$ such that $\nabla H(\boldsymbol{x}') = \mathbb{0}$.

1. If $\operatorname{rank}\{\boldsymbol{R}(\boldsymbol{x})\} = n$, then $\boldsymbol{R}(\boldsymbol{x}) \succ 0$ and thus

$$\left(\frac{\partial V_i}{\partial \boldsymbol{x}}\right)^{\mathsf{T}} \left(\boldsymbol{f}(\boldsymbol{x}) + \sum_{\substack{j=1\\ j\neq i}}^{N} \boldsymbol{G}_j(\boldsymbol{x}) \boldsymbol{u}_j \right)$$
$$= \left(\frac{\partial H}{\partial \boldsymbol{x}}\right)^{\mathsf{T}} (\boldsymbol{J}(\boldsymbol{x}) - \boldsymbol{R}(\boldsymbol{x})) \frac{\partial H}{\partial \boldsymbol{x}} - \sum_{\substack{j=1\\ j\neq i}}^{N} \left(\frac{\partial H}{\partial \boldsymbol{x}}\right)^{\mathsf{T}} \boldsymbol{G}_j(\boldsymbol{x})(\boldsymbol{S}_j(\boldsymbol{x}))^{-1} \boldsymbol{G}_j^{\mathsf{T}}(\boldsymbol{x}) \frac{\partial H}{\partial \boldsymbol{x}} \cdot \Upsilon_j$$
$$= -\underbrace{\left(\frac{\partial H}{\partial \boldsymbol{x}}\right)^{\mathsf{T}} \boldsymbol{R}(\boldsymbol{x}) \frac{\partial H}{\partial \boldsymbol{x}}}_{>0} - \sum_{\substack{j=1\\ j\neq i}}^{N} \underbrace{\left(\frac{\partial H}{\partial \boldsymbol{x}}\right)^{\mathsf{T}} \boldsymbol{G}_j(\boldsymbol{x})(\boldsymbol{S}_j(\boldsymbol{x}))^{-1} \boldsymbol{G}_j^{\mathsf{T}}(\boldsymbol{x}) \frac{\partial H}{\partial \boldsymbol{x}}}_{\geq 0} \cdot \underbrace{\Upsilon_j}_{>0} < 0 \tag{5.114}$$

   is always fulfilled for all $i \in \mathbb{P}$, which implies that (5.108) is fulfilled.

2. If at least two input matrices $\boldsymbol{G}_{i_1}(\boldsymbol{x}), \boldsymbol{G}_{i_2}(\boldsymbol{x})$ have full rank n, then for all $i \in \mathbb{P}$ there exists at least one $k \neq i$ such that $\boldsymbol{G}_k(\boldsymbol{x})(\boldsymbol{S}_k(\boldsymbol{x}))^{-1}\boldsymbol{G}_k^{\mathsf{T}}(\boldsymbol{x}) \succ 0$. Accordingly, for all $i \in \mathbb{P}$ the inequality

$$\left(\frac{\partial H}{\partial \boldsymbol{x}}\right)^{\mathsf{T}} (\boldsymbol{J}(\boldsymbol{x}) - \boldsymbol{R}(\boldsymbol{x})) \frac{\partial H}{\partial \boldsymbol{x}} - \sum_{\substack{j=1\\ j\neq i}}^{N} \left(\frac{\partial H}{\partial \boldsymbol{x}}\right)^{\mathsf{T}} \boldsymbol{G}_j(\boldsymbol{x})(\boldsymbol{S}_j(\boldsymbol{x}))^{-1} \boldsymbol{G}_j^{\mathsf{T}}(\boldsymbol{x}) \frac{\partial H}{\partial \boldsymbol{x}} \cdot \Upsilon_j$$
$$= -\underbrace{\left(\frac{\partial H}{\partial \boldsymbol{x}}\right)^{\mathsf{T}} \boldsymbol{R}(\boldsymbol{x}) \frac{\partial H}{\partial \boldsymbol{x}}}_{\geq 0} - \underbrace{\left(\frac{\partial H}{\partial \boldsymbol{x}}\right)^{\mathsf{T}} \boldsymbol{G}_k(\boldsymbol{x})(\boldsymbol{S}_k(\boldsymbol{x}))^{-1} \boldsymbol{G}_k^{\mathsf{T}}(\boldsymbol{x}) \frac{\partial H}{\partial \boldsymbol{x}} \cdot \Upsilon_k}_{>0}$$
$$- \sum_{\substack{j=1\\ j\neq i,k}}^{N} \underbrace{\left(\frac{\partial H}{\partial \boldsymbol{x}}\right)^{\mathsf{T}} \boldsymbol{G}_j(\boldsymbol{x})(\boldsymbol{S}_j(\boldsymbol{x}))^{-1} \boldsymbol{G}_j^{\mathsf{T}}(\boldsymbol{x}) \frac{\partial H}{\partial \boldsymbol{x}} \cdot \Upsilon_j}_{\geq 0} < 0 \tag{5.115}$$

   holds, which implies that (5.108) is fulfilled. □

In the following, Lemma 5.43 gives a necessary and sufficient condition for the special case of (5.70b) being a *linear* input-state-output PHS:

Lemma 5.43. *For the linear input-state-output PHS*

$$\dot{\boldsymbol{x}} = (\boldsymbol{J} - \boldsymbol{R})\boldsymbol{Q}\boldsymbol{x} + \sum_{i=1}^{N} \boldsymbol{G}_i \boldsymbol{u}_i \tag{5.116}$$

with $H(\boldsymbol{x}) = \frac{1}{2}\boldsymbol{x}^\mathsf{T}\boldsymbol{Q}\boldsymbol{x}$, $\boldsymbol{Q} \succ 0$, *the vector* $\boldsymbol{V}_0(\boldsymbol{x}) = \mathrm{col}_i\{H(\boldsymbol{x})\} = H(\boldsymbol{x}) \cdot \mathbb{1}$ *is an admissible vector of CLFs, if and only if*

$$\ker(\boldsymbol{R}) \cap \left(\bigcap_{i=1}^{N} \ker(\boldsymbol{G}_i^\mathsf{T})\right) \cap \left(\bigcup_{i=1}^{N} \bigcap_{\substack{j=1 \\ j\neq i}}^{N} \ker(\boldsymbol{G}_j \boldsymbol{S}_j^{-1} \boldsymbol{G}_j^\mathsf{T})\right) = \emptyset. \tag{5.117}$$

Proof. (see Appendix D.2). □

5.3.5 Adaptive Differential Game

From the previous subsection, it can be concluded that the application of the modified differential game (5.75) with $\boldsymbol{V}(\boldsymbol{x})$ being an admissible vector of CLFs always allows the derivation of stabilizing controllers $\boldsymbol{u}_i^\star(\boldsymbol{x})$ which constitute a Nash strategy with respect to (5.94) and (5.70b). However, since Υ_i depends on $\boldsymbol{V}(\boldsymbol{x})$, the condition $\Upsilon_i = \Upsilon_j = 1$ in Corollary 5.34 is in general not fulfilled for all $i \in \mathbb{P}$. Thus, according to Corollary 5.34, the modified differential game is not equal to the formulation (5.70).

To eliminate this discrepancy in the same spirit as for the single-player case, we thus introduce a vector of extended CLFs $\boldsymbol{V}(\boldsymbol{x}, \boldsymbol{w})$ with additional weighted sum of basis functions. The aim is to achieve $\Upsilon_i(t) \equiv 1$ for all $i \in \mathbb{P}$, so that according to Lemma 5.33, the scaled objective functions (5.94) being minimized by the resulting modified optimal controllers (5.76)–(5.77), are identical to the original objective functions (5.70a).

Adaptation using Extended CLFs

Each extended CLF $V_i(\boldsymbol{x}, \boldsymbol{w}_i)$ is supposed to be the sum of $H(\boldsymbol{x})$ and a weighted sum of $r_i \in \mathbb{N}$ basis functions

$$V_i(\boldsymbol{x}, \boldsymbol{w}_i) = H(\boldsymbol{x}) + \boldsymbol{w}_i^\mathsf{T}\boldsymbol{\Xi}_i(\boldsymbol{x}), \tag{5.118}$$

where $\boldsymbol{\Xi}_i : \mathbb{R}^n \to \mathbb{R}^{r_i}$ and $\boldsymbol{w}_i \in \mathbb{R}^{r_i}$. We assume that all basis functions $\boldsymbol{\Xi}_i(\boldsymbol{x})$ are smooth and "properly chosen" in the sense that the (optimal) value functions $V_i^\star(\boldsymbol{x})$, i.e. the solutions of the corresponding HJB equations

$$\mathscr{H}_i\left(\boldsymbol{x}, \boldsymbol{u}^\star, \frac{\partial V_i^\star}{\partial \boldsymbol{x}}\right) = 0, \tag{5.119}$$

can be parameterized via $\boldsymbol{\Xi}_i(\boldsymbol{x})$ and a vector $\boldsymbol{w}_i^\star$ of optimal weighting factors:

Assumption 5.44. *For each player* $i \in \mathbb{P}$, *there exists a* $\boldsymbol{w}_i^\star \in \mathbb{R}^{r_i}$ *such that*

$$V_i^\star(\boldsymbol{x}) = H(\boldsymbol{x}) + (\boldsymbol{w}_i^\star)^\mathsf{T}\boldsymbol{\Xi}_i(\boldsymbol{x}). \tag{5.120}$$

Analogously to Remark 5.16, Assumption 5.44 is justified by the Weierstraß approximation theorem whenever the number of basis functions is large (see also [VL11, p. 1559]).

Now the aim is to set up $\boldsymbol{w}_i$ and thus $V_i(\boldsymbol{x}, \boldsymbol{w}_i)$ in such a way that $\boldsymbol{w}_i = \boldsymbol{w}_i^\star$ in order to enable that $\boldsymbol{V}(\boldsymbol{x}, \boldsymbol{w}) = \mathrm{col}_i\{V_i(\boldsymbol{x}, \boldsymbol{w}_i)\}$ constitutes a vector of value functions for (5.70), i.e. a solution of the set of HJB equations (5.119) associated to the original problem (5.70). This property can be checked using the following corollary:

Corollary 5.45. *For all $i \in \mathbb{P}$, let $V_i(\boldsymbol{x}) = H(\boldsymbol{x}) + (\boldsymbol{w}_i^\diamond)^\mathsf{T}\boldsymbol{\Xi}_i(\boldsymbol{x})$ be a given CLF with $\boldsymbol{w}_i^\diamond \in \mathbb{R}^{r_i}$. Then $\boldsymbol{V}(\boldsymbol{x})$ is equivalent to the vector of value functions $\boldsymbol{V}^\star(\boldsymbol{x}) = \mathrm{col}_i\{V_i^\star(\boldsymbol{x})\}$ of (5.70), if and only if*

$$\forall i \in \mathbb{P},\ \boldsymbol{x} \in \mathbb{R}^n : \qquad (\boldsymbol{x}, \boldsymbol{w}_i^\diamond) \in \mathcal{Q}_i(\boldsymbol{x}, \boldsymbol{w}_i), \tag{5.121}$$

where

$$\mathcal{Q}_i(\boldsymbol{x}, \boldsymbol{w}) := \left\{(\boldsymbol{x}, \boldsymbol{w}_i) \in \mathbb{R}^n \times \mathbb{R}^{r_i} : \boldsymbol{w}_i^\mathsf{T}\boldsymbol{A}_i(\boldsymbol{x})\boldsymbol{w}_i + \boldsymbol{a}_i^\mathsf{T}(\boldsymbol{x})\boldsymbol{w}_i + a_i(\boldsymbol{x}) = 0\right\}, \tag{5.122}$$

$$\boldsymbol{A}_i(\boldsymbol{x}) := \frac{\partial \boldsymbol{\Xi}_i}{\partial \boldsymbol{x}}\boldsymbol{K}_i(\boldsymbol{x})\left(\frac{\partial \boldsymbol{\Xi}_i}{\partial \boldsymbol{x}}\right)^\mathsf{T}, \tag{5.123}$$

$$\begin{aligned}\boldsymbol{a}_i^\mathsf{T}(\boldsymbol{x}) := 2\Bigg(&\left(\frac{\partial H}{\partial \boldsymbol{x}}\right)^\mathsf{T}\boldsymbol{K}_i(\boldsymbol{x})\left(\frac{\partial \boldsymbol{\Xi}_i}{\partial \boldsymbol{x}}\right)^\mathsf{T} + \left(\frac{\partial H}{\partial \boldsymbol{x}}\right)^\mathsf{T}(\boldsymbol{J}(\boldsymbol{x}) + \boldsymbol{R}(\boldsymbol{x}))\left(\frac{\partial \boldsymbol{\Xi}_i}{\partial \boldsymbol{x}}\right)^\mathsf{T} \\ &- 2\sum_{\substack{j=1\\ j\neq i}} \boldsymbol{u}_j^\mathsf{T}\boldsymbol{G}_j^\mathsf{T}(\boldsymbol{x})\frac{\partial \boldsymbol{\Xi}_i}{\partial \boldsymbol{x}}\Bigg),\end{aligned} \tag{5.124}$$

$$\begin{aligned}a_i(\boldsymbol{x}) := &\left(\frac{\partial H}{\partial \boldsymbol{x}}\right)^\mathsf{T}\boldsymbol{K}_i(\boldsymbol{x})\frac{\partial H}{\partial \boldsymbol{x}} + 2\left(\frac{\partial H}{\partial \boldsymbol{x}}\right)^\mathsf{T}\boldsymbol{R}(\boldsymbol{x})\frac{\partial H}{\partial \boldsymbol{x}} - \ell_i(\boldsymbol{x}) \\ &- 4\sum_{\substack{j=1\\ j\neq i}} \boldsymbol{u}_j^\mathsf{T}\boldsymbol{G}_j^\mathsf{T}(\boldsymbol{x})\frac{\partial H}{\partial \boldsymbol{x}},\end{aligned} \tag{5.125}$$

$$\boldsymbol{K}_i(\boldsymbol{x}) := \boldsymbol{G}_i(\boldsymbol{x})(\boldsymbol{S}_i(\boldsymbol{x}))^{-1}\boldsymbol{G}_i^\mathsf{T}(\boldsymbol{x}). \tag{5.126}$$

Proof. (see Appendix D.2). □

Lemma 5.46. *For all $i \in \mathbb{P}$, let*

$$\mathsf{Q}_i(\boldsymbol{x}, \boldsymbol{w}_i) := \boldsymbol{w}_i^\mathsf{T}\boldsymbol{A}_i(\boldsymbol{x})\boldsymbol{w}_i + \boldsymbol{a}_i^\mathsf{T}(\boldsymbol{x})\boldsymbol{w}_i + a_i(\boldsymbol{x}) \tag{5.127}$$

with $\boldsymbol{A}_i(\boldsymbol{x})$, $\boldsymbol{a}_i(\boldsymbol{x})$ and $a_i(\boldsymbol{x})$ as in (5.123)–(5.125) be the corresponding quadratic function of the quadric $\mathcal{Q}_i(\boldsymbol{x}, \boldsymbol{w}_i)$ in (5.122). Then for each arbitrary but fixed $\boldsymbol{x} \in \mathbb{R}^n$, the objective *function*

$$\mathcal{J}_{\mathrm{w},i}(\boldsymbol{x}, \boldsymbol{w}_i) := \sum_{j=1}^{r_i} (\mathtt{Q}_i(\boldsymbol{x} + \boldsymbol{c}_{i,j}))^2, \tag{5.128}$$

consisting of a sum of squared quadratic functions $\mathtt{Q}_i$ *whose arguments are shifted by* $\boldsymbol{c}_{i,1}, \ldots, \boldsymbol{c}_{i,r_i} \in \mathbb{R}^N$, *has a global minimizer at* $\boldsymbol{w}_i = \boldsymbol{w}_i^\star$ *and is locally strictly convex in an open neighborhood* $\mathcal{M}_i$ *of* $\boldsymbol{w}_i^\star$, *if and only if the matrices*

$$\hat{\boldsymbol{A}}_i := \begin{bmatrix} 2(\boldsymbol{w}_i^\star)^\mathsf{T} \boldsymbol{A}_i^\mathsf{T}(\boldsymbol{x} + \boldsymbol{c}_{i,1}) + \boldsymbol{a}_i^\mathsf{T}(\boldsymbol{x} + \boldsymbol{c}_{i,1}) \\ \vdots \\ 2(\boldsymbol{w}_i^\star)^\mathsf{T} \boldsymbol{A}_i^\mathsf{T}(\boldsymbol{x} + \boldsymbol{c}_{i,r_i}) + \boldsymbol{a}_i^\mathsf{T}(\boldsymbol{x} + \boldsymbol{c}_{i,r_i}) \end{bmatrix} \tag{5.129}$$

have full rank r_i.

Proof. The proof is a direct consequence of the proof of Lemma 5.19. Here it has to hold for each player separately. □

Remark 5.47. *Let* $\boldsymbol{w} = \mathrm{col}_i\{\boldsymbol{w}_i\}$ *with* $\boldsymbol{w} \in \mathbb{R}^r$ *and* $r = r_1 + \cdots + r_N$ *denote the concatenation of all weights* $\boldsymbol{w}_i$, $i \in \mathbb{P}$ *and let* $\hat{\mathcal{J}}_\mathrm{w}(\boldsymbol{x}, \boldsymbol{w}) = \sum_{i \in \mathbb{P}} \mathcal{J}_{\mathrm{w},i}(\boldsymbol{x}, \boldsymbol{w}_i)$ *be the sum of all objective functions* $\mathcal{J}_{\mathrm{w},i}(\boldsymbol{x}, \boldsymbol{w})$. *Then*

$$\boldsymbol{w}^\star = \arg\min_{\boldsymbol{w}} \{\hat{\mathcal{J}}_\mathrm{w}(\boldsymbol{x}, \boldsymbol{w})\}. \tag{5.130}$$

◇

Remark 5.48. *For simplicity of notation, it is pointed out that the condition in Lemma 5.46 concerning local strict convexity of objective functions* $\mathcal{J}_{\mathrm{w},i}(\boldsymbol{w}_i)$ *in an open neighborhood* $\mathcal{M}_i$ *of* $\boldsymbol{w}_i^\star$ *is equivalent to the statement that the compound objective function* $\hat{\mathcal{J}}_\mathrm{w}(\boldsymbol{x}, \boldsymbol{w})$ *is locally strictly convex in an open neighborhood* $\mathcal{M} = \mathcal{M}_1 \times \cdots \times \mathcal{M}_N$ *of* $\boldsymbol{w}^\star$. ◇

With Remarks 5.47 and 5.48, we can again apply the unconstrained gradient flow to adapt $\boldsymbol{w}$, which finally leads to

$$\dot{\boldsymbol{w}}_i = -\alpha \cdot \frac{\partial \hat{\mathcal{J}}_\mathrm{w}(\boldsymbol{x}, \boldsymbol{w})}{\partial \boldsymbol{w}_i} = -\alpha \cdot \frac{\partial \mathcal{J}_{\mathrm{w},i}(\boldsymbol{x}, \boldsymbol{w}_i)}{\partial \boldsymbol{w}_i}, \tag{5.131}$$

where $\alpha > 0$ specifies the learning rate.

Closed-Loop System

The system dynamics (5.70b) together with the adaptation procedure (5.131) result in an extended system

$$\dot{\boldsymbol{x}} = (\boldsymbol{J}(\boldsymbol{x}) - \boldsymbol{R}(\boldsymbol{x}))\frac{\partial H(\boldsymbol{x})}{\partial \boldsymbol{x}} + \sum_{i=1}^{N} \boldsymbol{G}_i(\boldsymbol{x})\boldsymbol{u}_i, \tag{5.132a}$$

$$\dot{\boldsymbol{w}}_i = -\alpha \cdot \frac{\partial \mathcal{J}_{\mathrm{w},i}(\boldsymbol{x}, \boldsymbol{w}_i)}{\partial \boldsymbol{w}_i}, \qquad i \in \mathbb{P}. \tag{5.132b}$$

With (5.118), we obtain

$$\frac{\partial V_i(\boldsymbol{x}, \boldsymbol{w}_i)}{\partial \boldsymbol{x}} = \frac{\partial H}{\partial \boldsymbol{x}} + \left(\frac{\partial \boldsymbol{\Xi}_i}{\partial \boldsymbol{x}}\right)^{\mathsf{T}} \boldsymbol{w}_i \tag{5.133}$$

and accordingly the modified differential game for the extended system (5.132) reads

$$\forall i \in \mathbb{P}: \quad \min_{\boldsymbol{u}_i} \left\{ \frac{1}{2} \int_0^{t_{\mathrm{f}}} \ell_i(\boldsymbol{x}) + (\boldsymbol{u}_i)^{\mathsf{T}} \boldsymbol{S}_i(\boldsymbol{x}) \boldsymbol{u}_i \,\mathrm{d}t \right\} \tag{5.134a}$$

$$\text{s.t.} \quad \dot{V}_i(\boldsymbol{x}, \boldsymbol{w}) = \left(\frac{\partial V_i(\boldsymbol{x}, \boldsymbol{w})}{\partial \boldsymbol{x}}\right)^{\mathsf{T}} \dot{\boldsymbol{x}} + \sum_{i=1}^{N} \left(\frac{\partial V_i(\boldsymbol{x}, \boldsymbol{w})}{\partial \boldsymbol{w}_i}\right)^{\mathsf{T}} \dot{\boldsymbol{w}}_i. \tag{5.134b}$$

For this modified differential game, it is straightforward to obtain the explicit control law from (5.9)

$$\boldsymbol{u}_i^{\star} = -(\boldsymbol{S}_i(\boldsymbol{x}))^{-1} \boldsymbol{G}_i^{\mathsf{T}}(\boldsymbol{x}) \frac{\partial V_i}{\partial \boldsymbol{x}} \cdot \Upsilon_i(\boldsymbol{x}, \boldsymbol{w}_i), \tag{5.135}$$

where

$$\Upsilon_i(\boldsymbol{x}, \boldsymbol{w}_i) = \begin{cases} \dfrac{f'_{\Upsilon,i} \pm \sqrt{(f'_{\Upsilon,i})^2 + Q'_{\Upsilon,i} \cdot S'_{\Upsilon,i}}}{S'_{\Upsilon,i}}, & S'_{\Upsilon,i} \neq 0, \\ -\dfrac{Q'_{\Upsilon,i}}{2f'_{\Upsilon,i}}, & S'_{\Upsilon,i} = 0, \end{cases} \tag{5.136}$$

$$\begin{aligned} f'_{\Upsilon,i}(\boldsymbol{x}, \boldsymbol{w}) &= \left(\frac{\partial V_i}{\partial \boldsymbol{x}}\right)^{\mathsf{T}} (\boldsymbol{J}(\boldsymbol{x}) - \boldsymbol{R}(\boldsymbol{x}))\frac{\partial H}{\partial \boldsymbol{x}} - \frac{1}{2}\sum_{\substack{j=1 \\ j\neq i}}^{N} \left(\frac{\partial V_i}{\partial \boldsymbol{x}}\right)^{\mathsf{T}} \boldsymbol{K}_j(\boldsymbol{x})\frac{\partial V_j}{\partial \boldsymbol{x}} \cdot \Upsilon_j \\ &\quad - \left(\frac{\partial V_i}{\partial \boldsymbol{w}_i}\right)^{\mathsf{T}} \cdot \alpha \cdot \frac{\partial \mathcal{J}_{\mathrm{w},i}}{\partial \boldsymbol{w}_i} \\ &= \left(\left(\frac{\partial H}{\partial \boldsymbol{x}}\right)^{\mathsf{T}} + \boldsymbol{w}_i^{\mathsf{T}}\frac{\partial \Xi_i}{\partial \boldsymbol{x}}\right) \times \cdots \\ &\quad \cdots \times \left((\boldsymbol{J}(\boldsymbol{x}) - \boldsymbol{R}(\boldsymbol{x}))\frac{\partial H}{\partial \boldsymbol{x}} - \sum_{\substack{j=1 \\ j\neq i}}^{N} \frac{\Upsilon_j}{2}\boldsymbol{K}_j(\boldsymbol{x}) \left(\frac{\partial H}{\partial \boldsymbol{x}} + \left(\frac{\partial \Xi_j}{\partial \boldsymbol{x}}\right)^{\mathsf{T}} \boldsymbol{w}_j\right)\right) \\ &\quad - \left(\frac{\partial V_i}{\partial \boldsymbol{w}_i}\right)^{\mathsf{T}} \cdot \alpha \cdot \frac{\partial \mathcal{J}_{\mathrm{w},i}}{\partial \boldsymbol{w}_i} \qquad (5.137) \\ &= \left(\left(\frac{\partial H}{\partial \boldsymbol{x}}\right)^{\mathsf{T}} + \boldsymbol{w}_i^{\mathsf{T}}\frac{\partial \Xi_i}{\partial \boldsymbol{x}}\right) \left((\boldsymbol{J}(\boldsymbol{x}) - \boldsymbol{R}(\boldsymbol{x}))\frac{\partial H}{\partial \boldsymbol{x}} + \frac{1}{2}\sum_{\substack{j=1 \\ j\neq i}}^{N} \boldsymbol{G}_j(\boldsymbol{x})\boldsymbol{u}_j^{\star}\right) \\ &\quad - \left(\frac{\partial V_i}{\partial \boldsymbol{w}_i}\right)^{\mathsf{T}} \cdot \alpha \cdot \frac{\partial \mathcal{J}_{\mathrm{w},i}}{\partial \boldsymbol{w}_i}, \qquad (5.138) \\ S'_{\Upsilon,i}(\boldsymbol{x}, \boldsymbol{w}) &= \left(\frac{\partial V_i}{\partial \boldsymbol{x}}\right)^{\mathsf{T}} \boldsymbol{K}_i(\boldsymbol{x})\frac{\partial V_i}{\partial \boldsymbol{x}} \\ &= \boldsymbol{w}_i^{\mathsf{T}}\frac{\partial \Xi_i}{\partial \boldsymbol{x}}\boldsymbol{K}_i(\boldsymbol{x})\left(\frac{\partial \Xi_i}{\partial \boldsymbol{x}}\right)^{\mathsf{T}} \boldsymbol{w} + 2\left(\frac{\partial H}{\partial \boldsymbol{x}}\right)^{\mathsf{T}} \boldsymbol{K}_i(\boldsymbol{x})\left(\frac{\partial \Xi_i}{\partial \boldsymbol{x}}\right)^{\mathsf{T}} \boldsymbol{w}_i \\ &\quad + \left(\frac{\partial H}{\partial \boldsymbol{x}}\right)^{\mathsf{T}} \boldsymbol{K}_i(\boldsymbol{x})\frac{\partial H}{\partial \boldsymbol{x}}, \qquad (5.139) \\ Q'_{\Upsilon,i}(\boldsymbol{x}) &= \ell_i(\boldsymbol{x}), \qquad (5.140) \\ \boldsymbol{K}_i(\boldsymbol{x}) &= \boldsymbol{G}_i(\boldsymbol{x})\boldsymbol{S}_i^{-1}(\boldsymbol{x})\boldsymbol{G}_i^{\mathsf{T}}(\boldsymbol{x}). \qquad (5.141) \end{aligned}$$

Note that the "+" sign in the first row of (5.136) can be discarded with the same reasoning as in the proof of Lemma 5.37. Moreover, continuity of $\Upsilon_i(\boldsymbol{x}, \boldsymbol{w}_i)$ is still preserved (cf. Remark 5.38). Finally, (5.137) and (5.138) are equivalent as it follows from (5.135) that

$$\Upsilon_i \cdot \boldsymbol{K}_j(\boldsymbol{x}) \left(\frac{\partial H}{\partial \boldsymbol{x}} + \left(\frac{\partial \Xi_j}{\partial \boldsymbol{x}}\right)^{\mathsf{T}} \boldsymbol{w}_j\right) = -\boldsymbol{G}_j(\boldsymbol{x})\boldsymbol{u}_j. \tag{5.142}$$

Remark 5.49. *From the controller equations* (5.135)–(5.136), (5.138)–(5.141) *it becomes evident that neither* $\boldsymbol{\Xi}_j(\boldsymbol{x})$ *nor* $\boldsymbol{w}_j$ *of the other players* $j \neq i$ *must be known by* $i \in \mathbb{P}$. ♢

By combining the modified optimal control law and the adaptation procedure (5.131), we get the closed-loop system

$$\dot{\boldsymbol{x}} = (\boldsymbol{J}(\boldsymbol{x}) - \boldsymbol{R}(\boldsymbol{x}))\frac{\partial H}{\partial \boldsymbol{x}} - \sum_{i=1}^{N} \Upsilon_i \cdot \boldsymbol{G}_i(\boldsymbol{x})(\boldsymbol{S}_i(\boldsymbol{x}))^{-1}\boldsymbol{G}_i^{\mathsf{T}}(\boldsymbol{x})\frac{\partial V_i}{\partial \boldsymbol{x}}, \tag{5.143a}$$

$$\dot{\boldsymbol{w}}_i = -\alpha \cdot \frac{\partial \mathcal{J}_{\mathrm{w},i}}{\partial \boldsymbol{w}_i}, \tag{5.143b}$$

$$0 = q_i^0(\boldsymbol{x}) + \Upsilon_i \cdot q_i^1(\boldsymbol{x}, \boldsymbol{w}_i) + \Upsilon_i^2 q_i^2(\boldsymbol{x}, \boldsymbol{w}_i) + \Upsilon_i \sum_{\substack{j=1 \\ j \neq i}}^{N} q_{ij}^2(\boldsymbol{x}, \boldsymbol{w}_i, \boldsymbol{w}_j)\Upsilon_j, \quad i \in \mathbb{P}, \tag{5.143c}$$

$$\boldsymbol{x}(0) = \boldsymbol{x}_0, \tag{5.143d}$$

$$\boldsymbol{w}_i(0) = \mathbb{0}_{r_i}, \tag{5.143e}$$

where

$$q_i^0(\boldsymbol{x}) := \frac{1}{2}\ell_i(\boldsymbol{x}), \tag{5.144}$$

$$q_i^1(\boldsymbol{x}, \boldsymbol{w}_i) := q_i^1(\boldsymbol{x}) - \left(\frac{\partial V_i}{\partial \boldsymbol{w}_i}\right)^{\mathsf{T}} \cdot \alpha \cdot \frac{\partial \mathcal{J}_{\mathrm{w},i}}{\partial \boldsymbol{w}_i}, \tag{5.145}$$

$$q_i^2(\boldsymbol{x}, \boldsymbol{w}_i) := -\frac{1}{2}\left(\frac{\partial V_i}{\partial \boldsymbol{x}}\right)^{\mathsf{T}} \boldsymbol{G}_i(\boldsymbol{x})(\boldsymbol{S}_i(\boldsymbol{x}))^{-1}\boldsymbol{G}_i^{\mathsf{T}}(\boldsymbol{x})\frac{\partial V_i}{\partial \boldsymbol{x}}, \tag{5.146}$$

$$q_{ij}^2(\boldsymbol{x}, \boldsymbol{w}_i, \boldsymbol{w}_j) := -\left(\frac{\partial V_i}{\partial \boldsymbol{x}}\right)^{\mathsf{T}} \boldsymbol{G}_j(\boldsymbol{x})(\boldsymbol{S}_j(\boldsymbol{x}))^{-1}\boldsymbol{G}_j^{\mathsf{T}}(\boldsymbol{x})\frac{\partial V_j}{\partial \boldsymbol{x}}. \tag{5.147}$$

5.3.6 Stability of the Adaptive Differential Game

To perform a stability analysis of the equilibrium $(\mathbb{0}_n, \boldsymbol{w}^\star)$ of (5.143), we define the Lyapunov function candidate

$$\hat{V}(\boldsymbol{x}, \boldsymbol{w}) := \sum_{i=1}^{N} V_i(\boldsymbol{x}, \boldsymbol{w}) \tag{5.148}$$

and show that it is indeed a Lyapunov function for the closed-loop system (5.143).

Lemma 5.50. *Consider the closed-loop system* (5.143) *starting at* $(\boldsymbol{x}_0, \mathbb{0}_r)$. *The Lyapunov function candidate* $\hat{V}(\boldsymbol{x}, \boldsymbol{w})$ *is strictly monotonically decreasing over time unless* $(\boldsymbol{x}, \boldsymbol{w}) = (\mathbb{0}_n, \boldsymbol{w}^\star)$.

Proof. The time derivative of $V_i(\boldsymbol{x}, \boldsymbol{w}_i)$ equals

$$\begin{aligned}\dot{V}_i(\boldsymbol{x}, \boldsymbol{w}_i) &= \left(\frac{\partial V_i}{\partial \boldsymbol{x}}\right)^{\mathsf{T}} \dot{\boldsymbol{x}} + \left(\frac{\partial V_i}{\partial \boldsymbol{w}_i}\right)^{\mathsf{T}} \dot{\boldsymbol{w}}_i \\ &= \left(\frac{\partial V_i}{\partial \boldsymbol{x}}\right)^{\mathsf{T}} (\boldsymbol{J}(\boldsymbol{x}) - \boldsymbol{R}(\boldsymbol{x}))\frac{\partial H}{\partial \boldsymbol{x}} - \Upsilon_i \cdot \left(\frac{\partial V_i}{\partial \boldsymbol{x}}\right)^{\mathsf{T}} \boldsymbol{G}_i(\boldsymbol{x})(\boldsymbol{S}_i(\boldsymbol{x}))^{-1}\boldsymbol{G}_i^{\mathsf{T}}(\boldsymbol{x})\frac{\partial V_i}{\partial \boldsymbol{x}} \\ &\quad - \sum_{\substack{j=1\\ j\neq i}}^{N} \Upsilon_j \cdot \left(\frac{\partial V_j}{\partial \boldsymbol{x}}\right)^{\mathsf{T}} \boldsymbol{G}_j(\boldsymbol{x})(\boldsymbol{S}_j(\boldsymbol{x}))^{-1}\boldsymbol{G}_j^{\mathsf{T}}(\boldsymbol{x})\frac{\partial V_i}{\partial \boldsymbol{x}} + \left(\frac{\partial V_i}{\partial \boldsymbol{w}_i}\right)^{\mathsf{T}} \dot{\boldsymbol{w}}_i. \end{aligned} \tag{5.149}$$

Dividing (5.143c) by $\Upsilon_i > 0$ and inserting $\dot{\hat{V}}_i$ from (5.149) yields

$$\frac{1}{2}\frac{\ell_i(\boldsymbol{x})}{\Upsilon_i} + \dot{\hat{V}}_i + \frac{\Upsilon_i}{2} \cdot \left(\frac{\partial V_i}{\partial \boldsymbol{x}}\right)^{\mathsf{T}} \boldsymbol{G}_i(\boldsymbol{x})(\boldsymbol{S}_i(\boldsymbol{x}))^{-1}\boldsymbol{G}_i^{\mathsf{T}}(\boldsymbol{x})\frac{\partial V_i}{\partial \boldsymbol{x}} = 0, \tag{5.150}$$

which finally leads to

$$\dot{\hat{V}}_i = -\underbrace{\frac{1}{2}\frac{\ell_i(\boldsymbol{x})}{\Upsilon_i}}_{\succ 0} - \underbrace{\frac{\Upsilon_i}{2}}_{>0} \cdot \underbrace{\left(\frac{\partial V_i}{\partial \boldsymbol{x}}\right)^{\mathsf{T}} \boldsymbol{G}_i(\boldsymbol{x})(\boldsymbol{S}_i(\boldsymbol{x}))^{-1}\boldsymbol{G}_i^{\mathsf{T}}(\boldsymbol{x})\frac{\partial V_i}{\partial \boldsymbol{x}}}_{\succcurlyeq 0} \prec 0. \tag{5.151}$$

As a consequence, $\dot{\hat{V}} = \sum_{i=1}^{N} V_i(\boldsymbol{x}, \boldsymbol{w}_i) < 0$, unless $\boldsymbol{x} = \mathbb{0}_n$. □

Remark 5.51. *As already mentioned in Remark 5.39, it is sufficient for negative definiteness of* $\dot{\hat{V}}(\boldsymbol{x}, \boldsymbol{w})$ *(see* (5.151)*) and thus asymptotic stability of the closed-loop equilibrium to require that* $\ell_i(\boldsymbol{x}) \succ 0$, $\ell_{\neg i}(\boldsymbol{x}) \succcurlyeq 0$ *holds for some* $i \in \mathbb{P}$. ◇

Lemma 5.52. *Consider the closed-loop system* (5.143) *starting at* $(\boldsymbol{x}_0, \mathbb{0}_r)$. *Let the conditions of Lemma 5.46 hold with* $\boldsymbol{w}_0 \in \mathcal{M}$. *Then the Lyapunov function candidate* $\hat{V}(\boldsymbol{x}, \boldsymbol{w}) = \mathbb{1}^{\mathsf{T}}\boldsymbol{V}(\boldsymbol{x}, \boldsymbol{w})$ *as in* (5.148) *fulfills*

$$\forall\, t > 0, (\boldsymbol{x}, \boldsymbol{w}) \neq (\mathbb{0}_n, \boldsymbol{w}^\star): \qquad \hat{V}(\boldsymbol{x}(t), \boldsymbol{w}(t)) > 0. \tag{5.152}$$

Proof. $\hat{V}(\boldsymbol{x}, \boldsymbol{w})$ can be written in the form

$$\hat{V}(\boldsymbol{x}, \boldsymbol{w}) = N \cdot H(\boldsymbol{x}) + \boldsymbol{w}^{\mathsf{T}}\hat{\Xi}(\boldsymbol{x}), \tag{5.153}$$

where $\hat{\Xi}(\boldsymbol{x}) = \mathrm{col}_i\{\Xi_i(\boldsymbol{x})\}$. Due to the fact that both $\hat{V}(\boldsymbol{x}, \mathbb{0}_r) \succ 0$ as well as $\hat{V}(\boldsymbol{x}, \boldsymbol{w}^\star) \succ 0$ hold for all $\boldsymbol{x} \in \mathbb{R}^n$ and $\hat{\mathcal{J}}_{\mathrm{w}}(\boldsymbol{x}, \boldsymbol{w})$ is strictly convex, the conditions of Lemma 5.22 are fulfilled, which implies that

$$\lim_{t\to\infty} \|\boldsymbol{w}(t) - \boldsymbol{w}^\star\|_2 = 0. \tag{5.154}$$

Accordingly, from Lemma 5.23 it follows that $\hat{V}(\boldsymbol{x}(t), \boldsymbol{w}(t)) \succ 0$ for all $t > 0$. □

Theorem 5.53. *Consider the closed-loop system* (5.143) *starting at* $(\boldsymbol{x}_0, \mathbb{0}_r)$. *Let the conditions of Lemma 5.46 hold with* $\boldsymbol{w}_0 \in \mathcal{M}$. *Then* $\boldsymbol{x} = \mathbb{0}, \boldsymbol{w} = \boldsymbol{w}^\star$ *is a locally asymptotically stable equilibrium of* (5.143).

Proof. Define the shifted Lyapunov function candidate $\widetilde{V}(\boldsymbol{x}, \boldsymbol{w}) = \hat{V}(\boldsymbol{x}, \boldsymbol{w} - \boldsymbol{w}^\star)$. According to Lemma 5.52, $\widetilde{V}(\boldsymbol{x}, \boldsymbol{w})$ is positive definite and according to Lemma 5.50, $\dot{\widetilde{V}}(\boldsymbol{x}, \boldsymbol{w})$ is negative definite. As such, $\widetilde{V}(\boldsymbol{x}, \boldsymbol{w})$ is a strict Lyapunov function for the equilibrium $(\mathbb{0}, \boldsymbol{w}^\star)$ of (5.143), which is consequently a locally asymptotically stable equilibrium [Kha02, Theorem 4.1]. □

Example 8 (Nonlinear Differential Game) – ***cont'd***

Now the adaptation strategy (5.143) *is applied to the nonlinear differential game* (5.101)–(5.102). *As* $\boldsymbol{R}(\boldsymbol{x})$ *in* (5.102) *is positive definite, Lemma 5.42 states that* $\boldsymbol{V}_0(\boldsymbol{x}) = (H(\boldsymbol{x}), H(\boldsymbol{x}))^\mathsf{T}$ *is an admissible vector of CLFs.*

The basis functions for the extended CLFs are chosen to $\Xi_1(\boldsymbol{x}) = \Xi_2(\boldsymbol{x}) = (x_1^2, x_1x_2, x_2^2)^\mathsf{T}$, *which, after comparison of* (5.103) *and* (5.118), *implies that the optimal weights are* $\boldsymbol{w}_1^\star = (1.5, -2, 0.5)^\mathsf{T}$ *and* $\boldsymbol{w}_2^\star = (0, 0, 0.5)^\mathsf{T}$.

The system is initialized at $\boldsymbol{x}_0 = (1, 1)^\mathsf{T}$, $\boldsymbol{w}_0 = (0, 0)^\mathsf{T}$ *and the shifts in* $\mathcal{J}_{\mathrm{w},i}$ *(see* (5.128)*) are set to* $\mathbf{c}_{i,1} = (0, 0)^\mathsf{T}$, $\mathbf{c}_{i,2} = (1, 0)^\mathsf{T}$, $\mathbf{c}_{i,3} = (0, 1)^\mathsf{T}$, *and* $\mathbf{c}_{i,4} = (-1, -1)^\mathsf{T}$. *To investigate the behavior of the control system in case of an external excitation, we apply an additive disturbance input* $\mathbf{d} = (\delta(t - 10), \delta(t - 20))^\mathsf{T}$.

Fig. 5.10a shows the trajectories of $\boldsymbol{w}_1(t)$, $\boldsymbol{w}_2(t)$, *and* $\boldsymbol{\Upsilon}(t)$, *if the learning rate* α *is set to* 0.5, *with the dashed lines indicating the optimal weights* $\boldsymbol{w}_1^\star$ *and* $\boldsymbol{w}_2^\star$, *respectively. It can be seen that after* 2 s, $\boldsymbol{x}$ *has converged to its equilibrium value and both* $\boldsymbol{\Upsilon}$ *as well as* $\boldsymbol{w}_1$ *and* $\boldsymbol{w}_2$ *have converged to their optimal values. Moreover, their values remain identical once the learning process is completed, even after the additive disturbance at* 10 s *and* 20 s. *Thus, it can be deduced that* $V_1(\boldsymbol{x}, \boldsymbol{w}_1)$ *and* $V_2(\boldsymbol{x}, \boldsymbol{w}_2)$ *converge to the respective value functions* $V_1^\star(\boldsymbol{x})$ *and* $V_2^\star(\boldsymbol{x})$, *which means that the implemented control strategies* $u_1^\star(\boldsymbol{x}) = -\frac{1}{S_1} \cdot \boldsymbol{G}_1^\mathsf{T}(\boldsymbol{x})\frac{\partial V_1}{\partial \boldsymbol{x}}\Upsilon_1$ *and* $u_2^\star(\boldsymbol{x}) = -\frac{1}{S_2} \cdot \boldsymbol{G}_2^\mathsf{T}(\boldsymbol{x})\frac{\partial V_2}{\partial \boldsymbol{x}}\Upsilon_2$ *converge to feedback Nash strategies with regard to the original objective function* (5.70a).

Fig. 5.10b shows the trajectories, if α *is reduced to* 0.05. *In this case, the equilibrium values of* $\boldsymbol{w}_1$, $\boldsymbol{w}_2$, *and* $\boldsymbol{\Upsilon}$ *remain the same. However, convergence of* $\boldsymbol{\Upsilon}$, $\boldsymbol{w}_1$, *and* $\boldsymbol{w}_2$ *is considerably slower, with a time to convergence of around* 20 s. *The overshoot of* $\boldsymbol{\Upsilon}$ *is about ten times larger than in Fig. 5.10a. The disturbance at* $t = 10$ s *results in a slight displacement of* $\boldsymbol{\Upsilon}$, $\boldsymbol{w}_1$, *and* $\boldsymbol{w}_2$, *whereas there is virtually no displacement caused by the second disturbance at* $t = 20$ s.

This again shows that even after a disturbance, $\boldsymbol{w}_1$ and $\boldsymbol{w}_2$ remain at their optimal values once the learning process is completed.

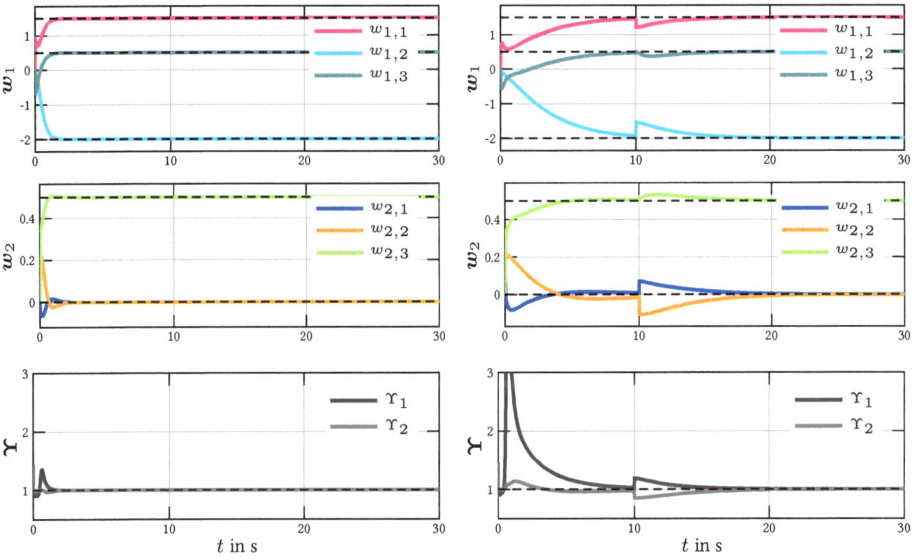

(a) $\alpha = 0.5$ (b) $\alpha = 0.05$

Figure 5.10: Adaptation results of the weights $\boldsymbol{w}$ and Lagrange multipliers $\boldsymbol{\Upsilon}$ depending on the chosen learning rate α.

5.3.7 Summary

In this section, a novel approach for finding feedback Nash equilibria in N-player noncooperative differential games for input-state-output PHSs has been presented. The proposed design procedure is based on a modified differential game which yields Sontag-type explicit control laws for all players, provided that an admissible vector of CLFs can be found. First, necessary and sufficient conditions have been stated under which the Hamiltonian of the input-state-output PHS accomplishes this purpose. Then, in order to ensure the optimal controllers for the modified differential game to coincide with the optimal controllers for the original problem, a simple adaptation strategy based on gradient flow is applied which is proven to be asymptotically stabilizing and able to approximate the value functions of the individual players (cf. Theorem 5.53).

5.4 Discussion

In this section, we restate the optimal balancing problem (4.37) from Subsection 4.4.2 and provide a conclusive statement about its explicit solution followed by a brief outline of prospective avenues of further research.

5.4.1 Application to Balance of Payments Restoration

Recall the optimal balancing problem from Subsection 4.4.2, yielding

$$\min_{\boldsymbol{u}} \quad \frac{1}{2}\int_0^{t_\mathrm{f}} \boldsymbol{x}_\sigma^\mathsf{T} \underbrace{\begin{bmatrix} \boldsymbol{Q}_1 & \mathbb{0} \\ \mathbb{0}^\mathsf{T} & Q_2 \end{bmatrix}}_{\boldsymbol{Q}} \boldsymbol{x}_\sigma + \boldsymbol{u}^\mathsf{T}\boldsymbol{S}\boldsymbol{u}\,\mathrm{d}t \tag{5.155a}$$

$$\text{subject to} \quad \dot{\boldsymbol{x}}_\sigma = (\boldsymbol{J}_\sigma - \boldsymbol{R}_\sigma)\frac{\partial H_\sigma(\boldsymbol{x}_\sigma)}{\partial \boldsymbol{x}_\sigma} + \boldsymbol{G}_\sigma \boldsymbol{u}. \tag{5.155b}$$

Note that $\boldsymbol{Q} \succcurlyeq 0$ and $\boldsymbol{S} \succ 0$ hold by definition. Provided that the (mild) technical assumption of nontrivial zonal power injections $\hat{\boldsymbol{p}}_\mathrm{g} \neq \mathbb{0}$ holds, the following corollary gives the asymptotic stability of the resulting (closed-loop) equilibrium governed by the PHS dynamics (5.155b) and the adaptive optimal controller (5.55b), (5.55c).

Corollary 5.54. *Let $V_\sigma^\star(\boldsymbol{x}_\sigma) = H_\sigma(\boldsymbol{x}_\sigma) + (\boldsymbol{w}^\star)^\mathsf{T}\boldsymbol{\Xi}(\boldsymbol{x}_\sigma)$ be the value function of optimization problem* (5.155) *and let the conditions of Lemmas 5.19 and 4.19 hold with $\boldsymbol{w}_0 \in \mathcal{M}$. Then $\boldsymbol{x}_\sigma = \mathbb{0}, \boldsymbol{w} = \boldsymbol{w}^\star$ is an asymptotically stable equilibrium of* (5.155b), (5.55b), (5.55c).

Proof. If the conditions of Lemma 5.19 hold, then Theorem 5.24 can be applied to state that (5.155b), (5.55b), (5.55c) is at least (non-asymptotically) stable around $(\boldsymbol{x}_\sigma = \mathbb{0}, \boldsymbol{w} = \boldsymbol{w}^\star)$. Moreover, if the conditions of Lemma 4.19 hold, then this lemma states that the (open-loop) PHS dynamics (5.155b) are globally asymptotically stable. Thus, asymptotic stability of the closed-loop system (5.155b), (5.55b), (5.55c) around $(\boldsymbol{x}_\sigma = \mathbb{0}, \boldsymbol{w} = \boldsymbol{w}^\star)$ follows by Theorem 5.24. □

5.4.2 Possible Extensions

The novel methodology for ADP-based solutions of optimal control problems and differential games with PHS dynamics is founded on the "Lyapunov-like" properties of the Hamiltonian $H(\boldsymbol{x})$, which allow to obviate some common challenges in ADP literature, such as the need for initial stabilizing weights. However, for didactical reasons, a number of simplifying yet not necessarily essential assumptions are made in the context of this chapter. In the following we thus briefly discuss worthwhile extensions and generalizations of the previous findings.

Extension to Weak CLFs

A remarkable insight following from Corollary 5.54 is that the CLF property of $H(\boldsymbol{x})$ is *not* a necessary condition for asymptotic stability of the closed-loop system. In particular, although $H_\sigma(\boldsymbol{x}_\sigma)$ is only a *weak CLF*[70], the special structure of $H_\sigma(\boldsymbol{x}_\sigma)$ ensures that the largest invariant set contained in $\dot{H}_\sigma(\boldsymbol{x}_\sigma)$ is still equal to the origin, such that asymptotic stability can be addressed by invoking LaSalles invariance principle. In general, note that $H(\boldsymbol{x})$ being a weak CLF is also sufficient for non-asymptotic stability of the closed-loop system for both the single- and multi-player case[71]. However, for asymptotic stability, it always has to be investigated "by hand" whether the largest invariant set contained in $\{\boldsymbol{x} \neq \mathbb{0} : \dot{V}(\boldsymbol{x}, \boldsymbol{w}) = 0\}$ is equal to the origin.

Extension to Passive Input-Output Systems

Although all considerations of this chapter focus on optimization problems in which system dynamics can be described by an input-state-output PHS, they can seamlessly be extended to the more general system class of *passive input-output systems* (cf. Definition A.15 in Appendix A.4). In particular, all stability statements made so far still hold as long as the storage function $V(\boldsymbol{x})$ of the passive input-output system has the property of a (weak) CLF (cf. Definition 5.26 and Definition 5.30). A very practical benefit of PHSs, however, is that the Hamiltonian obtained there as a direct outcome of the energy-based modeling procedure always reflects the actual energy stored in the system. Therefore, $H(\boldsymbol{x})$ highly qualifies as a natural CLF candidate, whereas in (generic) passive input-output systems a suitable storage function has to be found artificially by means of educated guessing and thus often has no intrinsic physical interpretability.

Extension of Gradient Flow

As stated in Remark 3.28, the gradient flows in the adaptation schemes in (5.55c), (5.131) can alternatively be replaced by the Newton gradient flows

$$\dot{\boldsymbol{w}}_i = -\alpha \cdot \left(\frac{\partial^2 \mathcal{J}_{\mathrm{w},i}}{\partial \boldsymbol{w}_i^2}\right)^{-1} \frac{\partial \mathcal{J}_{\mathrm{w},i}}{\partial \boldsymbol{w}_i}, \qquad \alpha > 0. \tag{5.156}$$

If ill-conditioning of the Hessian in (5.156) hinders the numerical calculation of its inverse, there is a broad literature on alternative formulations of the Newton descent direction, such as *regularized Newton's method* [Pol09b] or the pseudo-inverse formulation [GK65]

$$\dot{\boldsymbol{w}}_i = -\alpha \cdot \left(\frac{\partial^2 \mathcal{J}_{\mathrm{w},i}}{\partial \boldsymbol{w}_i^2}\right)^{\dagger} \frac{\partial \mathcal{J}_{\mathrm{w},i}}{\partial \boldsymbol{w}_i}, \qquad \alpha > 0. \tag{5.157}$$

---

[70] $H(\boldsymbol{x})$ is said to be a weak CLF if the requirement "< 0" in (5.6) is relaxed to "≤ 0".

[71] Provided that the conditions of Lemma 5.19 (or Lemma 5.46 for the N-player case) with $\boldsymbol{w}_0 \in \mathcal{M}$ continue to apply.

Besides the standard least-squares pseudoinverse (5.157), there exists a large number of advanced approaches based on singular value decomposition such as truncated pseudoinverse or damped least-squares pseudoinverse, see [MCA95] for a discussion of alternative formulations.

6 Case Studies

The aim of this chapter is to demonstrate the capability of the overall control scheme shown in Fig. 4.7 in its entirety. Section 6.1 introduces the employed benchmark simulation model, which is based on the IEEE 57-bus system. Section 6.2 demonstrates the performance of the cell-based frequency and voltage controller from Chapter 3 using a three-cell scenario and showcases the benefits of the WoC scheme compared to an islanded MG operation (*Case Study I*). Section 6.3 evaluates the efficiency of the overall control scheme by means of a ten-cell scenario (*Case Study II*). The latter Case Study Incorporates the frequency and voltage controller from Chapter 3, the price-based controllers for congestion management and BoP containment from Chapter 4, and the BoP restoration strategy developed in Chapter 5. Section 6.4 draws a conclusive statement about the main findings of both case studies.

6.1 Benchmark Model Setup

The simulation is carried out on a modified IEEE 57-bus system depicted in Fig. 6.1. As part of the *IEEE power system test case* benchmark systems [IEE21] as well as the MATPOWER package [ZMST11], the IEEE 57-bus system was originally derived from a segment of a real-world power system localized in the midwestern US. It has already gained notable popularity and application in numerous works from different research directions. For instance, [JW16] employs the IEEE 57-bus test case to test a probabilistic OPF algorithm considering both uncertain infeed and consumption. In [ZC06], the same test system is used to demonstrate the effectiveness and performance of a new interline power flow controller approach. In [KPS15], a price-based optimization algorithm which minimizes the total cost of active power generation by minimizing the sum of active power losses in the network is tested on the IEEE 57-bus system.

Due to its complex structure with 80 power lines, a high mesh degree[72] and widely diverse line admittances, the IEEE 57-bus system is supposed to be a pertinent basis for the purpose of thoroughly validating the controller concepts derived in this dissertation. Yet, we have conducted the following modifications and amendments to the original model [IEE21] in order to provide a more insightful playground with regard to all possible network connectors and participants established in sections 3.2 and 3.3:

- Originally equipped with only 7 SGs and 42 loads, we have connected 38 controllable sources to the network (19 each of $\mathcal{S}$ and $\mathcal{I}$ type) as well as one (uncontrollable) load $(p_{\ell,i}, q_{\ell,i})$ per bus $i \in \mathcal{V}$, modeling the connection to local low-voltage feeders. The

[72] The associated graph $\mathscr{G}_{\mathrm{IEEE}}$ has an average degree of 2.81.

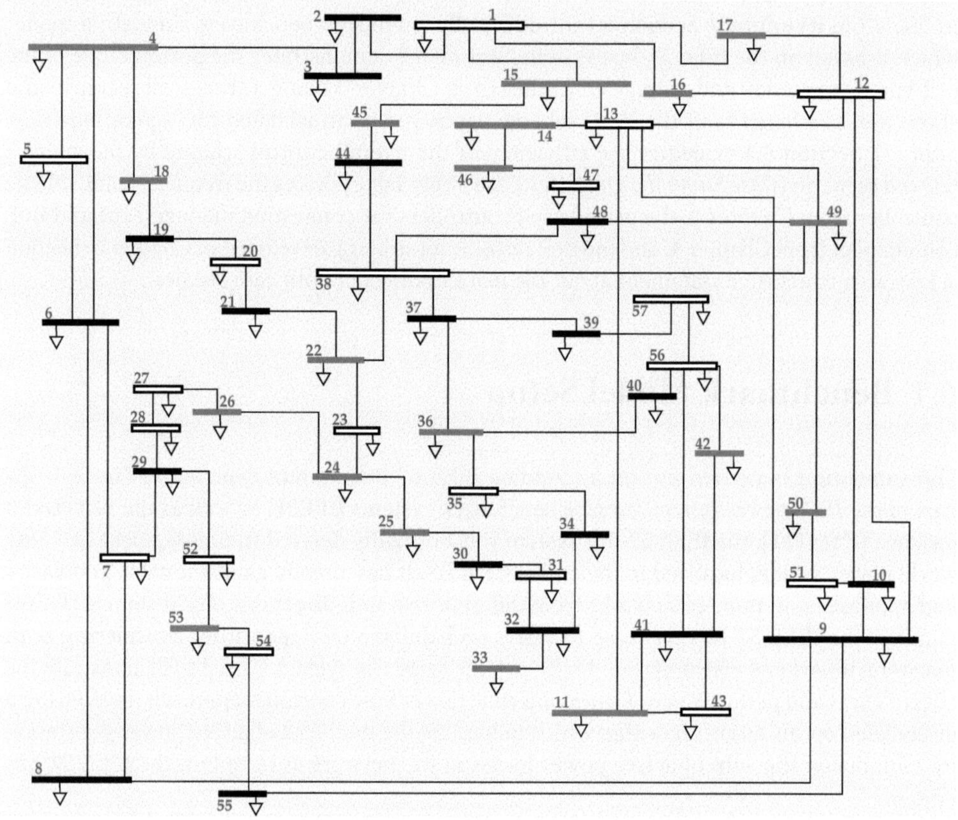

Figure 6.1: Stylized representation of the IEEE 57-bus system [IEE21]. $\mathcal{S}$ nodes are depicted in black, $\mathcal{I}$ nodes are depicted in gray, and $\mathcal{L}$ nodes are depicted in white. Each bus is equipped with an uncontrollable load modelled by (uncontrollable) active and reactive power consumption.

Table 6.1: Node parameters.

Type	A_i	Γ_i	$X_{d,i} - X'_{d,i}$	$\tau_{U,i}$
$\mathcal{S}$	$[1.2, 1.7]$	$[20, 27]$	$[0.12, 0.19]$	$[6.4, 7.7]$
$\mathcal{I}$	$[1.2, 1.7]$	$[4, 5.5]$	---	---
$\mathcal{L}$	$[1.2, 1.7]$	---	---	---

assignment of buses to node types is illustrated in Fig. 6.1. A detailed list is provided by Table E.1 in Appendix E.1.

- Since the node parameters are not specified in [IEE21], they are chosen to be randomly distributed within the specific intervals shown in Table 6.1. The parameter values for $\mathcal{S}$ nodes, $\mathcal{L}$ nodes and transmission lines are based on those provided in [TBD16, KBKH19]. The parameter values of $\mathcal{I}$ nodes base upon [MDS+18]. A full list of all numerical values is given in Table E.1 in Appendix E.1.

Throughout this chapter, all of the following numerical values are given in p.u. with $U_{\text{base}} = 135\,\text{kV}$, $S_{\text{base}} = 100\,\text{MVA}$, and $\mathtt{C}_{\text{base}} = 1\,\text{MU}/S_{\text{base}}$.

All simulations conducted in this chapter are implemented using the software system `Wolfram Mathematica` (Version 12.0.0) with the numerical solver `NDSolve`. In general, the program components developed in the context of this dissertation exhibit a high degree of modularity, so that changes to the plant model and variations or partial replacements of the employed controllers are straightforward. This enables to conveniently select and evaluate a wide variety of combinations of network topologies, parameter ranges, and control strategies.

The numerical computations are conducted on a machine with an Intel Core i7-6600U and 12 GB of RAM.

6.2 Case Study I: Three Cells

The objective of the first Case Study Is to investigate the performance of the frequency and voltage controllers developed in Chapter 3 and to evaluate the economic efficiency of the obtained results by examining different cell topologies.

6.2.1 Overview and Stimulation Signal

The IEEE 57-bus system is divided into three cells which are each interconnected via multiple boundary lines (see Fig. 6.2). Let $\boldsymbol{D}_{\text{IEEE}}$ denote the incidence matrix of the IEEE 57-bus system and $\boldsymbol{D}^3_{\text{IEEE}}$ denote the respective incidence matrix if all $|\hat{\mathcal{E}}_{\text{p}}| = 8$ boundary edges of adjacent cells (colored red in Fig. 6.2) are removed. In the following, we investigate the role of connectivity of $\mathcal{G}_{\text{p}}$ and $\mathcal{G}_{\text{c}}$ as well as the influence of $\boldsymbol{\kappa}$ by examining the following four different topological scenarios:

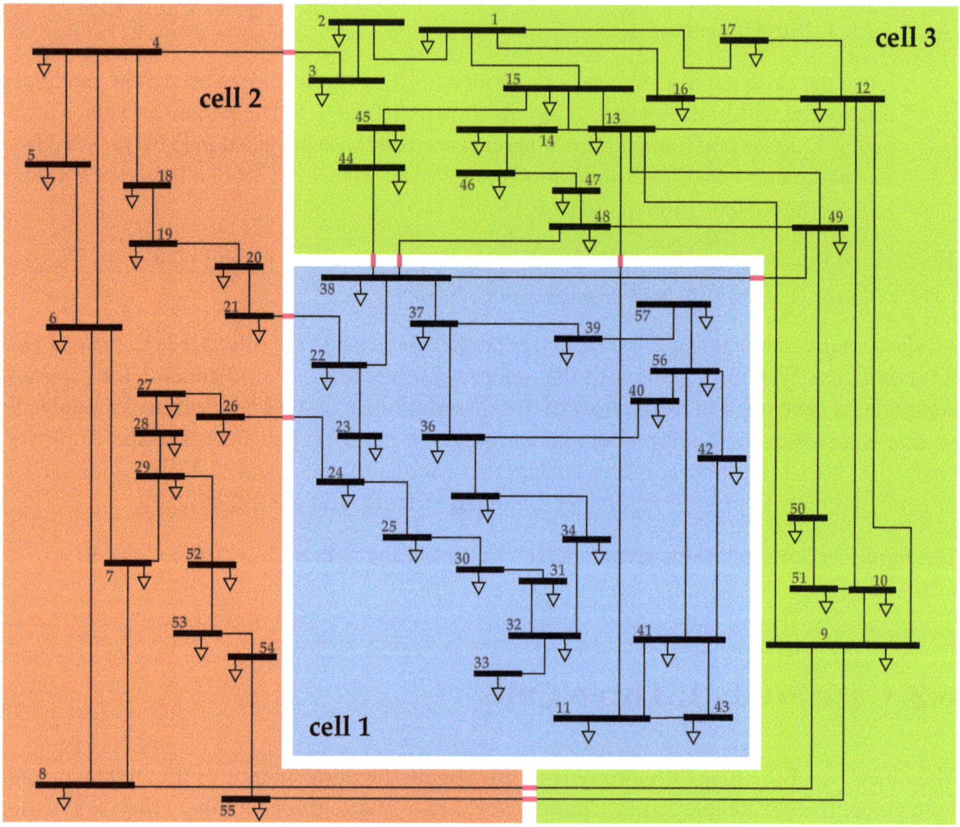

Figure 6.2: IEEE 57-bus system divided into three cells.

Table 6.2: List of scenarios elaborated in Chapter 6 (CM = congestion management).

Scenario	D_p	D_c	Zonal Pricing	BoP
I-a)	D^3_{IEEE}	D^3_{IEEE}	–	–
I-b)	D_{IEEE}	D^3_{IEEE}	–	–
I-c)	D_{IEEE}	D_{IEEE}	–	–
I-d)	D_{IEEE}	D^3_{IEEE}	$\kappa'' = (1, 3, 9)^\mathsf{T}$	–
II-a)	D_{IEEE}	D^{10}_{IEEE}	CM without integral action	containment
II-b)	D_{IEEE}	D^{10}_{IEEE}	CM with integral action	containment + restoration

I-a) *Islanded scenario:* All physical connections between different cells via boundary lines as well as well as communication links between CCs are cut such that both $\mathscr{G}_p$ and $\mathscr{G}_c$ are not (weakly) connected. In particular, we choose $D_p = D_c = D^3_{IEEE}$. This scenario is intended to show the behavior of three autonomous MGs, each of which are operated in islanded mode and controlled separately by the price-based frequency and voltage controller as derived in Chapter 3.

I-b) *Independent price zones:* The cells are physically connected, but the communication networks of CCs are separated from each other, i.e. $D_p = D_{IEEE}$ and $D_c = D^3_{IEEE}$. This corresponds to the "core scenario" of Section 3.4 where there exists no interconnection of price zones.

I-c) *Uniform pricing:* Both the physical power lines and the communication networks of cells are connected by choosing $D_p = D_c = D_{IEEE}$, which results in one large cell with one centralized CC.

I-d) *Constant participation factors:* The cells are physically connected as in Scenario I-b) with $D_p = D_{IEEE}$ and $D_c = D^3_{IEEE}$. In addition, CCs coordinate the relationship between specific prices based on (3.145) with participation factors κ'' set to a constant value of $\kappa'' = (1, 3, 9)^\mathsf{T}$. This portrays the idea of interconnected zonal prices pursued in Subsection 3.5. Since no further means of monetary balancing mechanism is active, κ is set equal to κ''.

A comparison chart summarizing the different scenarios of Case Study I is provided in the upper half of Table 6.2.

Controller Parameters

The controller parameters in τ_g from (3.110a) are set to 0.1, while the controller parameters in $\tau_{(\cdot)}$ from (3.110b)–(3.110k) are set to 0.01. The lower and upper bounds of $p_{g,i}$ are set to $\underline{p}_{g,i} = -0.002$ and $\overline{p}_{g,i} = 0.003$ respectively, and the voltage limits are set to $\underline{U}_i = 0.98$ and $\overline{U}_i = 1.02$.

Initialization

At $t = 0$, the system is in synchronous mode with $\boldsymbol{L} = \mathbb{0}$. All controller variables $\boldsymbol{p}_{\mathrm{g}}$, $\boldsymbol{\nu}$, $\boldsymbol{\lambda}$, $\boldsymbol{\mu}_{(\cdot)}$ are initialized to zero. Without loss of generality, we set $\mathcal{W} \equiv \mathcal{Z}$ and choose the cost functions to

$$\mathrm{C}_{\pi}(\boldsymbol{p}_{\pi}) = \frac{1}{2} \cdot \Bigg(\sum_{i \in \mathcal{V}_{\mathcal{P},\pi}} \frac{1}{\varpi_i} \cdot p_{\mathrm{g},i}^2 \Bigg), \qquad \pi \in \mathcal{W}, \tag{6.1}$$

where $\varpi_i = 1 + 0.04 \cdot (i - 1)$. The initial values of the voltage angle deviations ϑ_{ij} are chosen to be randomly distributed within the interval $[-0.04, 0.014]$ and all voltages U_i are chosen to be randomly distributed within the interval $[0.98, 1.02]$.

Stimulation Signal

The active and reactive power consumptions $\boldsymbol{p}_{\ell}$ and $\boldsymbol{q}_{\ell}$ are set piecewise constant in order to simulate a step change in generation or consumption in specific areas of the network. To this end, the initial values of $\boldsymbol{p}_{\ell}$ and $\boldsymbol{q}_{\ell}$ are chosen such that the AC power flow equations (3.8) are satisfied. Subsequently, step-wise load changes are applied at equal intervals of 5 min at the six $\mathcal{L}$ nodes 12, 13, 20, 27, 28, and 43. The selection of these nodes was made to ideally cover a wide range of load concentrations and thus to provoke diverse and reverted power flows. Nodes 12 and 13 as well as 27 and 28 belong to the same cell and are direct neighbors. Bus 20 also belongs to the same cell as 12 and 13, but is no direct neighbor of the latter. Bus 43 belongs to cell 2 and is situated more remotely. A total simulated time of $t = 30\,\mathrm{min}$ is chosen, which incorporates the following five load jumps:

- At $t_1 = 5\,\mathrm{min}$, the active power consumption $p_{\ell,28}$ in cell 3 increases by 0.015,
- at $t_2 = 10\,\mathrm{min}$, the reactive power consumption $q_{\ell,28}$ at the same node increases by 0.015,
- at $t_3 = 15\,\mathrm{min}$, the active and reactive power consumption at two additional nodes, 20 and 27, which are also located in cell 3, increase by 0.0075,
- at $t_4 = 20\,\mathrm{min}$, the power consumptions at nodes 20, 27, and 28 are reset to their initial values. At the same time, active and reactive power consumption at load nodes 12, 13, and 43 in cells 1 and 2 increase by 0.0075 each,
- at $t_5 = 25\,\mathrm{min}$, the active and reactive power consumptions of all above-mentioned nodes are multiplied by -1 to simulate a reversed load flow.

A graphical representation of the resulting active and reactive power consumptions is given in Fig. 6.3.

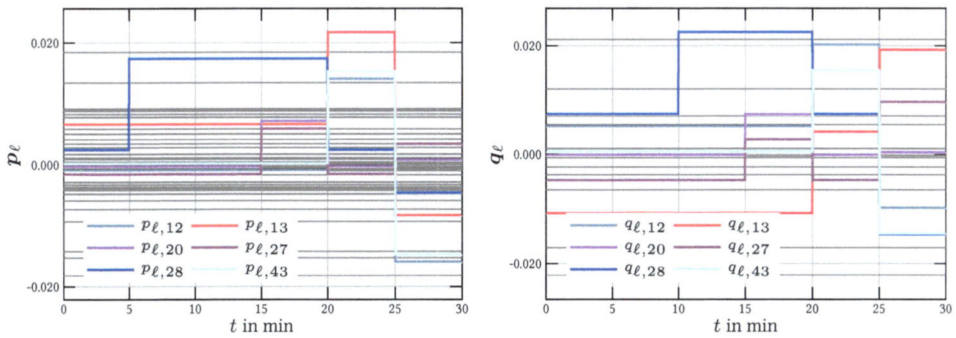

(a) Active power consumption

(b) Reactive power consumption

Figure 6.3: Step-wise jumps in power consumption $(\boldsymbol{p}_\ell, \boldsymbol{q}_\ell)$ for Case Study I. The thin gray lines indicate the (constant) power consumption at all other nodes.

6.2.2 Numerical Results

Figs. 6.4 and 6.6 show the resulting nodal prices $\boldsymbol{\lambda}$, active power generations $\boldsymbol{p}_{\mathrm{g}}$, nodal frequencies $\boldsymbol{f}$ and voltage magnitudes $\boldsymbol{U}$ for Scenarios I-a)–I-d), respectively.

In each of the four scenarios, no nodal price differences within each cell $k \in \{1, 2, 3\}$ can be detected. A close-up plot of the nodal prices for Scenario I-a) after t_1 and t_4 (given in Fig. 6.5) shows that the nodal prices converge to a common (cell-specific) price Λ_k in less than 10 s.

Prices and Active Power Generation

Scenario I-a) (islanded cells) It can be seen in the upper panel of Fig. 6.4a that due to physical decoupling, the loads steps in cell 3 at t_1, t_2, and t_3 only affect the price in cell 3 itself, while prices in other cells remain unaffected. Likewise, the load jump in cells 1 and 2 at t_5 only affects the prices in cells 1 and 2 itself, while the price in cell 3 retains its initial value. With regard to the active power generations (see the upper panel of Fig. 6.4b), it can be seen that the upper and lower limits $\overline{\boldsymbol{p}}_{\mathrm{g}}$ and $\underline{\boldsymbol{p}}_{\mathrm{g}}$ are respected except after t_3 in cell 3 and after t_5 in cell 2, where we see a significant oscillation of all generations around the lower limit. Moreover, the active power generations in cell 1 do not converge within the interval between t_4 and t_5.

Scenario I-b) (free zonal prices) The resulting cell-specific prices shown in the second upper panel in Fig. 6.4a predominantly follow a similar course as in Scenario I-a). In particular, the load jumps in cell 3 at t_1, t_2, and t_3 lead to an increase in the prices in cell 3. Likewise, the load jumps in cells 1 and 2 at t_4 lead to an increase in the prices in cells 1 and 2. However, in contrast to Scenario I-a), load jumps in one cell also affect the prices in the other cells, signifying the effect of physical interconnection via inter-cell lines. Moreover, the spread between zonal prices is lower than in Scenario I-a). Similar to the price curves, the resulting active power generations in different cells are affected by each other and exhibit a lower

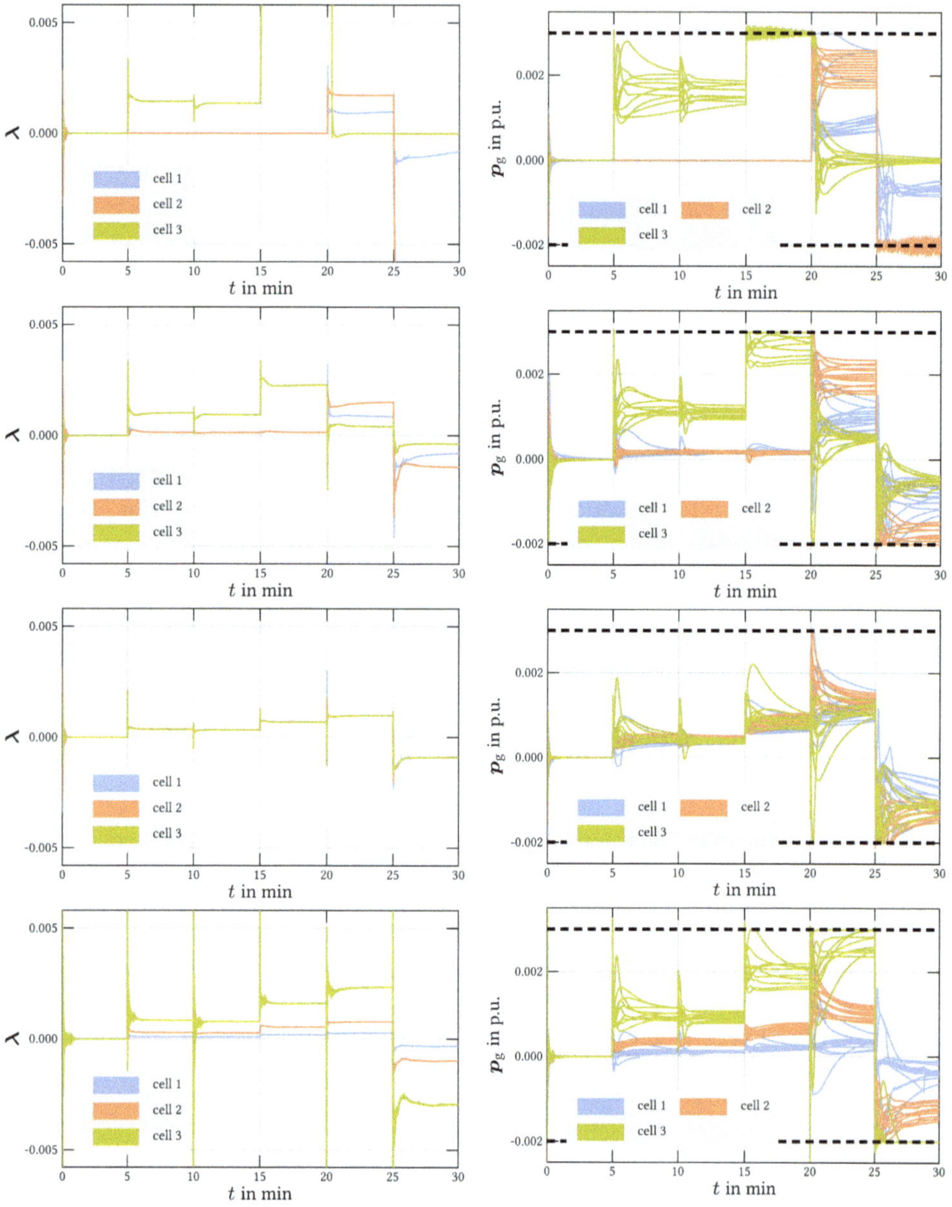

(a) Nodal prices

(b) Active power generation

Figure 6.4: Nodal prices and active power generation for Scenarios I-a)–I-d).

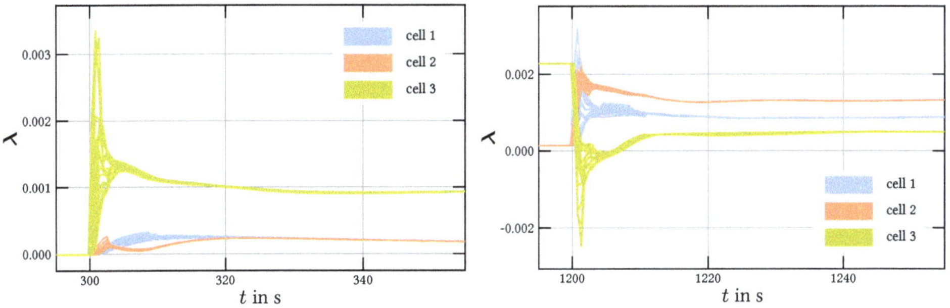

(a) load jump at t_1

(b) load jump at t_4

Figure 6.5: Synchronization of nodal prices towards a zonal price.

spread than in Scenario I-a) (see the second upper panel in Fig. 6.4b). This illustrates that the other cells contribute to eliminating the imbalance in a specific cell.

Scenario I-c) (uniform pricing) The second lower panel in Fig. 6.4a reveals that in this scenario, all prices quickly synchronize to a common value after each load jump. Accordingly, load jumps in one specific cell impact the active power generations, nodal frequencies, and voltage magnitues in all cells. This corresponds to the individual active power generations being closer to each other, barely reaching their specific upper or lower limits, see the second lower panel in Fig. 6.4b.

Scenario I-d) (constant participation factors) As depicted in the lower panel of Fig. 6.4a, the cell-specific prices Λ_k converge to distinct steady-state values, which invariably follow a ratio of 1:3:9 regardless of their absolute values. This is in line with the specified participation factors $\boldsymbol{\kappa} = (1, 3, 9)^\mathsf{T}$. Accordingly, the individual share of active power generation for each cell (see the lower panel in Fig. 6.4b) is distributed proportially to *both* the weighting factors ϖ_i in (6.1) *and* the cell-specific participation factors $\boldsymbol{\kappa}$, while always respecting the specified constraints $(\underline{\boldsymbol{p}}_\mathrm{g}, \overline{\boldsymbol{p}}_\mathrm{g})$ on active power generation as observed in the previous two scenarios. For the formation of prices (and thus the resulting active power generations), it is irrelevant in which region of the network the specific load jump occurs.

Frequency and Voltage Control

The resulting frequency and voltage curves show significant differences between Scenario I-a) and the other scenarios (see Fig. 6.6). As evident in the upper panel of Fig. 6.6a, in Scenario I-a), each load jump causes temporary deviations of the nodal frequencies in the involved cells, which are subsequently regulated back to their nominal value. The maximum overshoot during the entire simulated time is $+40\,\mathrm{mHz}$. After the steps at t_4 and t_5, the performance of the frequency regulation in cell 1 is poor. By contrast, frequency regulation is maintained

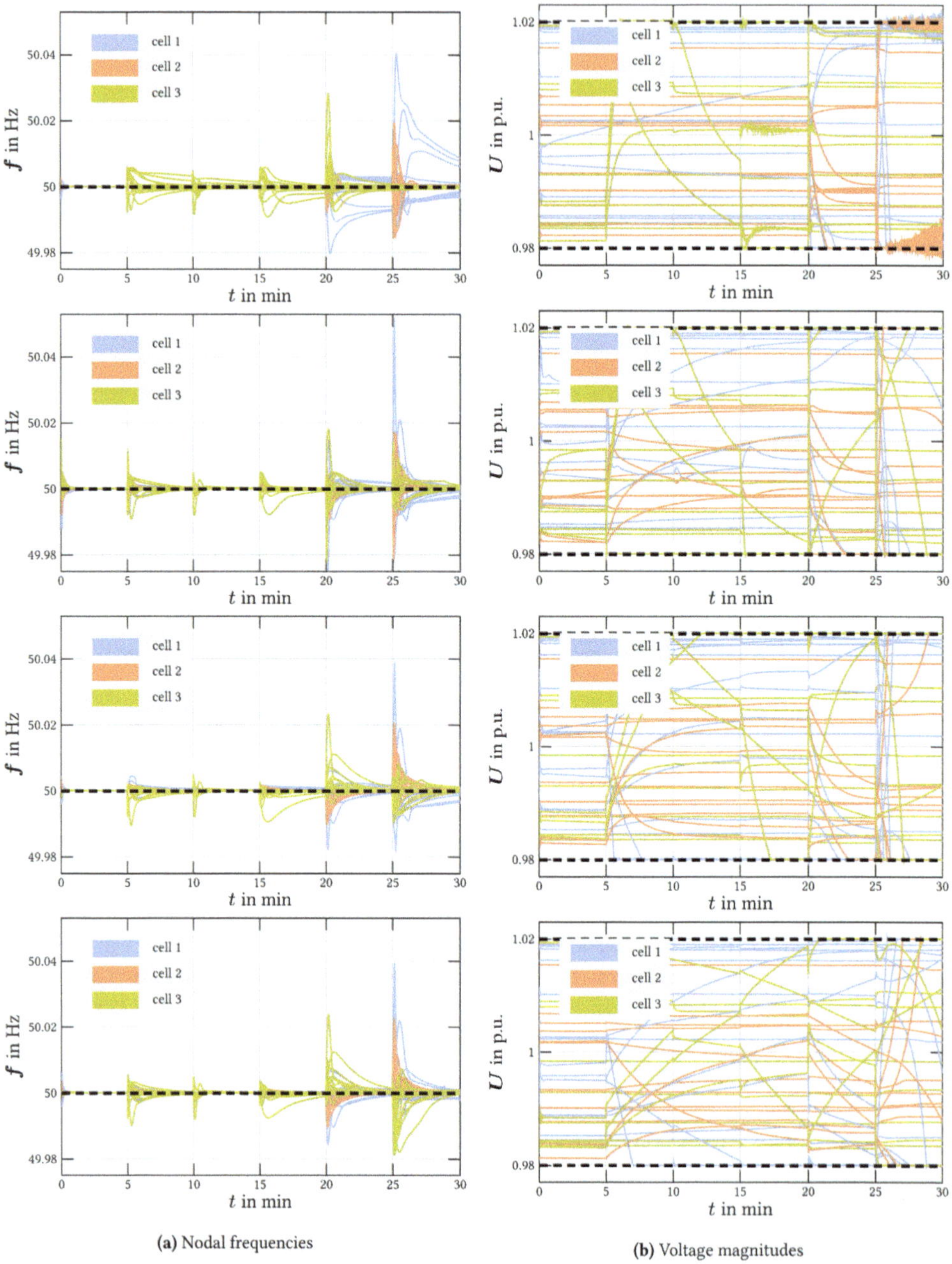

Figure 6.6: Nodal frequencies and voltage magnitudes for Scenarios I-a)–I-d).

without exception for Scenarios I-b)–I-d), with maximum overshoots of $+50\,\mathrm{mHz}$ (see the other subplots in Fig. 6.6a).

Fig. 6.6b shows the resulting voltage magnitudes in all cells. In Scenario I-a) (upper panel in Fig. 6.6b), all upper and lower limits are met except after t_5, where most of the voltage magnitudes in cell 2 oscillate around their upper or lower limits. As with nodal frequencies, the issues with voltage magnitudes are limited to Scenario I-a), while in the other scenarios voltage regulation is consistently provided, i.e. voltage magnitudes are kept within their respective limits without any oscillations and/or constraint violations.

6.2.3 Summary of Case Study I

The presented case study demonstrates the general applicability of the proposed control scheme for the automatic regulation of frequency and voltage by means of price signals. However, the detailed results for Scenario I-a) show that WPs in cell 2 are unable to autonomously meet the local active and reactive power demand without violating the constraints (3.83b)–(3.83c). This result exemplifies that the operation as autonomous MGs creates *unnnecessarily reduced robustness*, unless each MG is over-equipped with instantaneous generation capacity at any point in time based on peak demand considerations. As discussed at the beginning of Subsection 2.2.1, the latter would be highly cost-ineffective. It can be seen in Fig. 6.4a that with islanded MG operation, price jumps only occur in cells where the load changes actually occur, and that these price jumps are far more severe than in the other scenarios. On the contrary, the WoC-based infrastructure with a physically interconnected overall network helps to indirectly couple producers and consumers from different regions and thus to increase the robustness of the grid against sudden load changes. Interestingly, deploying the new WoC infrastructure comes at almost no additional fixed cost compared to autonomous operation, since the (few) additional coupling lines necessary for the interconnection of cells are either already in place or can be retrofitted between nearby buses with little effort.

The cell-specific ratios of the free zonal prices resulting in Scenario I-b) provide a mixture of the results in islanded MG operation (Scenario I-a)) and those in uniform pricing (Scenario I-c)), but, as discussed in Remark 3.34, do not yet follow an overarching goal. Furthermore, the results of Scenario I-d) stress that in general, any given relationships between zonal prices can be specified via κ. In the subsequent Case Study II, such "non-academic" adjustments of κ as discussed in Chapter 4, will be examined in detail.

6.3 Case Study II: Ten Cells

This case study considers the full control scheme as proposed in this dissertation, with zonal pricing through $\kappa(t)$ along with frequency and voltage control. Special focus is put on real-time congestion management as well as on the attainment of BoP.

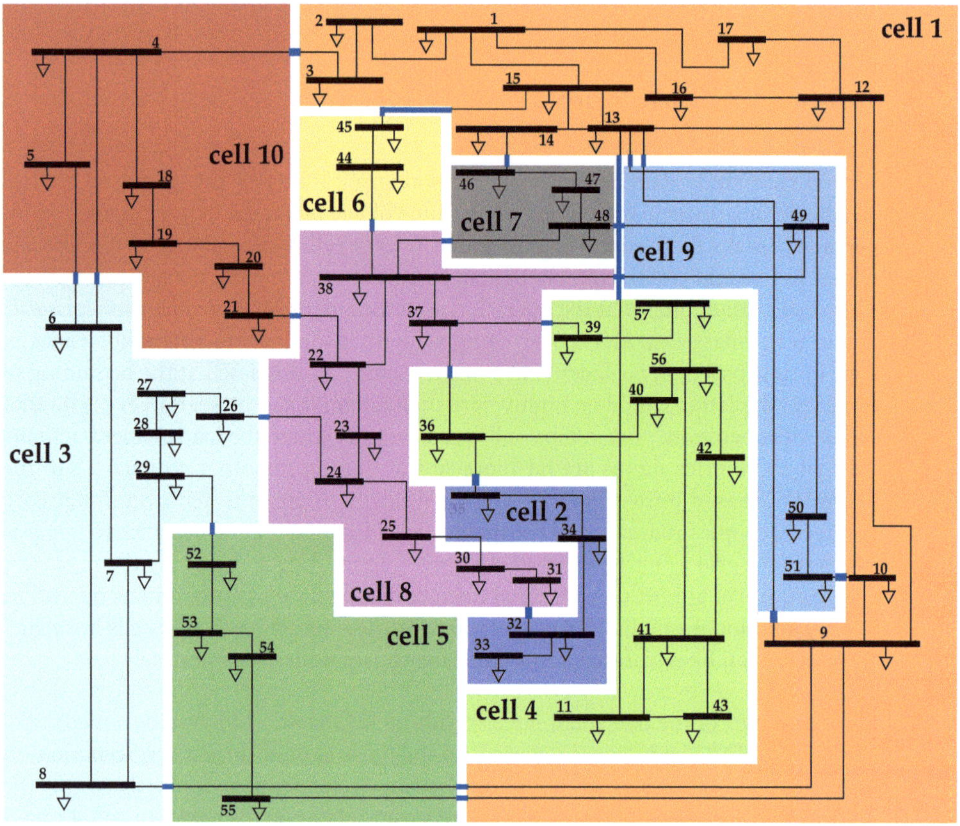

Figure 6.7: IEEE 57-bus system divided into ten cells.

6.3.1 Overview and Stimulation Signal

The modified IEEE 57-bus system is now divided into ten cells as illustrated in Fig. 6.7 whereby the assignment of buses to the node types and all physical parameters (see Table E.1) are left identical to Case Study I.

The communication networks represented by $\boldsymbol{D}_\mathrm{c}$ and $\hat{\boldsymbol{D}}_\mathrm{c}$ are chosen identically[73] to the physical interconnection structure, i.e. $\boldsymbol{D}_\mathrm{c} = \boldsymbol{D}_\mathrm{p}$ and $\hat{\boldsymbol{D}}_\mathrm{c} = \hat{\boldsymbol{D}}_\mathrm{p}$. We examine the following two different real-time pricing scenarios:

II-a) *BoP containment and congestion management without integral action*: The consensus-based controller for BoP containment developed in Subsection 4.4.1 is applied with $\hat{\mathscr{G}}_\mathrm{c} = \hat{\mathscr{G}}_\mathrm{p}$. Moreover, the congestion controller (4.14) from Subsection 4.3.1 is applied to enforce congestion management without transient constraint violation.

II-b) *BoP restoration and congestion management with integral action*: In addition to the controller for BoP containment, which is left identical to Scenario II-a), BoP restoration via the adaptive optimal controller from Subsection 5.4.1 is activated. Moreover, congestion controller (4.14) is replaced by congestion controller (4.15) with integral action.

A comparison chart summarizing the scenarios of Case Study II is provided in the lower half of Table 6.2 in Subsection 6.2.1.

Controller Parameters

To ensure sufficient timescale separation (Fig. 4.7), the controller parameters are chosen as follows: The frequency and voltage controller (3.110) is given the fastest response times by setting all parameters $\tau_{(\cdot)}$ in (3.110) to 0.01. For the "higher-level" balancing (4.29) and pricing distribution controllers (4.11), we choose the components in $\boldsymbol{\tau}_a$ and $\boldsymbol{\tau}_\phi$ to 0.1, and the components in $\boldsymbol{\tau}_{\nu_a}$ to 10. Finally, we set the components in $\boldsymbol{\tau}_{\mho}$ and $\boldsymbol{\tau}_{U^\mathrm{set}}$ to 100 for congestion controller (4.14) without integral action, and in $\boldsymbol{\tau}_{\mho}^-$ and $\boldsymbol{\tau}_{\mho}^+$ to 1000 for congestion controller (4.15) with integral action.

The lower and upper bounds of $p_{\mathrm{g},i}$ are set to $\underline{p}_{\mathrm{g},i} = 0$ and $\overline{p}_{\mathrm{g},i} = 0.5$ respectively, and the voltage limits are set to $\underline{U}_i = 0.98$ and $\overline{U}_i = 1.02$. The maximum admissible voltage setpoint deviation is chosen to $\overline{U}^\Delta = 0.015$. For both congestion controllers (4.14) and (4.15), the thresholds C_m^min, $m \in \hat{\mathcal{E}}_\mathrm{p}$ are chosen as $C_m^\mathrm{min} = 0.8$. The maximum permissible power flow of line $35 \to 35$ is set to $S_{35\to 36}^\mathrm{max} = 0.2$. For the other lines, it is set to $S_{i\to j}^\mathrm{max} = 0.1$. As state and controller weight matrices for BoP restoration, we choose $\boldsymbol{Q}_1 = 4 \cdot \boldsymbol{I}$, $Q_2 = 20$, and $\boldsymbol{S} = 4 \cdot \boldsymbol{I}$,

[73] As shown in further simulative studies in [KBKH19], the concrete choice of $\boldsymbol{D}_\mathrm{c}$ or $\hat{\boldsymbol{D}}_\mathrm{c}$ has no influence on the resulting steady state, yet only effects on the transient behavior, provided that the respective communication graphs $\mathscr{G}_\mathrm{c}$ or $\hat{\mathscr{G}}_\mathrm{c}$ are connected.

Table 6.3: Step-wise load changes for Case Study II.

timestep	cell k	node i	$\Delta p_{\ell,i}$	$\Delta q_{\ell,i}$
$t_1 = 15\,\text{min}$	1	1	$+0.05$	$+0.01$
$t_2 = 30\,\text{min}$	2	35	$+0.03$	$+0.001$
$t_3 = 45\,\text{min}$	3	28	$+0.01$	$+0.01$
$t_4 = 60\,\text{min}$	3	28	-0.01	-0.01
$t_5 = 75\,\text{min}$	2	35	-0.03	-0.001
$t_6 = 90\,\text{min}$	1	1	-0.05	-0.01

respectively. The basis functions for BoP restoration are chosen to

$$\Xi(\boldsymbol{x}_\sigma) = \begin{bmatrix} (\epsilon_1^\kappa)^2 \\ \vdots \\ (\epsilon_{10}^\kappa)^2 \\ \sigma^2 \\ \prod_{k=1}^{10} \epsilon_k^\kappa \\ \sigma \cdot \left(\prod_{k=1}^{10} \epsilon_k^\kappa\right) \end{bmatrix}, \tag{6.2}$$

and as objective function for the adaptation (5.55c), we use

$$\mathcal{J}_\mathrm{w}(\boldsymbol{x}_\sigma, \boldsymbol{w}) = (\mathtt{Q}(\boldsymbol{x}_\sigma, \boldsymbol{w}))^2 + \sum_{k=1}^{N} (\mathtt{Q}(\boldsymbol{x}_\sigma + \mathbf{e}_k, \boldsymbol{w}))^2 + \sum_{k=1}^{N} (\mathtt{Q}(\boldsymbol{x}_\sigma - \mathbf{e}_k, \boldsymbol{w}))^2 \tag{6.3}$$

with a learning rate of $\alpha = 0.5$.

Initialization and Stimulation Signal

The initial values of all controller states are set to 0, except for $\hat{\boldsymbol{U}}^{\text{set}}(0) = \mathbb{1}$. We set $\mathcal{W} = \mathcal{Z}$ and choose the cost functions to

$$\mathtt{C}_\pi(\boldsymbol{p}_{\mathrm{g},\pi}) = \frac{1}{2} \left(\sum_{i \in \mathcal{V}_{\mathcal{P},\pi}} \frac{1}{\varpi_i} p_{\mathrm{g},i}^2 \right), \qquad \pi \in \mathcal{W}, \tag{6.4}$$

where $\varpi = 1 + 0.01 \cdot (i-1)$.

The external disturbance is given by step-wise load jumps at the three nodes 1, 28, 35 from different zones in the network as listed in Table 6.3 and illustrated in Fig. 6.8. In order to investigate the path dependence of the resulting solutions, the load jumps occurring at t_1, t_2 and t_3 are compensated by reversed load jumps at t_4, t_5, and t_6 (in descending order with respect to the order of occurence).

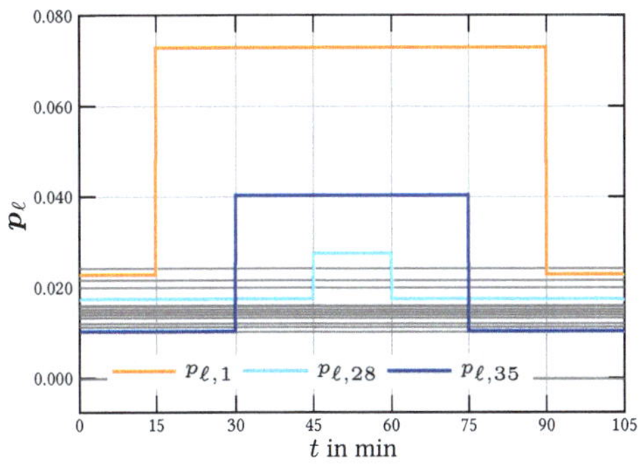

(a) Active power consumption

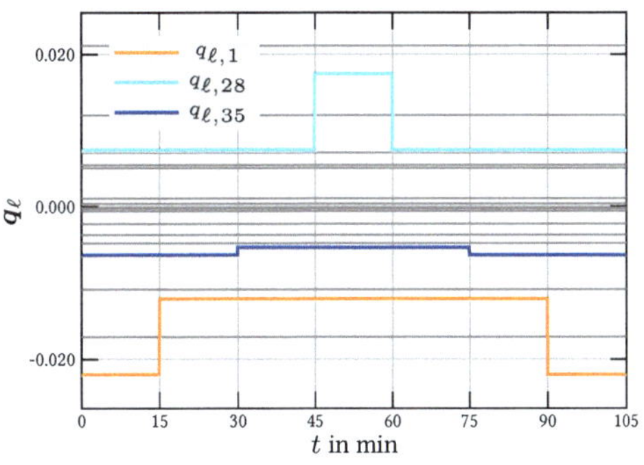

(b) Reactive power consumption

Figure 6.8: Step-wise jumps in power consumption $(\boldsymbol{p}_\ell, \boldsymbol{q}_\ell)$ for Case Study II. The thin gray lines indicate the (constant) power consumption at all other nodes.

6.3.2 Numerical Results

Triggered by the step-wise load jumps in $\boldsymbol{p}_\ell$ and $\boldsymbol{q}_\ell$, respectively, all voltage magnitudes, active power generations, and nodal prices attain a new steady state during each of the 15 min intervals (see Fig. 6.9). Notably, the jumps at t_1 and t_5 as well as those at t_2 and t_3 yield *exactly the same* steady state for each of the quantities shown in Fig. 6.9. All nodal frequencies converge to the nominal frequency 50 Hz within 10 min and exhibit a maximum deviation of -38 mHz after the step at t_6. These overshoots of nodal frequencies tend to be higher in amplitude if the load jump is closer to the respective node. E.g. after t_1, the load jump at node 1 in cell 1 causes a much more significant transient frequency deviation for the nodes in cell 1 than for those in the other cells. Furthermore, all specified voltage limits are maintained during the entire simulation.

Similarly to Case Study I, the nodal prices λ_i show a very fast convergence of less than one second towards a zonal price Λ_k. At steady state, the cell-specific active power generations are equidistant to each other, while their absolute values depend on all current active and reactive power consumptions. These results indicate that active power sharing (cf. Definition 3.38) is given across the entire network. whereby the weighting factors ϖ_i in (6.4) are identical to those in (3.116).

Congestion Management

Fig. 6.10 shows the resulting apparent power flows over all boundary lines $m \in \hat{\mathcal{E}}_\text{p}$. It can be clearly seen that in both scenarios all power flow limits S_m^max are respected at steady state. In Scenario II-a) (see Fig. 6.10a), there is no transient violation of any of the line flow limits. The participation factors $\boldsymbol{\kappa}$ synchronize within 2 min after initialization to a common value of less than 0.87 (see Fig. 6.11a). Subsequently, with the apparent power of line 38 $\rightarrow$ 48 approaching its specified maximum absolute value, the participation factors split over a range of 0.1, where the zonal pricing factor in cell 7 takes the highest value and the zonal pricing

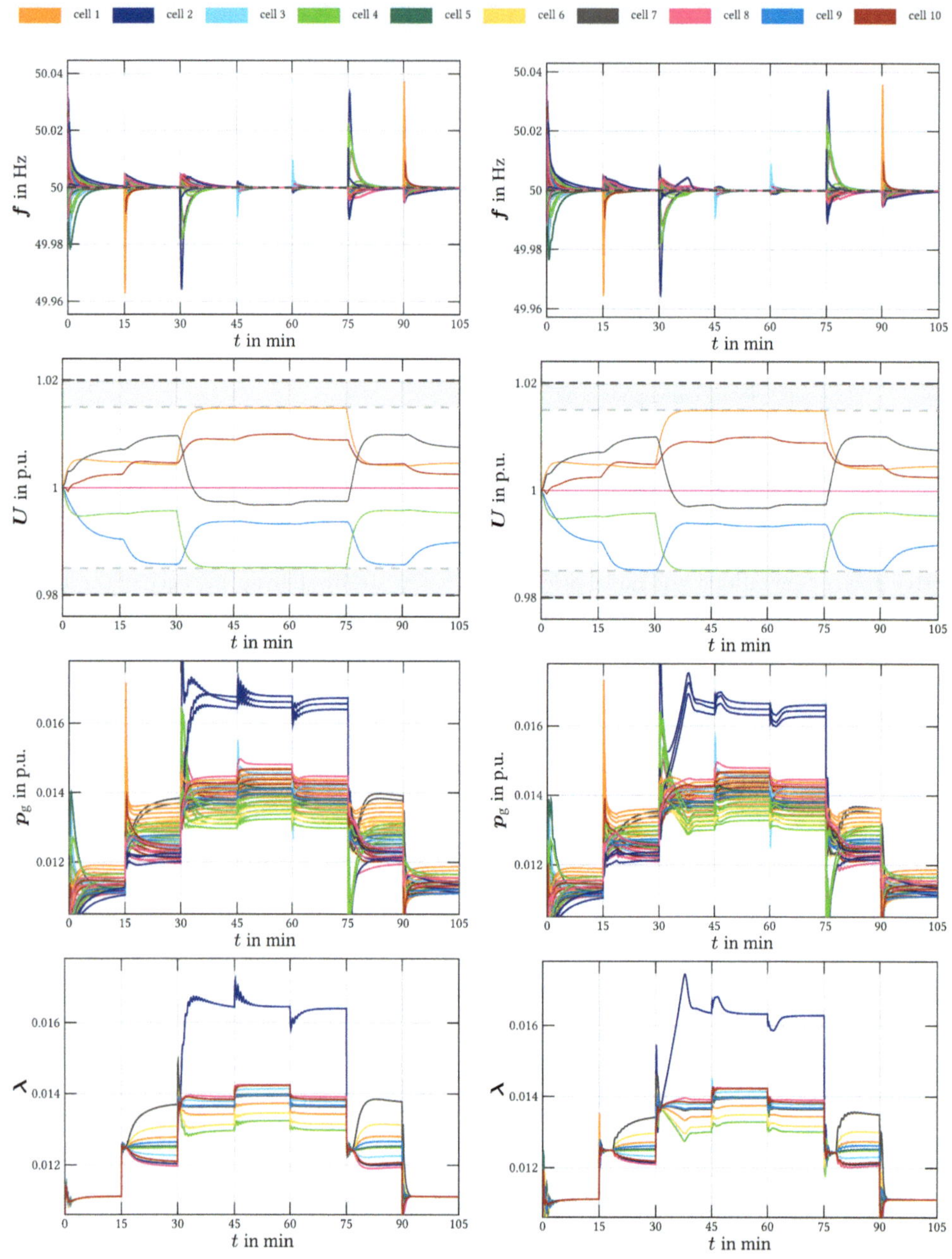

(a) Scenario II-a) (b) Scenario II-b)

Figure 6.9: Nodal frequencies, voltage magnitudes, active power generation, and nodal prices.

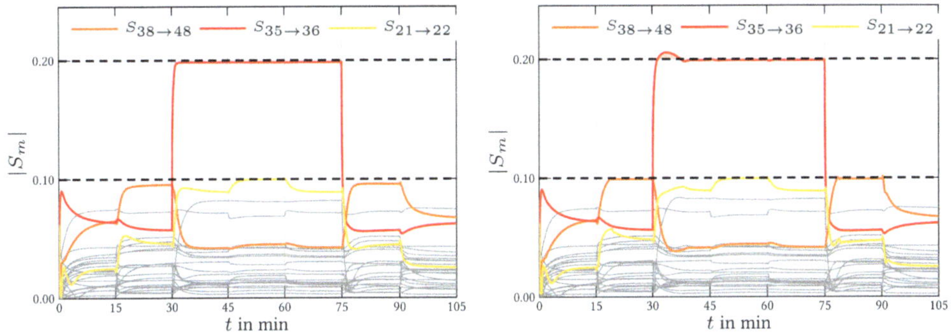

(a) Scenario II-a) **(b)** Scenario II-b)

Figure 6.10: Apparent power flows over inter-cell lines $\hat{\mathcal{E}}_\text{p}$.

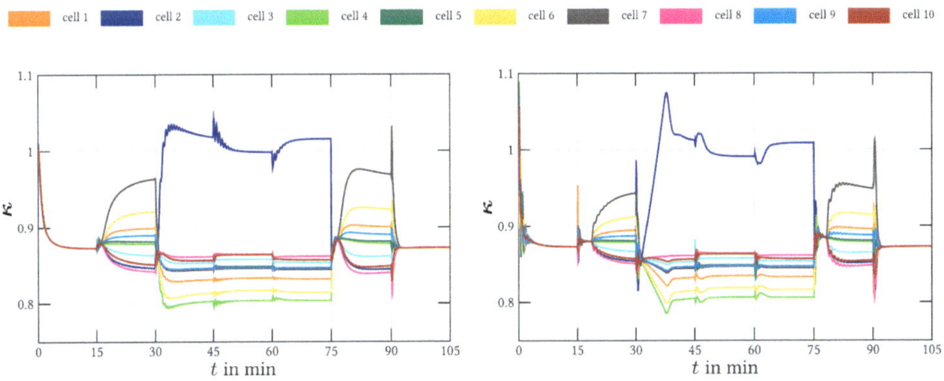

(a) Scenario II-a) **(b)** Scenario II-b)

Figure 6.11: Participation factors $\boldsymbol{\kappa}$.

factors in cells 8 and 10 take the lowest value. After the second jump at t_2, the line between node 35 in cell 2 and node 36 in cell 4 reaches its limit of $S_{35\to36}^{\max} = 0.2$. Subsequently, κ_2 increases significantly, while κ_4 decreases. After the third jump at t_3, also the power line between node 21 in cell 10 and node 22 in cell 8 reaches its limit. At this point κ_2 slightly recedes, while κ_{10} increases. After the occurence of the opposite jumps at t_4 and t_5, the former steady states are recovered, before all participation factors $\boldsymbol{\kappa}$ converge to their initial value 0.87 after the last jump at t_6.

Fig. 6.11a reveals that in Scenario II-a), the individual participation factors are overlaid by a ripple. By contrast, the individual participation factors in Scenario II-b) resulting from the congestion controller with integral action move more slowly and show almost no ripple during the transient process, but exhibit higher transient overshoots (see Fig. 6.11b). Accordingly, the resulting apparent power flows in Fig. 6.10b exhibit temporary constraint violations of a maximum duration of $10\,\text{min}$. Remarkably, the steady-state apparent power flows $S_{38\to48}$, $S_{35\to36}$, and $S_{21\to22}$ of Scenario II-b) are exactly at their specified maximum limits, while in

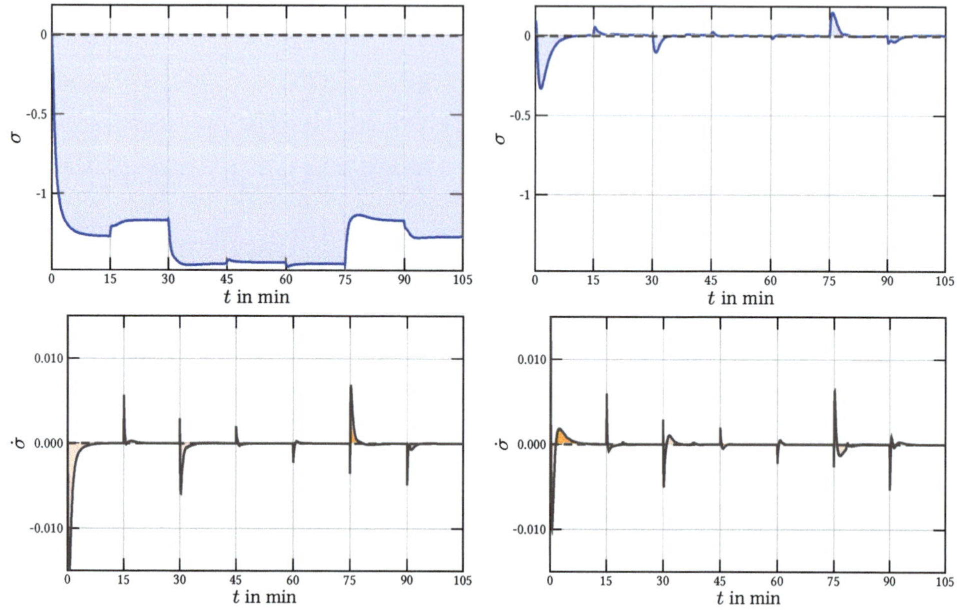

(a) Scenario II-a)

(b) Scenario II-b)

Figure 6.12: BoP containment and restoration.

Scenario II-a) they only approach these respective limits. This can be elucidated by the fact that the barrier functions $\mho_m$ used for the zonal pricing controller for congestion management (4.14) without integral action do not allow for an exact convergence to the threshold $\mathcal{C}_m^{\min}$ (and thus $|S_m| = S_m^{\max}$), while for the congestion controller (4.15) with integral action, the barrier function $\mho_m$ only requires that the *integrated constraint violation* has to be finite.

Balancing of Payments

The trajectories of σ indicating the BoP error are depicted in the upper panels of Fig. 6.12. In Scenario II-a), σ settles to a new steady-state value between -1.18 and -1.43 after each load jump. The resulting bias of σ depends on the magnitude of the applied load jump, although there is no clear relationship between the sign of the load jump and its impact on σ. However, BoP containment is given, which is shown by the bottom panel in Figure 6.12a. Although the associated time derived BoP deviations $\dot{\sigma}$ (see ibid.) never exceed 0.016, their accumulation over a period of several minutes always causes a net deficit of more than 1. Thus, with respect to the sign convention in (4.34), we see that the overall payments of RPs exceed the WPs' overall revenues, thereby resulting in an unwanted accumulation of capital. By contrast, in Scenario II-b), σ recovers the desired steady-state value of $\sigma^\star = 0$ no later than $15\,\text{min}$ after each load jump, thus BoP restoration (and hence of course also BoP containment) is accomplished.

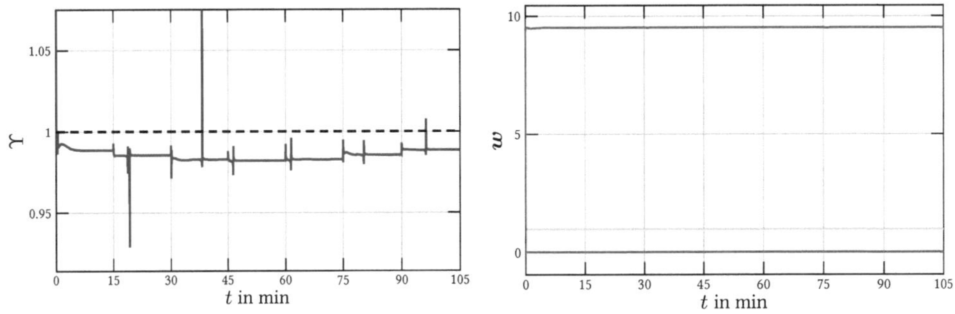

(a) Lagrange multiplier Υ

(b) Weights $\boldsymbol{w}$ of adaptation procedure

Figure 6.13: Learning-based optimal controller for BoP restoration in Scenario II-b).

Fig. 6.13 shows the progression of the Lagrange multiplier $\Upsilon(t)$ of the employed adaptive optimal controller from Chapter 5 as well as the resulting weighting factors $\boldsymbol{w}(t)$ of the adaptation procedure (5.55c). The weighting factors $w_1, \ldots, w_{10}$ associated to the basis functions $\Xi_1 = (\epsilon_1^\kappa)^2$ to $\Xi_{10} = (\epsilon_{10}^\kappa)^2$ converge to the steady-state value $w_1^\star = \cdots = w_{10}^\star = 0.964$ within 5 s after starting the simulation, and w_{11} belonging to $\Xi_{11} = \sigma^2$ converges to $w_{11}^\star = 9.4999$. The weighting factors associated to the remaining basis functions converge to zero. After the adaptation is accomplished, the value of $\boldsymbol{w}(t)$ is no longer affected by subsequent load jump. Yet, at the same time, $\Upsilon(t)$ is slightly affected by each load jump, while always converging to a new steady-state value within 10 s which stays in a corridor between 0.978 and 0.985. These results indicate that the underlying HJB equation (5.24) is approximately (but not exactly) fulfilled and thus the learned, extended CLF $V(\boldsymbol{x}) = H_\sigma(\boldsymbol{x}_\sigma) + \boldsymbol{w}^\star \Xi(\boldsymbol{x}_\sigma)$ is a good approximation of the value function. Occasionally, after $\Upsilon(t)$ has reached a steady state, the graph of $\Upsilon(t)$ reveals steep-flanked spikes of $\Upsilon(t)$ with a magnitude up to 0.12 from time to time. This is caused by a near-zero denominator in (5.55b), leading to a division by (almost) zero.

Zonal Pricing

Fig. 6.14 compares the resulting (cell-specific) wholesale prices $\boldsymbol{\lambda}(t)$ and the (cell-independent) retail price $\Lambda_\mathrm{R}(t)$. Similar to Case Study I, nodal prices $\boldsymbol{\lambda}$ in each cell $k \in \mathcal{Z}$ converge to a cell-specific wholesale price Λ_k within 1 s, such that no differences between nodes can be observed in Fig. 6.14. The retail price increases with increasing total demand at the $\mathcal{L}$ nodes, and vice versa. In line with the trajectories of $\boldsymbol{\kappa}(t)$ in Fig. 6.11b, there are *identical* wholesale prices, if none of the power flows along the boundary lines reaches its limit, and different wholesale prices, if at least one of the power lines reaches its limit (see Fig. 6.10). Interestingly, the retail price is hardly affected in cases where $\boldsymbol{\Lambda} = \mathrm{col}_{k\in\mathcal{Z}}\{\Lambda_k\}$ is after a line flow limit has been reached, such as after 19 min through the congestion of line 38 $\rightarrow$ 48 or after 37 min triggered by line 35 $\rightarrow$ 36. Most of the time, each of the wholesale prices is *below* the retail price. This is caused by the BoP containment, which enforces the RPs to collectively compensate for the momentary nonzero resistive losses in addition to their own

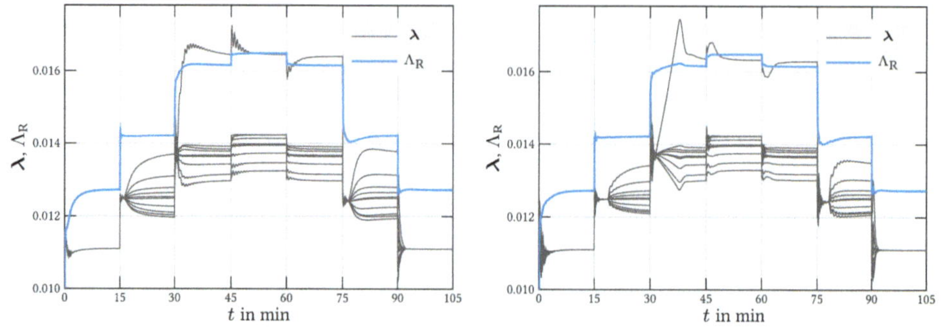

(a) Scenario II-a) (b) Scenario II-b)

Figure 6.14: Comparison between wholesale prices $\boldsymbol{\lambda}$ and retail price Λ_R.

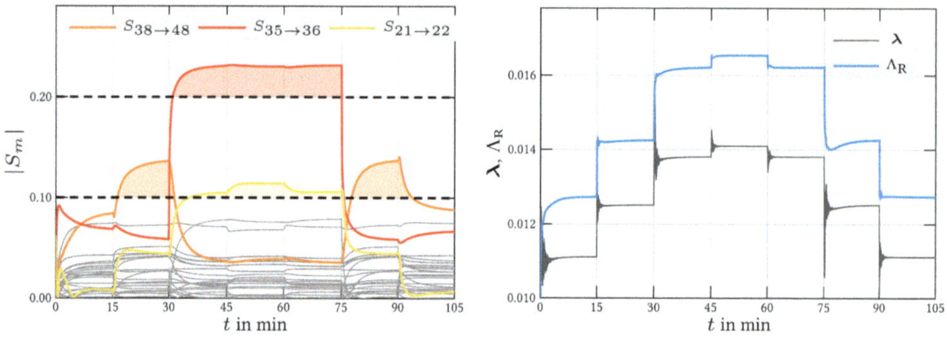

(a) Apparent power flows (b) Wholesale and retail prices

Figure 6.15: Apparent power flows and corresponding wholesale and retail prices in case of relaxed power flow limits for Scenario II-a).

power consumption. However, after the load jumps at t_3 and t_5, the spread between wholesale prices is high enough that wholesale price Λ_2 of cell 2 even exceeds the retail price. This can be explained by the fact that cell 2 is located directly at the *critical line* $35 \rightarrow 36$, thus the zonal pricing generates a heavy increase in κ_2 (and hence Λ_2) in order to combat congestion locally by means of incentivizing increased infeed by cell 2.

For comparison, Fig. 6.15 shows the resulting apparent power flows and price curves in case the congestion controller of Scenario II-b) is deactivated. In this case, there is a significant overload of the lines $38 \rightarrow 48$, $35 \rightarrow 36$, and $21 \rightarrow 22$ (see Fig. 6.15a) by up to 18%. Accordingly, all wholesale prices are equal and always lie below the retail price, as can be seen in Fig. 6.15b.

6.3.3 Summary of Case Study II

The zonal pricing controller used for congestion management in Case Study II causes a demand-oriented, incentive-driven shift of power generation to regions that are "more favorable" from a global grid perspective, whenever the purely merit-order-based network operation (with uniform prices) is no longer feasible. Compared to the alternative mechanism of cost-based redispatch, however, two major improvements are evident here: First of all, congestion is not alleviated "unidimensionally" by simply affecting two participants pairwise. Instead, a convenient and automatic adjustment of *all* cells is executed, as can be seen e.g. in Fig. 6.11 after the jump at t_1. Secondly, the presented congestion controller is able to handle multiple congested lines at the same time, as e.g. demonstrated in Fig. 6.10 after the second and third jump. By contrast, cost-based redispatch strategies quickly become challenging in case of multiple congested lines due to the possibly very large number of combinations of increase/decrease pairs.

Case Study II also shows that the selected pricing strategy has only a minor influence on the retail prices, as can be seen by comparison between Λ_R in Fig. 6.14 and Λ_R in Fig. 6.15b. The reason for this is the appropriate settlement of *regularized* wholesale and retail prices through distributed balancing. In combination with the pricing strategy, it provides for additional competition between WPs, since, loosely speaking, an increasing κ_k in one cell is automatically compensated by a decreasing κ_k in some other cell. From this perspective, the automatic settlement of zonal price differences resulting from congested lines can be interpreted as a *dynamic rearrangement of the (copper plate-based) merit order, triggered by network utility considerations.*

While the presented schemes for frequency and voltage control, congestion management, and BoP containment are fully distributed, the developed adaptive optimal controller for BoP restoration is still operated in a centralized manner, since both the complete system dynamics as well as all inputs have to be known. Especially, if the state vector of the underlying system is very large, this considerably hinders the determination of suitable basis functions. Although the trajectories of $\Upsilon(t)$ and $\boldsymbol{w}(t)$ can be employed to qualitatively evaluate whether or not the value function candidate $V(\boldsymbol{x}, \boldsymbol{w})$ obtained with a *given* set of basis functions is a sufficiently good approximation of the value function, so far, there exists no constructive approach how to reasonably choose the basis functions. Therefore, further research is advisable on how to usefully incorporate existing a priori model knowledge into the choice of basis functions.

6.4 Discussion

In this chapter, the main capabilities of the developed dynamic zonal pricing controller with combined frequency and voltage regulation have been systematically investigated using an IEEE benchmark model. Table 6.4 summarizes the respective control objectives substantiated in each of the simulation scenarios. Case Study I in Subsection 6.2 showcases the benefits of cell-based approaches for frequency and voltage control of large-scale systems. This is in addition to Example 4, which already demonstrates the integrated price-based frequency controller in its main features incorporating the classical primary, secondary, and tertiary control layers. Finally, Case Study II in Subsection 6.3 demonstrates the effectiveness of incentive-based dynamic pricing using participation factors for real-time control of line congestion. This unification of the classical aspects of power system stabilization and incentive-compatible, real-time dispatch represents a new, holistic framework for assigning a concrete, monetary

Table 6.4: Scenario chart for Case Studies I and II based on and in supplement to the literature comparison chart in Table 2.2. For abbreviations not listed, see ibid.

NCs	ED	CM	FC	VC	BoP♠	
	distributed controllers / competitive agents	thermal model / active + reactive power / meshed topology	lossy networks / lossless networks	secondary/tertiary / primary	containment/restoration	
$\mathcal{SIL}$	●●	– – –	●●	– –	–	Example 4
$\mathcal{SIL}$	●●	– – –	●●	●●	–	Case Study I
$\mathcal{SIL}$	●●	●● –	●●	●●	◐	Case Study II-a)
$\mathcal{SIL}$	●●	●●●	●●	●●	●	Case Study II-b)

♠ BoP: ● = BoP restoration, ◐ = BoP containment, – = not provided.

Table 6.5: Comparison of simulation times T_{CPU}.

Scenario	T_{sim}	T_{CPU}
I-a)	1800	278.65
I-b)	1800	102.40
I-c)	1800	67.86
I-d)	1800	70.72
II-a)	6300	73.95
II-b)	6300	389.59

value to *grid support* in addition to a pure remuneration for power provision itself—something that has not been encountered in the literature so far.

Table 6.5 lists the computation times T_{CPU} required for each of the simulated scenarios. It is evident that all computation times are far below the simulated time T_{sim} in each case, which shows the *real-time capability* of the overall control scheme. The individual durations shown in Table 6.5 along with further observations from the accompanying simulation experiments suggest that the simulation time is highly correlated with the number of active constraints. This correlation is plausible taking into account that the `NDSolve` routine, which is specialized for solving differential-algebraic equations, is a *variable step* solver which tends to a reduction of the step sizes (and consequently an increased computation time t_{CPU}) whenever the simulated system becomes more stressed due to higher fluctuations of the state variables close to the active constraints. However, since the proposed control framework is distributed, in real-time applications, separate controllers with significantly lower computational complexity along with fixed-step solvers will be operated in parallel at every single node, thereby resulting in significantly lower computation times.

7 Conclusion and Outlook

7.1 Conclusion

Modern power systems exhibit a remarkable paradox: On one hand, increasing RES-based intermittent power generation requires close cooperation among all resources for efficiency and synergy reasons, in order to incorporate a “physical awareness” among all network participants about what actions are deemed most ideal from a global network perspective. On the other hand, the unbundling principle along with antitrust considerations impose, for good reason, that the exchange of information both between system operators and competitive prosumers, as well as between prosumers themselves, should be kept to a minimum.

The dissertation at hand provides a contribution towards bridging this gap by proposing an incentive-compatible, market-integrated control scheme with cell-based price consensus. Relying on and complying with the classical division into regulated system operators and profit-driven prosumers, the proposed brownfield framework allows each self-serving WP to autonomously make decisions in order to maximize its profit. In contrast to a copper plate-like market which does not consider the location of buyers and sellers [RS19], in the proposed framework any information about the temporal and spatial network state is indicated by the node distinct real-time price set by the local CCs which act as the network’s neutral market facilitators. Inefficient allocations and normative add-on mechanisms such as cost-based redispatch or the BRP role, which are both critically discussed in Subsection 2.1.2, can thus be omitted.

Since all power system stability issues and physical constraints are already settled at market clearing, there is no further need of a capacity-based BM as compared to the baseline scenario. Thus, potential strategic market behavior as discussed in Subsection 2.1.2 is prevented. Consequently, there are no after-market costs to charge to the RPs[74]. In addition to the proposed real-time mechanism, the WEM may still be in place, which all network participants may use for their own risk hedging or for long-term agreements (cf. Fig. 3.7). If the proposed framework is adopted in practice, the only change for market participants is the way in which their own real-time prices for electricity generation or consumption have to be calculated.

For each of the individual network participants and for each of the technical and economic layers in Fig. 4.7, this dissertation has derived dedicated control schemes, with a special focus on optimization-based controller design and distributed controller communication. The major contributions of this dissertation and their relevance to the research gaps addressed in Section 2.3 are depicted in Fig. 7.1 and briefly summarized below.

[74] Remaining grid tariffs, if any, now only apply for construction and maintenance of the grid infrastructure itself.

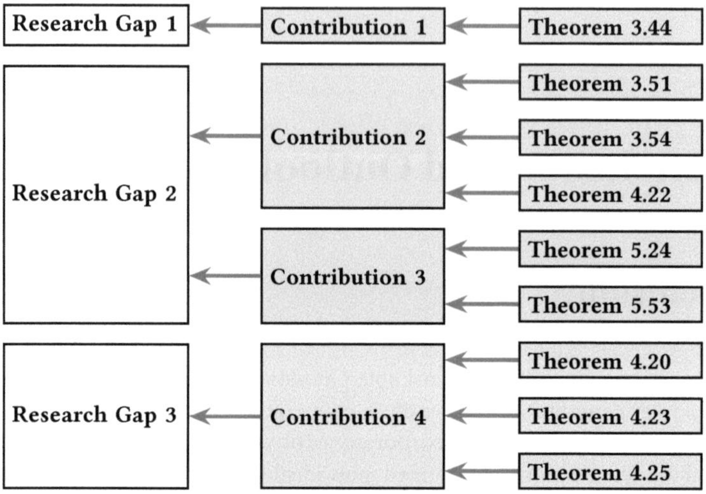

Figure 7.1: Overview of main contributions.

In Chapter 3, we have worked towards a market-based integration of power system stability by developing a distributed, optimization-based frequency and voltage controller such that the actions of the local WPs are stimulated solely by their individual nodal price. By thoroughly accounting for network parameters in the CCs' calculations, the price calculated by CCs also covers all accompanying line losses, thereby closing Research Gap 1. A sufficient condition for stability of the closed-loop system has been provided in Theorem 3.44.

The rigorous pursuit of an economically efficient interplay between stakeholders has constituted a major focus of all endeavors. By introducing a novel, vector-valued zonal participation factor $\boldsymbol{\kappa}$ along with a distributed price consensus protocol, the equilibrium set of the resulting closed-loop system is restricted to the set of Pareto efficient allocations, which has been elaborated in Theorem 3.51 and Theorem 3.54. The dynamic *regularization* of all participation factors by means of distributed balancing, which has been discussed in Theorem 4.22, thereby forces only those solutions that provide BoP containment, turning the entire power infrastructure into a *monetarily self-contained system* without any accumulation of capital.

The methodological foundation for efficiently regulating the remaining steady-state deviations of aggregate net payments has been laid by introducing a new adaptive framework for optimal control of (generic) PHSs in Theorem 5.24. Since the class of PHSs has proven to be a universally applicable modeling tool for a variety of multi-physics or techno-economic systems, possible applications of the developed method reach far beyond the particular area of power system control applications presented earlier in this work. In Theorem 5.53, the developed methodology has been further extended to the case of N-player noncooperative differential games.

Thanks to distributed balancing via the time-dependent regularization factor $\boldsymbol{a}$, all of the remaining $|\mathcal{Z}| - 1$ degrees of freedom when choosing $\boldsymbol{\kappa}$ could be exploited to implement higher-level goals. Namely, control laws with and without integral action have been proposed in the context of this dissertation on how to usefully impose "incentive-signaling" spatial differences among zonal prices, while also respecting the obligation of treating all RPs equally (cf. Claim 4.5). Theoretical conclusions about the resulting closed-loop system have been stated in Theorems 4.20, 4.23 and 4.25, and the practicality of the overall control framework in automatic congestion control has been demonstrated in Case Study II of Chapter 6, finally closing Research Gap 3.

Since all individual controllers are systematically derived from optimization problems, there are no remaining degrees of freedom in the controller design except for the (inverse) controller gains $\tau_{(\cdot)}$, which, however, do not affect the value of the closed-loop equilibrium. Reasonably, the controller gains can be used to provide the necessary timescale separation between the individual control layers. The overall controller thus features low efforts for parameter tuning.

By automatic control of the participation factors $\boldsymbol{\kappa}$ based on continuous measurements of the state of the network, this work has given rise to allocating marginal production in specific regions of the power system in real time. In contrast to a pure merit order system, the cell-specific multiplication by $\boldsymbol{\kappa}$ provides a real-time incentive mechanism that encourages profit-driven network participants with power sources and flexible loads to strategically shift their capacities to regions with higher average prices and thus to enhance the reliability of the system by inducing an additional *geographical awareness* among WPs.

As discussed in Subsection 2.1.2, all disruptions caused by the increasing RES infeed have so far been charged to the end customer in an intransparent manner via ever-increasing grid tariffs. Through the concept of spatially differentiated prices, which cluster around the (virtual) uniform WP price Λ^0, the presented work marks a *paradigm shift towards a "polluter-pays" principle.* Exemplarily shown in Case Study II, where the social optimum gained by uniform pricing would fail to alleviate congestion at market clearing, WPs in unfavorable locations are subject to significantly lower prices through the dynamic pricing mechanism, which in turn pushes them out of the market in the long run. This creates an *ongoing market consolidation* that is favorable from an overall perspective, along with preferential investment in regions that are most desirable by virtue of the global pricing policy.

Remarkably, the proposed control framework does not require any internal model of the WPs involved (apart from the minor assumption of rationally chosen cost functions). This not only protects the privacy of the stakeholders involved, but also enables the integration of an arbitrary number of small-scale flexible prosumers, fully in line with the future vision of power networks as transactive *energy internet* (EI) platforms with minimum entry barriers [MKR+20, PCK+21].

7.2 Outlook

This work advocates for a careful separation between *sovereignty over the grid* (held by CCs) and *sovereignty over electricity generation* (held by WPs) to maximize fairness without restricting the benefits of a free market. Nonetheless, there will always be the need for a certain normative framework set by policy makers, e.g. rules that ensure compliance with

Assumption 3.17 or the resolution of overriding long-term goals, in order to keep the results of the market mechanism within the bounds of what is socially desirable [NBM+20].

A cell-based infrastructure *without* cutting the physical network has proven to be advantageous compared to the MG concept, as it unites the merits of informational and operational unbundling with those of non-discriminatory power flows. However, the question of most effective cell size and appropriate definition of cell boundaries, which is still open and controversially discussed in academia, is not explicitly addressed in the context of this dissertation. On one hand, as it becomes apparent from the application to congestion control, a finer cellularization, especially along lines that may be vulnerable in terms of congestion, basically contributes to a more elaborate influence of specific line flows. On the other hand, a too fine cellularization counteracts the "no market power" imperative (cf. Assumption 3.17), as it opens up unwanted exploitation opportunities for WPs located at critical points in the network to affect prices unilaterally, leading to reduced competition in the overall network. Hence, an "optimal" cell partitioning must always trade-off between both requirements [Sto97]. While in this dissertation, the determination of appropriate price zones is done manually, based on a priori knowledge and in a fit-and-forget manner, recent lines of research advocate *dynamic* cell partitioning that, in contrast to static approaches, adaptively adjusts to the instantaneous needs of the network [XZK+16, p. 135ff.; Wan15, p. 97]. In [HDA21], a first practical algorithm is presented where the dynamic formation of *local energy markets* is adjusted on a daily basis to gather and reward the flexibility of end users.

The explicit control methods derived in this dissertation can all be applied regardless of the eventual (static or dynamic) cell partitioning and therefore constitute a practical and easily implementable toolbox for any superordinate cellularization and pricing strategy to conveniently realize the intended higher-level goals.

The explicit concepts for real-time pricing elaborated in this dissertation open up a broad field for extensions and amendments. For instance, by adequately extending κ'' (see Subsection 4.3.3), additional summands could not only incorporate generic long-term goals, but also account for *network effects anticipated for the future* right before their actual occurrence. This could contribute to an internalization of follow-up costs as current costs and therefore pave the way towards a *proactive energy infrastructure* [MS21]. In particular, the distributed balancing mechanism from Subsection 4.4.1, if active, always ensures that any of these additional price components which incorporate future tendencies are already fairly allocated among consumers as of today.

Especially for large-scale power infrastructures, aiming for a distributed controller implementation with neighbor-to-neighbor communication, as pursued in this work, is advantageous both in terms of computational complexity and cyber security. However, compared to the other control schemes presented in this dissertation, the novel adaptive methodology for optimal control of single- and multi-player PHSs introduced in Chapter 5 still requires a complete model of the physical plant system (along with perfect measurement information on the other players' control actions in the multi-player case) and therefore lacks a distributed implementation. Paving the way for the application to large-scale systems, future research should thus focus on optimal controllers and noncooperative differential games for PHSs with *limited information sharing* and non-exact model knowledge. In addition, further emphasis could be put on more sophisticated adaptation strategies for the individual weighting factors to increase the speed of convergence while still guaranteeing asymptotic stability.

The major stumbling block towards a *comprehensive stability analysis of the overall system* dynamics is the general absence of Lyapunov-based a priori statements in case of lossy power networks (see also the discussion in Subsection 3.4.4 on ongoing research efforts). For this reason, the existing literature on PHS-based control of AC power systems is almost without exception confined to the lossless case. In the context of this work, a sufficient condition for stability has been presented by Theorem 4.25, which, however, contains an impractical criterion that requires a priori knowledge of the numerical steady-state value. This in turn requires the solution of an intractable (and possibly high-dimensional) nonlinear system of power flow equations, which in general is only given through approximation (cf. [Tay15]). A trend in recent research on power system stability and control is therefore towards the design of versatile controllers that preserve *equilibrium-independent passivity* of the overall system [SNM+21]. Future advances in this field may allow to derive controllers whose stabilizing properties do not depend upon a condition that is difficult to prove in advance—yet at the price of a conservative controller design with limited performance and/or the abandonment of potential degrees of freedom.

Another requirement to be critically investigated by further research is the dynamic security requirement of "sufficient timescale separation" (cf. Section 4.5), which was necessary in the context of this work, but is generally difficult to certify. For this purpose, an even more rigorous *first-principles modeling* e.g. by means of a thorough port-Hamiltonian modeling of the physical plant system and all controller parts bears great potential towards a more native understanding of the (generalized) power flows of the individual subsystems along with their particular impact on overall system stability. Such a generalized stability analysis could eventually obviate the need for any artificial security assumptions regarding timescale separation.

Yet the goals of the energy transition can only be achieved if all parties are enabled in equal measure to be seen as part of the solution. Due to its variability over time, the retail price Λ_R already introduced as part of this dissertation provides a thriving linkage point for research and innovation on demand-side oriented participation. Suitable price incentives and low-entry optimization tools for personal use might thus also encourage RPs to play a much more active role in future electricity markets. Potential approaches currently being piloted range from smart micro-storages at household level [MGR+18], widespread installation of every household with a smart meter infrastructure [BM17, CGM19, YLGW19], and app-based optimal scheduling to exploit *smart home* flexibilities [ATM+20], to the deployment of completely decentralized EI platforms for automated settlement of *smart contracts* among micro-scale energy resources, e.g. on the Ethereum blockchain [ACRK19, BAA+21, SKSJ21]. However, two factors are decisive to the success of such concepts: First, there must be a political debate about how much flexibility is tolerable for the individual and how much individual inflexibility is tolerable for society. Secondly, any innovation that enables the participation of RPs must be made accessible to everybody with lowest possible entry barriers. If unbureaucratic ways to involve end users can be enabled on a large scale and kept accessible for everyone, there is huge potential in turning the narrative of energy transition from a bare necessity to a collaborative societal project where everyone willing can participate and benefit.

A Mathematical Supplements

A.1 Notation

Throughout this dissertation, vectors and matrices are written in boldface. All vectors are column vectors $\boldsymbol{a} = \text{col}_i\{a_i\} = \text{col}\{a_1, a_2, \ldots\} = (a_1, a_2, \ldots)^\mathsf{T}$ with elements a_i, $i = 1, 2, \ldots$. The notation $\boldsymbol{A} = \text{diag}\{\boldsymbol{a}\} = \text{diag}_i\{a_i\}$ means that $\boldsymbol{A}$ is a diagonal matrix with the entries of vector $\boldsymbol{a}$ on its diagonal. The all-zeros and all-ones vectors of dimension n are denoted by $\mathbb{0}_n$ and $\mathbb{1}_n$, respectively. The $(n \times n)$-identity matrix is denoted by $\boldsymbol{I}_n$. The unit vector of the k-th axis of a Cartesian coordinate system is denoted by $\mathbf{e}_k$. For vectors $\boldsymbol{a}, \boldsymbol{b}$ of the same size, $\boldsymbol{a} \geq \boldsymbol{b}$ denotes that each component in $\boldsymbol{a}$ is greater than or equal to the corresponding component in $\boldsymbol{b}$. Positive semidefinite and positive definite matrices or functions are denoted by $\succcurlyeq 0$ and $\succ 0$, respectively. Moreover, we write $V(\boldsymbol{x}) \succ 0$ if $V(\boldsymbol{x} \neq \mathbb{0}) > 0$ and $V(\mathbb{0}) = 0$, and $V(\boldsymbol{x}) \succcurlyeq 0$ if $V(\boldsymbol{x} \neq \mathbb{0}) \geq 0$ and $V(\mathbb{0}) = 0$.

Optimizers or equilibria are marked with a star $\square^\star$ and shifted values with respect to the optimizer or equilibrium are marked with a tilde $\widetilde{\square}$, i.e. $\widetilde{\boldsymbol{x}}(t) = \boldsymbol{x}(t) - \boldsymbol{x}^\star$. Upper and lower bounds of a variable are denoted by $\overline{\square}$ and $\underline{\square}$, respectively. Phasor variables are marked with an arrow $\vec{\square}$ and the complex conjugate of a variable is marked with an asterisk $\square^*$.

To allow distinction from the Hamiltonian H of the input-state-output PHS (cf. Definition A.13), the Hamiltonian function of the optimization problem (also called *Pontryagin function*) is denoted by $\mathscr{H}$. With the set of players $\mathbb{P}$, the vector $\boldsymbol{u}_{\neg i}$ refers to the vector of all inputs $j \in \mathbb{P}\backslash\{i\}$. Where appropriate, we also the subscript index $\neg i$ to denote *all elements from a given index set except* i, if the index set becomes clear from context. For clarity of presentation, the time dependence (t) of the variables is not explicitly mentioned, unless it is essential for transparency of the statements. If no dimension is specified for a certain vector or matrix, it can be derived from context.

For scalars $a, b \in \mathbb{R}$, we define

$$\langle a \rangle_b^+ = \begin{cases} a, & a > 0 \text{ or } b > 0, \\ 0, & \text{otherwise}, \end{cases} \tag{A.1}$$

and for vectors $\boldsymbol{a}, \boldsymbol{b} \in \mathbb{R}^n$, (A.1) is applied component-wise, i.e. $\langle \boldsymbol{a} \rangle_{\boldsymbol{b}}^+ = \text{col}\{\langle a_1 \rangle_{b_1}^+, \ldots, \langle a_n \rangle_{b_n}^+\}$. For a scalar $a \in \mathbb{R}$, we define the saturation operator

$$\langle a \rangle_{-1}^1 = \begin{cases} 1, & a > 1, \\ a, & a \in [-1, 1], \\ -1, & a < -1, \end{cases} \tag{A.2}$$

and for vectors $\boldsymbol{a} \in \mathbb{R}^n$, (A.2) is applied component wise, i.e. $\langle \boldsymbol{a} \rangle_{-1}^{1} = \mathrm{col}(\langle a_1 \rangle_{-1}^{1}, \ldots, \langle a_n \rangle_{-1}^{1})$.

The operations absolute value $|\cdot|$, natural logarithm $\ln(\cdot)$, and exponential function $\exp(\cdot)$ when applied to vectors $\boldsymbol{a} \in \mathbb{R}^n$ are applied component wise, e.g. $|\boldsymbol{a}| = \mathrm{col}\{|a_1|, \ldots, |a_n|\}$ etc. The Hadamard product of two vectors is denoted by $\circ$, i.e. $\boldsymbol{a} \circ \boldsymbol{b} = \mathrm{col}\{a_1 b_1, \ldots, a_n b_n\}$ for each $\boldsymbol{a}, \boldsymbol{b} \in \mathbb{R}^n$.

The gradient of a function $f\colon \mathbb{R}^n \to \mathbb{R}$ with respect to $\boldsymbol{x}$ is denoted by $\nabla f(\boldsymbol{x})$ and is the column vector of all partial derivatives of $f(\boldsymbol{x})$, i.e. $\nabla f(\boldsymbol{x}) = \mathrm{col}\{\frac{\partial f(\boldsymbol{x})}{\partial x_1}, \ldots, \frac{\partial f(\boldsymbol{x})}{\partial x_n}\}$. The Jacobian of $\boldsymbol{\Phi} : \mathbb{R}^n \to \mathbb{R}^r$ with $\boldsymbol{x} \mapsto \boldsymbol{\Phi}(\boldsymbol{x})$ is defined as $\frac{\partial \boldsymbol{\Phi}}{\partial \boldsymbol{x}} = (\nabla \Phi_1(\boldsymbol{x}), \ldots, \nabla \Phi_r(\boldsymbol{x}))^\mathsf{T}$.

A.2 Basic Concepts of Game Theory

The conceptual framework for the investigation of decision-making situations among multiple network participants with different, possibly conflicting goals is provided by game theory. In recent decades, game theory has been established as a lingua franca for almost all research on competitive energy markets [ANHMIT20, p. 86; NBM+20, p. 2], since it not only offers modeling tools for conflicting situations, but also explicit instructions for strategic actions of the participants as well as a normative assessment of desired and undesired operating points from participants' viewpoints. After first conceptual studies in the 1970s [MD77, BHK78], game-theoretic concepts modeling the interaction of competitive network participants have been developed at every timescale, from optimal day-ahead scheduling on an hourly basis [AOS+13b] to real-time strategies using differential games[75] [LLL+19]. Moreover, a wide variety of use cases and network participants have been considered, from optimal generator dispatch, demand-side management and demand response[76] [MWY+19], to optimal balancing of interconnected distribution systems and optimal utilization of shared resources [MMM20]. An early review on the applications of game theory for power systems is conducted in [SHPB12]. For more recent survey papers, see [CY19, ANHMIT20, NBM+20].

In the context of game theory, each network participant is modeled as element $i \in \mathbb{P}$ of a set of players. It aims at maximizing its own profit P_i by choosing a strategy $\boldsymbol{a}_i$.

Definition A.1 (Game [BO99; Bau16, p. 4f.]). *A* game *is a tuple* $(\mathbb{P}, (\mathbb{A}_i)_{i \in \mathbb{P}}, (\mathsf{P}_i)_{i \in \mathbb{P}})$, *where* $\mathbb{P} = \{1, \ldots, N\}$ *is the set of players,* $\mathbb{A}_i$ *is the set of actions of player* $i \in \mathbb{P}$, *and*

$$\mathsf{P}_i : \mathbb{A}_1 \times \cdots \times \mathbb{A}_N \to \mathbb{R} \tag{A.3}$$

is the profit function of player $i \in \mathbb{P}$.

For convenience of notation, let $\mathbb{A} = \mathbb{A}_1 \times \cdots \times \mathbb{A}_N$ denote the set of *action profiles*, where an action profile $\boldsymbol{a} \in \mathbb{A}$ is the collection of all actions $\boldsymbol{a}_i$ of the players. If P_i depends not

75 (cf. Definition A.2 in Appendix A.2)

76 The term "demand-side management" encompasses all (long- and short-term) control actions at consumer side, while "demand response" specifically refers to the provision of short-term incentives for price-elastic consumers.

only on the actions $\boldsymbol{a}_i \in \mathbb{A}_i$ but also on the state $\boldsymbol{x}(t)$ of a dynamical system, this leads to the notion of a differential game:

Definition A.2 (Differential Game [SH69; BO99; Bau16, p. 90]). *A* differential game *is a tuple* $(\mathbb{P}, \mathbb{S}, (\mathbb{A}_i)_{i\in\mathbb{P}}, (\mathrm{P}_i)_{i\in\mathbb{P}})$*, where* $\mathbb{P} = \{1, \ldots, N\}$ *is the set of players,* $\mathbb{S}$ *is a controlled dynamical system*

$$\dot{\boldsymbol{x}}(t) = \boldsymbol{f}(\boldsymbol{x}(t), \boldsymbol{u}(t)), \tag{A.4a}$$
$$\boldsymbol{x}(0) = \boldsymbol{x}_0, \tag{A.4b}$$

where $\boldsymbol{f} : \mathbb{R}^n \times \mathbb{R}^{p_1} \times \cdots \times \mathbb{R}^{p_N}$ *is Lipschitz continuous with state* $\boldsymbol{x} \in \mathcal{X} \subseteq \mathbb{R}^n$ *and inputs* $\boldsymbol{u}_i \in \mathcal{U}_i \subseteq \mathbb{R}^{p_i}$. $\mathbb{A}_i$ *is the set of actions of player* $i \in \mathbb{P}$ *with* $\mathbb{A}_i = \mathcal{U}_i$*, and*

$$\mathrm{P}_i : \mathbb{A} \times \mathcal{X} \to \mathbb{R} \tag{A.5}$$

is the profit function of player $i \in \mathbb{P}$.

Remark A.3. *Independent to the notion of differential games, the term* dynamic game *encompasses those games where the players make their decisions* $\mathbb{A}_i$ sequentially, *in contrast to* static *games, where players act simultaneously.* ◇

A common assumption in classical game theory is that all players are subject to the following basic behavioral pattern:

Assumption A.4 (Rationality and Common Knowledge of Rationality [Aum76, AB95, Bic04]). *Each player acts* rationally, *i.e. strives to maximize its own profit* P_i. *Moreover, each player takes into account that the other players act rationally as well.*

An important implication of Assumption A.4 is that especially those strategies are preferable, where no player can achieve a higher profit without at least one other player achieving a lower profit. Such strategies are said to be *Pareto efficient* or *Pareto optimal.* A formal definition is given below:

Definition A.5 (Pareto Efficiency [Bau16]). *The action profile* $\boldsymbol{a}^\star = (\boldsymbol{a}_1^\star, \ldots, \boldsymbol{a}_i^\star, \ldots, \boldsymbol{a}_N^\star)^t ransp$ *is said to be* Pareto efficient *if there exists no other* $\boldsymbol{a} \in \mathbb{A}$ *such that*

$$\exists i \in \mathbb{P} : \qquad \mathrm{P}_i(\boldsymbol{a}) > \mathrm{P}_i(\boldsymbol{a}^\star) \qquad \wedge \qquad \mathrm{P}_{\neg i}(\boldsymbol{a}) \geq \mathrm{P}_{\neg i}(\boldsymbol{a}^\star). \tag{A.6}$$

If, instead, there exists an $\boldsymbol{a} \in \mathbb{A}$ *fulfilling* (A.6)*, then* $\boldsymbol{a}^\star$ *is said to be a* Pareto inefficient *or* Pareto dominated *strategy.*

For existence theorems on Pareto efficient solutions, see e.g. [Meh85].

Another prominent solution strategy arising from Assumption A.4 is to find *Nash equilibria* among all possible action profiles $\boldsymbol{a} \in \mathbb{A}$:

Definition A.6 (Nash Equilibrium [Bau16, p. 8]). *The action profile* $\boldsymbol{a}^\star = (\boldsymbol{a}_1^\star, \dots, \boldsymbol{a}_N^\star)$ *is said to be a* Nash equilibrium, *if none of the players can make a higher profit by unilaterally deviating from its own action profile, i.e., if*

$$\forall\, \boldsymbol{a}_i \in \mathbb{A}_i : \qquad \mathrm{P}_i(\boldsymbol{a}_i, \boldsymbol{a}_{\neg i}) \geq \mathrm{P}_i(\boldsymbol{a}_i^\star, \boldsymbol{a}_{\neg i}^\star) \tag{A.7}$$

holds for all $i \in \mathbb{P}$.

Depending on the *information structure* of the given differential game, we distinguish between *open-loop strategies*, where each player only has knowledge of the initial state $\boldsymbol{x}_0$ of (A.4), and *closed-loop strategies*, where each player is able to observe the actual state $\boldsymbol{x}(t)$ of (A.4).

If all profit functions P_i are Lagrange-type, implicit solutions exists for both open-loop and closed-loop Nash strategies.

Definition A.7 (Lagrange-Type Performance Index). *A profit function is said to be a* Lagrange-type *performance index, if it has the following form.*

$$\mathrm{P}_i(\boldsymbol{x}) = -\int_0^{t_\mathrm{f}} \mathrm{L}_i(\boldsymbol{x}(t), \boldsymbol{u}_1(t), \dots, \boldsymbol{u}_N(t))\,\mathrm{d}t. \tag{A.8}$$

In the open-loop case, the Nash strategy of each player is determined by solving a boundary value problem which follows from *Pontryagin's maximum principle* [Bre11]. The resulting open-loop control laws depends only on time, thus a disturbance leads most likely to suboptimal operation.

In the context of feedback control, where the actors base their strategic decisions on continuous measurements of the state vector and have limited information about other actors' objectives, emphasis is placed on finding closed-loop Nash strategies (also called feedback Nash strategies). Here, the players are able to react to disturbances by immediately noticing undesired deviations from the desired state. An implicit solution is given by a set of coupled Hamilton-Jacobi-Bellman (HJB) equations:

Lemma A.8 (Sufficient Condition for Feedback Nash Equilibrium [Bre11, HKZ12]). *For the differential game from Definition A.2 with fixed duration* $[0, t_\mathrm{f}]$, *the action profile* $\boldsymbol{a}^\star = (\boldsymbol{u}_i^\star(t), \dots, \boldsymbol{u}_N^\star(t))^\mathsf{T}$ *provides a feedback Nash equilibrium, if there exist continuously differentiable* value functions $V_i^\star : [0, T] \times \mathcal{X} \to \mathbb{R}$ *for each* $i \in \mathbb{P}$, *such that the HJB (partial differential) equation*

$$-\frac{\partial V_i^\star(t, \boldsymbol{x})}{\partial t} = \max_{\boldsymbol{u}_i \in \mathcal{U}_i} \left\{ \left(\frac{\partial V_i^\star(t, \boldsymbol{x})}{\partial \boldsymbol{x}} \right)^\mathsf{T} \boldsymbol{f}(t, \boldsymbol{x}, \boldsymbol{u}_i, \boldsymbol{u}_{\neg i}) - \mathrm{L}_i(t, \boldsymbol{x}, \boldsymbol{u}_i, \boldsymbol{u}_{\neg i}^\star) \right\} \tag{A.9}$$

and the terminal condition

$$V_i^{\star}(t_\mathrm{f}, \boldsymbol{x}) = 0 \tag{A.10}$$

are satisfied for all $(t, \boldsymbol{x}) \in [0, T] \times \mathcal{X}$.

Note that for $N = 1$, the above differential game with Lagrange-type performance index is equivalent to the classical optimal control problem. For existence theorems for open-loop and feedback Nash equilibria, see [Bre11, p. 360ff.].

A.3 Algebraic Graph Theory

Definition A.9 (Graph, Directed Graph, Connected Graph, Complete Graph [Bap14, p. 10; HKK⁺05, p. 9; BM76, p. 4]). *A* graph $\mathcal{G}$ *is a pair* $(\mathcal{V}, \mathcal{E})$*, where* $\mathcal{V}$ *is the set of nodes and* $\mathcal{E}$ *is the set of edges. The graph is said to be* directed *if the edges* $e \in \mathcal{E}$ *are ordered pairs* $e = (v_1, v_2)$*, where* $v_1 \in \mathcal{V}$ *represents the tail of* e *and* $v_2 \in \mathcal{V}$ *represents the head of* e*. The graph is said to be* (weakly) connected*, if there exists an undirected path connecting every pair of nodes in* $\mathcal{V}$*. The graph is said to be* complete*, if there exists an edge between each pair of nodes.*

Definition A.10 (Incidence Matrix [Bap14, p. 13]). *The* incidence matrix $\boldsymbol{D} \in \mathbb{R}^{n \times m}$ *of the graph* $\mathcal{G}$ *is defined by*

$$[\boldsymbol{D}]_{i,j} = \begin{cases} 1, & i \text{ is the head of } e_j, \\ -1, & i \text{ is the tail of } e_j, \\ 0, & \text{otherwise.} \end{cases} \tag{A.11}$$

If $\mathcal{G}$ *is a weighted graph, then the non-zero entries in* (A.11) *are multiplied by the weights* $w_{ij} \in \mathbb{R}$ *of the respective edge* $(i, j) \in \mathcal{E}$.

Definition A.11 (Laplacian Matrix [Bap14, p. 49]). *The* Laplacian matrix $\boldsymbol{\mathcal{A}} \in \mathbb{R}^{n \times n}$ *of the graph* $\mathcal{G}$ *is defined by*

$$\boldsymbol{\mathcal{A}} = \boldsymbol{D}\boldsymbol{D}^\top. \tag{A.12}$$

From Definition A.11 it follows that $[\boldsymbol{\mathcal{A}}]_{i,j} = [\boldsymbol{\mathcal{A}}]_{j,i} = -1$, if nodes i and j are incident, and $[\boldsymbol{\mathcal{A}}]_{i,i}$ being equal to the *degree* of node $i \in \mathcal{V}$ (i.e. the total number of edges connected to node i). The incidence and Laplacian matrix of a connected graph have the following properties:

Lemma A.12. *If a graph* $\mathscr{G} = (\mathcal{V}, \mathcal{E})$ *with* $|\mathcal{V}| = n$ *is connected, then the following statements hold:*

(1) $\operatorname{rank}(\boldsymbol{D}) = \operatorname{rank}(\boldsymbol{\mathcal{A}}) = n - 1$,

(2) the all-ones vector $\mathbb{1}$ *is a basis for the kernel of both* $\boldsymbol{D}^\top$ *and* $\boldsymbol{\mathcal{A}}$,

(3) the Laplacian matrix $\boldsymbol{\mathcal{A}}$ *is positive semidefinite and has exactly one zero eigenvector. The remaining eigenvectors of* $\boldsymbol{\mathcal{A}}$ *are strictly positive.*

Proof. (see [Bap14, pp. 14, 50f.]) □

A.4 Port-Hamiltonian Systems

In recent years, systematic modeling of dynamical multi-physics systems[77] in port-Hamiltonian form has become increasingly popular in a wide range of applications such as acoustics [FH16], aerospace [ACRMA19], robotics [GSC17, MMS07, MMS09], power electronics [BDST17, CGB+19], material science [LSM+20], or energy systems [FZO+13, SOA+14, CT14, HMM+20]. Among the broad class of implicit or explicit PHS representations (see e.g. [CGHM13] for a comprehensive overview), the state-space representation via *input-state-output PHSs*, in which the system dynamics are affected by the gradient field of a smooth, positive definite *Hamiltonian* $H(\boldsymbol{x})$ representing the overall energy stored in the system, is very familiar for modeling and control purposes, as it allows direct access to the inputs $\boldsymbol{u} \in \mathbb{R}^p$ and outputs $\boldsymbol{y} \in \mathbb{R}^p$ of the system.

Definition A.13 (Input-State-Output PHS [van17, Definition 6.1.1]). *Let* $\boldsymbol{x} \in \mathbb{R}^n$ *be a state vector. The state-space system*

$$\dot{\boldsymbol{x}} = (\boldsymbol{J}(\boldsymbol{x}) - \boldsymbol{R}(\boldsymbol{x}))\frac{\partial H(\boldsymbol{x})}{\partial \boldsymbol{x}} + \boldsymbol{G}(\boldsymbol{x})\boldsymbol{u}, \tag{A.13a}$$

$$\boldsymbol{y} = \boldsymbol{G}^\top(\boldsymbol{x})\frac{\partial H(\boldsymbol{x})}{\partial \boldsymbol{x}} \tag{A.13b}$$

is said to be an input-state-output PHS (without feedthrough), *if* $\boldsymbol{J}(\boldsymbol{x}) = -\boldsymbol{J}^\top(\boldsymbol{x})$ *and* $\boldsymbol{R}(\boldsymbol{x}) = \boldsymbol{R}^\top(\boldsymbol{x}) \succcurlyeq 0$. *The gradient* $\boldsymbol{z} := \nabla H(\boldsymbol{x})$ *is said to be the vector of* co-states *or* generalized efforts *of* (A.13).

The above input-state-output representation clearly visualizes a separation between the interconnection topology of the system and the constitutive relations of its components, which is an inherent property of PHSs [DBMS09]. The input-state-output representation of

[77] The PHS representation can be applied whenever energy plays a major role in the dynamical system to be modeled. Indeed, this modeling paradigm can also be abstracted to non-physical systems, e.g. by conceiving the momentary value of the objective function of gradient flow (cf. Subsection 3.4.1) as generalized *cyber-energy* stored in the system [JOv17, p. 109].

PHSs is thus particularly suitable for modeling of large-scale power systems with a repetitive structure. A generalization of input-state-output PHSs is given by the following definition:

Definition A.14 (Input-State-Output PHS with Nonlinear Resistive Structure [van17, Definition 6.1.4]). *Let $\boldsymbol{x} \in \mathbb{R}^n$ be a state vector. The state-space system*

$$\dot{\boldsymbol{x}} = \boldsymbol{J}(\boldsymbol{x})\frac{\partial H(\boldsymbol{x})}{\partial \boldsymbol{x}} - \boldsymbol{\mathcal{R}}(\boldsymbol{x}) + \boldsymbol{G}(\boldsymbol{x})\boldsymbol{u}, \quad \text{(A.14a)}$$

$$\boldsymbol{y} = \boldsymbol{G}^\mathsf{T}(\boldsymbol{x})\frac{\partial H(\boldsymbol{x})}{\partial \boldsymbol{x}} \quad \text{(A.14b)}$$

is said to be an input-state-output PHS with nonlinear resistive structure, *if $\boldsymbol{J}(\boldsymbol{x}) = -\boldsymbol{J}^\mathsf{T}(\boldsymbol{x})$ and $\boldsymbol{z}^\mathsf{T}\boldsymbol{\mathcal{R}}(\boldsymbol{x}) \geq 0$.*

Definition A.15 (Passivity, Shifted Passivity [van17, p. 94ff.]). *A nonlinear state-space system*

$$\dot{\boldsymbol{x}} = \boldsymbol{f}(\boldsymbol{x}, \boldsymbol{u}), \quad \text{(A.15a)}$$

$$\boldsymbol{y} = \boldsymbol{g}(\boldsymbol{x}, \boldsymbol{u}), \quad \text{(A.15b)}$$

where $\boldsymbol{f} : \mathbb{R}^n \times \mathbb{R}^p \to \mathbb{R}^n$ is locally Lipschitz, $\boldsymbol{g} : \mathbb{R}^n \times \mathbb{R}^p \to \mathbb{R}^p$ is continuous, $\boldsymbol{f}(\mathbb{0}, \mathbb{0}) = \mathbb{0}$ and $\boldsymbol{g}(\mathbb{0}, \mathbb{0}) = \mathbb{0}$ is said to be passive *if there exists a positive semidefinite* storage function *$V(\boldsymbol{x})$ such that*

$$\dot{V} = \left(\frac{\partial V}{\partial \boldsymbol{x}}\right)^\mathsf{T} \boldsymbol{f}(\boldsymbol{x}, \boldsymbol{u}) \leq \boldsymbol{u}^\mathsf{T}\boldsymbol{y} \quad \text{(A.16)}$$

holds for all $(\boldsymbol{x}, \boldsymbol{u}) \in \mathbb{R}^n \times \mathbb{R}^p$.

Let $(\boldsymbol{x}^\star, \boldsymbol{u}^\star)$ with $\boldsymbol{f}(\boldsymbol{x}^\star, \boldsymbol{u}^\star) = \mathbb{0}$ denote an equilibrium of (A.15). *Then* (A.15) *is said to be* shifted passive *with respect to $(\boldsymbol{x}^\star, \boldsymbol{u}^\star)$ if there exists a positive semidefinite storage function $V(\boldsymbol{x})$ with*

$$\dot{V} = \left(\frac{\partial V}{\partial \boldsymbol{x}}\right)^\mathsf{T} \boldsymbol{f}(\boldsymbol{x}, \boldsymbol{u}) \leq (\boldsymbol{u} - \boldsymbol{u}^\star)^\mathsf{T}(\boldsymbol{y} - \boldsymbol{y}^\star). \quad \text{(A.17)}$$

The following corollary trivially follows by comparison of Definition A.15 with Definition A.13 and Definition A.14.

Corollary A.16. *The input-state-output PHSs* (A.13) *or* (A.14) *are passive, if and only if their Hamiltonians $H(\boldsymbol{x})$ are bounded from below.*

Remark A.17. *Depending on the application, different sortings of the state vector $\boldsymbol{x}$ are chosen in the context of the dissertation. However, the PHS property* is invariant *with respect to these permutations. Consider e.g. the state transformation $\boldsymbol{x}' = \boldsymbol{P}\boldsymbol{x}$, where $\boldsymbol{P} \in \mathbb{R}^{n\times n}$ denotes an arbitrary permutation matrix. Then it can be directly inferred that the transformed input-state-output PHS*

$$\dot{\boldsymbol{x}}' = (\underbrace{\boldsymbol{P}\boldsymbol{J}(\boldsymbol{x})\boldsymbol{P}^{-1}}_{\boldsymbol{J}(\boldsymbol{x}')} - \underbrace{\boldsymbol{P}\boldsymbol{R}(\boldsymbol{x})\boldsymbol{P}^{-1}}_{\boldsymbol{R}(\boldsymbol{x}')})\frac{\partial H(\boldsymbol{x})}{\partial \boldsymbol{x}'} + \boldsymbol{P}\boldsymbol{G}(\boldsymbol{x})\boldsymbol{u}, \tag{A.18a}$$

$$\boldsymbol{y} = (\boldsymbol{P}\boldsymbol{G}(\boldsymbol{x}))^{\mathsf{T}}\frac{\partial H(\boldsymbol{x})}{\partial \boldsymbol{x}} \tag{A.18b}$$

fulfills $\boldsymbol{J}(\boldsymbol{x}') = -\boldsymbol{J}^{\mathsf{T}}(\boldsymbol{x}')$ and $\boldsymbol{R}(\boldsymbol{x}') = \boldsymbol{R}^{\mathsf{T}}(\boldsymbol{x}') \succcurlyeq 0$. This property also holds for the general case of input-state-output PHSs with nonlinear resistive structure. ◇

A.5 Convex Optimization

Definition A.18 (Optimization Problem in Standard Form).

$$\begin{aligned} &\min && f_0(\boldsymbol{x}) && \text{(A.19a)}\\ &\textit{subject to} && f_i(\boldsymbol{x}) \le 0, \quad i = 1,\dots,p && \text{(A.19b)}\\ & && h_j(\boldsymbol{x}) = 0, \quad j = 1,\dots,m && \text{(A.19c)} \end{aligned}$$

Definition A.19 (Convex Optimization Problem). *The optimization problem* (A.19) *is said to be a* convex optimization problem *if $f_i(\boldsymbol{x})$, $i = 0,\dots,p$ are convex functions and $h_j(\boldsymbol{x})$, $j = 1,\dots,m$ are affine functions.*

Definition A.20 (Karush-Kuhn-Tucker (KKT) Point [BV15, p. 243ff.]). *Consider the optimization problem* (A.19)*. Each $(\boldsymbol{x}^\star, \boldsymbol{\lambda}^\star, \boldsymbol{\mu}^\star)$ is said to be a* Karush-Kuhn-Tucker point, *if*

$$\frac{\partial \mathcal{L}}{\partial \boldsymbol{x}}(\boldsymbol{x}^\star, \boldsymbol{\lambda}^\star, \boldsymbol{\mu}^\star) = \mathbb{0}, \tag{A.20a}$$

$$\frac{\partial \mathcal{L}}{\partial \boldsymbol{\lambda}}(\boldsymbol{x}^\star, \boldsymbol{\lambda}^\star, \boldsymbol{\mu}^\star) = \mathbb{0}, \tag{A.20b}$$

$$\mu_i^\star f_i(\boldsymbol{x}^\star) = 0, \qquad i = 1,\dots,p, \tag{A.20c}$$

$$\boldsymbol{\mu}^\star \ge \mathbb{0}, \tag{A.20d}$$

where

$$\mathscr{L}(\boldsymbol{x}, \boldsymbol{\lambda}, \boldsymbol{\mu}) := f_0(\boldsymbol{x}) + \sum_{i=1}^{p} \mu_i f_i(\boldsymbol{x}) + \sum_{j=1}^{m} \lambda_j h_j(\boldsymbol{x}) \tag{A.21}$$

denotes the Lagrangian *of* (A.19).

Definition A.21 (Slater's Condition [BV15, p. 226f.]). *The convex optimization problem* (A.19) *is said to fulfill* Slater's condition, *if there exists some $\boldsymbol{x}'$ which satisfies* (A.19b)–(A.19c) *and $f_i(\boldsymbol{x}') < 0$ for all* non-affine *functions f_i.*

Lemma A.22. *Consider the the convex optimization problem* (A.19) *and assume that Slater's condition holds. Then $\boldsymbol{x}$ is a global optimizer of* (A.19), *if and only if $(\boldsymbol{x}^\star, \boldsymbol{\lambda}^\star, \boldsymbol{\mu}^\star)$ is a KKT point.*

Proof. (see e.g. [BV15, p. 226]) □

A.6 Basic Principles of Welfare Economics

Definition A.23 (Market Equilibrium, Market-Clearing Price, Market-Clearing Quantity [Deb59, Yu20]). *A* market equilibrium *is a condition where a market price is settled through competition such that the amount of produced goods equals the amount of demanded goods. The price reached at market equilibrium is said to be the* market-clearing price *or* competitive price, *and the quantity reached at market equilibrium is said to be the* market-clearing quantity *or* competitive quantity.

Definition A.24 (Perfect Market, Perfect Competition [Bor93, Ley18, Yu20]). *A market structure is said to be a* perfect market *or* perfect competition, *if the following idealized conditions are fulfilled:*

1. *The traded goods are completely homogeneous (i.e. indistinguishable).*
2. *The market participants act as price takers.*
3. *All market participants are fully informed about current market conditions.*
4. *The traded goods can be divided arbitrarily.*
5. *There are no transaction costs incurring.*
6. *All market participants act infinitely fast.*

7. *There is a large number of suppliers and demanders interacting with each other, so that none of them can individually influence market prices.*

A perfect market has the following feature.

Theorem A.25 (First Fundamental Theorem of Welfare Economics [Moo06, p. 150]). *In a perfect market, each market equilibrium leads to a Pareto efficient allocation of resources.*

In *monopolies* or *oligpolies*, individual suppliers have a considerable influence on prices[78]. However, if there are a sufficient number of players (*polypoly*), the assumption of price-taking behavior is justified, as a single player has only negligible influence on the overall price.

A simple but powerful welfare metric[79] is the sum of aggregate profits (*utilitarian* or *Benthamian welfare function*) [Var10, p. 635], which in our case equals

$$\mathtt{W} = \sum_{\pi \in \mathcal{W}} \mathtt{P}_\pi + \sum_{k \in \mathcal{Z}} \mathtt{P}_k - \sum_{k \in \mathcal{Z}} \mathtt{C}_k^\ell. \tag{A.22}$$

The usefulness of $\mathtt{W}$ becomes clear in the following result:

Theorem A.26 (Pareto Efficiency of Market Equilibria under Perfect Competition [Var10, p. 636ff.]). *$\mathtt{W}$ attains its maximum in an equilibrium of a perfect market.*

Remark A.27. *Theorem A.26 states that in a perfect market, the point of social optimum (i.e.* $\arg\max \mathtt{W}$*) and the point of individual optimum (in the sense of individual Pareto efficiency) coincide. From a welfare point of view, it is therefore desirable that the conditions of Definition A.24 are satisfied to the best possible extent.* ◇

78 Oligopolies with few participants are said to be *tight*, and those with many participants are said to be *wide*. The wider the oligopoly, the closer the market equilibrium is to the social optimum.

79 Another welfare measure which is sometimes encountered in the context of energy economics is the *maximin* (or Rawlsian) welfare function [Var10, p. 636] $\mathtt{W}' = \min_{i \in \mathbb{P}}\{\mathtt{P}_i\}$, where only the profit of the "worst-off" market participant is relevant. However, since the profits of the other market participants are completely disregarded and furthermore $\mathtt{W}'$ does not conform to the chosen profit functions for the case where WPs have different nominal outputs, this type of welfare functions is not considered further.

B Supplementary Material to *Optimization-based Control of Cellular Power Networks* (Chapter 3)

B.1 Synchronous Machine-Type Nodes: Connection with the Power Network

Fig. B.1 shows a vector diagram of the stator voltages U_i' and $U_{g,i}$ and their relationship to the external nodal voltage U_i. For convenience, we define a synchronous rotating reference frame and denote the absolute rotor angle with respect to the same reference frame by δ_i. φ_i defines the *load angle* between stator current and nodal voltage[80], $\vartheta_{g,i}$ is the *rotor angle* between $\vec{U}_{g,i}$ and $\vec{U}_i'$, and $\vartheta_{n,i}$ is the (pseudo-) rotor angle with respect to the synchronous rotating reference frame of $\vec{U}_i$. The current and voltage phasors in synchronous rotating reference frame can be characterized by $\vec{U}_i' = U_i' \exp \jmath\delta_i = U_i' \exp \jmath(\theta_i + \vartheta_{n,i})$, $\vec{U}_{g,i} = U_{g,i} \exp(\jmath(\theta_i + \vartheta_{g,i}))$, $\vec{U}_i = U_i \exp(\jmath\theta_i)$, and $\vec{I}_i = I_i \exp \jmath(\theta_i - \varphi_i) = I_i \exp \jmath(\theta_i + \vartheta_{n,i} - \varphi_i')$. From the right half of Fig. 3.4, it follows that

$$\vec{U}_i' - \vec{U}_{g,i} = \jmath X_{d,i}\vec{I}_{d,i} + \jmath X_{q,i}\vec{I}_{q,i}. \tag{B.1}$$

Moreover, it follows that $U_{gd,i} - U_{d,i} = \jmath X_{T,i} I_{q,i} + \jmath X_{n,i} I_{q,i}$ and $U_{gq,i} - U_{q,i} = \jmath X_{T,i} I_{d,i} + \jmath X_{n,i} I_{d,i}$, hence

$$\vec{U}_i' - \vec{U}_i = \jmath X_{d,i}\vec{I}_{d,i} + \jmath X_{q,i}\vec{I}_{q,i} + \jmath(X_{T,i} + X_{n,i})(\vec{I}_{d,i} + \vec{I}_{q,i}). \tag{B.2}$$

With the geometric relationships from Fig. B.1

$$I_{d,i} = I_i \sin(\varphi_i'), \tag{B.3a}$$
$$I_{q,i} = I_i \cos(\varphi_i'), \tag{B.3b}$$
$$U_{q,i} = U_i \sin(\vartheta_{n,i}), \tag{B.3c}$$
$$U_{d,i} = U_i \cos(\vartheta_{n,i}), \tag{B.3d}$$

we can calculate the active and reactive power injections

$$\begin{aligned} p_{inj,i} &= \Re(\vec{U}_i \vec{I}_i^*) = U_i I_i \cos(\varphi_i) = U_i I_i \cos(\varphi_i' - \vartheta_{n,i}) \\ &= U_i I_i \sin(\varphi_i') \sin(\vartheta_{n,i}) + U_i I_i \cos(\varphi_i') \cos(\vartheta_{n,i}), \end{aligned} \tag{B.4}$$

$$\begin{aligned} q_{inj,i} &= \Im(\vec{U}_i \vec{I}_i^*) = U_i I_i \sin(\varphi_i) = U_i I_i \sin(\varphi_i' - \vartheta_{n,i}) \\ &= U_i I_i \sin(\varphi_i') \cos(\vartheta_{n,i}) - U_i I_i \cos(\varphi_i') \sin(\vartheta_{n,i}). \end{aligned} \tag{B.5}$$

[80] The SM acts as generator if $\vartheta_{n,i} > 0$ and as motor if $\vartheta_{n,i} < 0$.

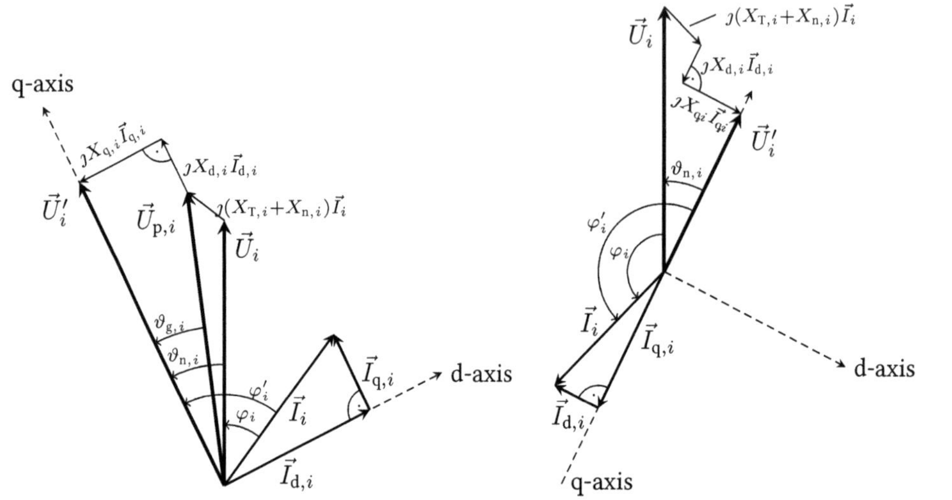

(a) Synchronous generator ($\vartheta_{\mathrm{n},i} > 0$) **(b)** Synchronous motor ($\vartheta_{\mathrm{n},i} < 0$)

Figure B.1: Vector diagram of a SM.

Inserting (B.3) into (B.4) and (B.5) yields

$$p_{\mathrm{inj},i} = U_{\mathrm{d},i}I_{\mathrm{d},i} + U_{\mathrm{q},i}I_{\mathrm{q},i}, \tag{B.6}$$
$$q_{\mathrm{inj},i} = U_{\mathrm{q},i}I_{\mathrm{d},i} - U_{\mathrm{d},i}I_{\mathrm{q},i}. \tag{B.7}$$

Since $\vec{I}_{\mathrm{d},i}$ and $\vec{I}_{\mathrm{q},i}$ are orthogonal, (B.2) can be written separately for both d- and q-axis, which yields

$$U'_{\mathrm{q},i} - U_{\mathrm{q},i} = I_{\mathrm{d},i}(X_{\mathrm{d},i} + X_{\mathrm{T},i} + X_{\mathrm{n},i}) \tag{B.8}$$
$$U'_{\mathrm{d},i} - U_{\mathrm{d},i} = I_{\mathrm{q},i}(X_{\mathrm{q},i} + X_{\mathrm{T},i} + X_{\mathrm{n},i}) \tag{B.9}$$

With Assumption 3.7b), we have $U'_{\mathrm{d},i} = 0$. Moreover, we define the auxiliary reactances $X_{\mathrm{dn},i} := X_{\mathrm{d},i} + X_{\mathrm{T},i} + X_{\mathrm{n},i}$ and $X_{\mathrm{qn},i} := X_{\mathrm{q},i} + X_{\mathrm{T},i} + X_{\mathrm{n},i}$ to get

$$I_{\mathrm{d},i} = \frac{U'_i - U_i\cos(\vartheta_{\mathrm{n},i})}{X_{\mathrm{dn},i}}, \tag{B.10a}$$
$$I_{\mathrm{q},i} = \frac{U_i\sin(\vartheta_{\mathrm{n},i})}{X_{\mathrm{qn},i}}. \tag{B.10b}$$

Substituting (B.10) into (B.4) and (B.7) yields

$$p_{\mathrm{inj},i} = \frac{U'_iU_i}{X_{\mathrm{dn},i}}\sin(\vartheta_{\mathrm{n},i}) + \frac{U_i^2}{2}\frac{X_{\mathrm{dn},i} - X_{\mathrm{qn},i}}{X_{\mathrm{dn},i}X_{\mathrm{qn},i}}\sin(2\vartheta_{\mathrm{n},i}), \tag{B.11}$$
$$q_{\mathrm{inj},i} = \frac{U'_iU_i}{X_{\mathrm{dn},i}}\cos(\vartheta_{\mathrm{n},i}) - \frac{U_i^2}{X_{\mathrm{dn},i}X_{\mathrm{qn},i}}\left(X_{\mathrm{dn},i}\sin^2(\vartheta_{\mathrm{n},i}) + X_{\mathrm{qn},i}\cos^2(\vartheta_{\mathrm{n},i})\right). \tag{B.12}$$

B.2 Distributed Consensus-Based Control

In the area of networked systems, there often arise control laws of the *separable type*

$$\dot{u} = \sum_{i=1}^{n} f_i(u), \tag{B.13}$$

where the calculation of the global variable u is composed of a finite amount of additive components $f_i(u)$ with $i = 1, \ldots, n$, representing "local" calculations, e.g. relying on parameters which are only locally available. For the local calculation of $f_i(u)$, however, each local agent needs instantaneous knowledge of the global variable u. Additionally, in order to calculate $\dot{u}$, all $f_i(u)$ need to be known globally.

The idea of *consensus-based control* [OSFM07] is to design local controllers for agent i which achieve the same equilibrium as (B.13), but without requiring knowledge of u or $f_{\neg i}(u)$. Instead, the agents communicate local variables in a distributed fashion via m neighbor-to-neighbor communication links. The communication structure is represented by a weakly connected, directed graph with incidence matrix $\boldsymbol{D} \in \mathbb{R}^{n \times n}$.

Lemma B.1. *Consider the centralized controller* (B.13) *and let the steady state* $u^\star$ *of* (B.13) *be unique. Then, each equilibrium* $(\boldsymbol{u}^\star, \boldsymbol{\nu}^\star)$ *of the distributed* consensus-based controller

$$\boldsymbol{\tau}_{\mathrm{u}} \dot{\boldsymbol{u}} = \begin{bmatrix} f_1(u_1) \\ \vdots \\ f_n(u_n) \end{bmatrix} + \boldsymbol{D}\boldsymbol{\nu}, \tag{B.14a}$$

$$\boldsymbol{\tau}_{\nu} \dot{\boldsymbol{\nu}} = -\boldsymbol{D}^\mathsf{T} \boldsymbol{u} \tag{B.14b}$$

with $\boldsymbol{\tau}_{\mathrm{u}}, \boldsymbol{\tau}_{\nu} \succ 0$ *is unique. Moreover, it holds that* $\boldsymbol{u}^\star = u^\star \cdot \mathbb{1}$.

Proof. The equilibrium of (B.14) is characterized by

$$\mathbb{0} = \begin{bmatrix} f_1(u_1^\star) \\ \vdots \\ f_n(u_n^\star) \end{bmatrix} + \boldsymbol{D}\boldsymbol{\nu}^\star, \tag{B.15a}$$

$$\mathbb{0} = -\boldsymbol{D}^\mathsf{T} \boldsymbol{u}^\star. \tag{B.15b}$$

Since $\boldsymbol{D}$ is the incidence matrix of a connected graph, (B.15b) implies $u_1^\star = \ldots = u_n^\star =: \hat{u}^\star$. Premultiplying (B.15a) with $\mathbb{1}^\mathsf{T}$ yields

$$0 = \mathbb{1}^\mathsf{T} \begin{bmatrix} f_1(\hat{u}^\star) \\ \vdots \\ f_n(\hat{u}^\star) \end{bmatrix} + \underbrace{\mathbb{1}^\mathsf{T} \boldsymbol{D}}_{=0} \boldsymbol{\nu}^\star = \sum_{i=1}^{n} f_i(\hat{u}^\star), \tag{B.16}$$

which is equivalent to (B.13), implying that $\hat{u}^\star$ is equal to $u^\star$. To prove uniqueness of $\boldsymbol{\nu}^\star$, define $\boldsymbol{\nu} =: \boldsymbol{\nu}_\mathrm{I} + \boldsymbol{\nu}_\mathrm{K}$, where $\boldsymbol{\nu}_\mathrm{I} \in \mathrm{im}(\boldsymbol{D}\boldsymbol{\tau}_\nu^{-1})$ and $\boldsymbol{\nu}_\mathrm{K} \in \ker(\boldsymbol{D}\boldsymbol{\tau}_\nu^{-1})$. We show that both $\boldsymbol{\nu}_\mathrm{I}^\star$ and $\boldsymbol{\nu}_\mathrm{K}^\star$ are unique, which implies that $\nu^\star$ is unique as well.

With regard to $\boldsymbol{\nu}_\mathrm{K}^\star$, note that for each $\boldsymbol{b} \in \ker(\boldsymbol{D}\boldsymbol{\tau}_\nu^{-1})$ it holds that $\boldsymbol{b}^\mathsf{T}\dot{\boldsymbol{\nu}} = 0$ and thus $\boldsymbol{b}^\mathsf{T}\boldsymbol{\nu}(t) = \boldsymbol{b}^\mathsf{T}\boldsymbol{\nu}(0)$ for all $t > 0$. Moreover, with $\boldsymbol{b}^\mathsf{T}\nu(t) = \boldsymbol{b}^\mathsf{T}\boldsymbol{\nu}_\mathrm{K}(t)$, it follows that $\boldsymbol{\nu}_\mathrm{K}^\star = \boldsymbol{\nu}_\mathrm{K}(0)$, thus $\boldsymbol{\nu}_\mathrm{K}^\star$ is unique. With regard to $\boldsymbol{\nu}_\mathrm{I}^\star$, note that $\dim(\mathrm{im}(\boldsymbol{D}\boldsymbol{\tau}_\nu^{-1})) = \mathrm{rank}(\boldsymbol{D})$ since $\boldsymbol{\tau}_\nu$ has full rank. With (B.15a), we state that $\boldsymbol{\nu}_\mathrm{I}^\star$ is the unique solution of

$$-\begin{bmatrix} f_1(\hat{u}^\star) \\ \vdots \\ f_n(\hat{u}^\star) \end{bmatrix} - \boldsymbol{D}\boldsymbol{\nu}_\mathrm{K}^\star = \boldsymbol{D}\boldsymbol{\nu}_\mathrm{I}^\star. \tag{B.17}$$

This completes the proof. □

In summary, the distributed consensus-based controller (B.14) enables each agent to calculate a local estimator of the global variable u, which is communicated among agents in a distributed sense by using the communication structure represented by $\boldsymbol{D}$.

Remark B.2. *If* (B.13) *has* $v > 0$ *equilibria, then the distributed controller* (B.14) *also has* v *equilibria. This fact can be proven by formulationg the proof of Lemma B.1 separately for each equilibrium* $u_1^\star, \ldots, u_n^\star$ *of* (B.13). ◇

Remark B.3. *The results of Lemma B.1 remain the same if the incidence matrix* $\boldsymbol{D}$ *in* (B.14) *is replaced by an arbitrary Laplacian matrix* $\mathcal{L}$ *(cf. Definition A.11).* ◇

B.3 Extension to Nonconvex Cost Functions

Strict convexity of $\mathtt{C}_{\mathrm{g},i}$ is an assumption that is very common in the literatur for simplification and regularity reasons. However, with a closer look, certain types of power plants exhibit a more delicate cost structure, whereby the cost function can be divided into three major areas which are caused by different levels of efficiency, see solid line in Fig. B.2a: The first area is characterized by relatively low efficiency, which is reflected by a rather steep increase of $\mathtt{C}_{\mathrm{g},i}(p_{\mathrm{g},i})$ per increment of power. What follows is a region of high efficiency, which causes a flattening of $\mathtt{C}_{\mathrm{g},i}(p_{\mathrm{g},i})$. At even higher levels of power, it becomes increasingly difficult (that is, increasingly expensive) to supply further amounts of power, with the consequence that $\mathtt{C}_{\mathrm{g},i}(p_{\mathrm{g},i})$ eventually rises with increasing steepness.

Let the (unique) inflection point of $\mathtt{C}_{\mathrm{g},i}(p_{\mathrm{g},i})$ be denoted by $p_{\mathrm{g},i}^\circ$[81]. Then, also $p_{\mathrm{g},i}^\circ$ is an inflection point of $\mathtt{P}_{\mathrm{g},i}(p_{\mathrm{g},i})$. Now assume that the current price λ_i fulfills $\lambda_i \geq \nabla\mathtt{C}_{\mathrm{g},i}(p_{\mathrm{g},i}^\circ)$ and define by $p_{\mathrm{g},i}^\sharp := \arg\max \mathtt{P}_{\mathrm{g},i}$ under the current price λ_i. Then, it is rational for each profit-oriented

[81] This point corresponds to the operating point of the power plant with highest efficiency.

WP to apply a *virtual cost function* $\breve{\mathtt{C}}_{\mathrm{g},i}$, which is strictly convex around $p^{\sharp}_{\mathrm{g},i}$ in order to guarantee global convergence to the own profit maximum. If however $\lambda_i \geq \nabla \mathtt{C}_{\mathrm{g},i}(p^{\circ}_{\mathrm{g},i})$, it is advisable for WPs to leave the market, since otherwise any nonzero power supply would result in a negative profit.

To obtain a continuously differentiable, strictly convex upper envelope of $\mathtt{C}_{\mathrm{g},i}$, it is thus a reasonable choice to choose the virtual cost function equal to

$$\breve{\mathtt{C}}_{\mathrm{g},i} = \begin{cases} \mathtt{C}_{\mathrm{g},i}, & p_{\mathrm{g},i} \geq p^{\circ}_{\mathrm{g},i}, \\ p^2_{\mathrm{g},i} - (p^{\circ}_{\mathrm{g},i})^2 - \frac{m^{\circ}}{2p^{\circ}_{\mathrm{g},i}}(p_{\mathrm{g},i} - p^{\circ}_{\mathrm{g},i}) + \mathtt{C}^{\circ}_{\mathrm{g},i}, & p_{\mathrm{g},i} < p^{\circ}_{\mathrm{g},i}, \end{cases} \tag{B.18}$$

where $m^{\circ} = \nabla \mathtt{C}_{\mathrm{g},i}(p^{\circ}_{\mathrm{g},i})$ denotes the steepness of $\mathtt{C}_{\mathrm{g},i}(p_{\mathrm{g},i})$ at the inflection point. We motivate this choice by a small example.

Example 9

Consider the cost function

$$\mathtt{C}_{\mathrm{g},i}(p_{\mathrm{g},i}) = \frac{1}{10} \cdot \left((p_{\mathrm{g},i} - 2)^3 + (p_{\mathrm{g},i} - 2)^2 + 2(p_{\mathrm{g},i} - 2) + 8\right). \tag{B.19}$$

Here, $p^{\circ}_{\mathrm{g},i} = \frac{5}{3}$, $\mathtt{C}^{\circ}_{\mathrm{g},i} = \frac{27}{20}$, *and* $m^{\circ}_i = \frac{1}{6}$. *Fig. B.2 shows the curves of* $\mathtt{C}_{\mathrm{g},i}$ *and* $\breve{\mathtt{C}}_{\mathrm{g},i}$ *as well as the resulting profit functions* $\mathtt{P}_{\mathrm{g},i}$ *for different prices* $\lambda_i \in [5, 25]$. *It can be seen that the replacement of* $\mathtt{C}_{\mathrm{g},i}$ *by the strictly convex, continuously differentiable function*

$$\breve{\mathtt{C}}_{\mathrm{g},i} = \begin{cases} \mathtt{C}_{\mathrm{g},i}, & p_{\mathrm{g},i} \geq \frac{10}{6}, \\ p^2_{\mathrm{g},i} - \frac{1}{20}p_{\mathrm{g},i} + \frac{509}{108}, & p_{\mathrm{g},i} < \frac{10}{6} \end{cases} \tag{B.20}$$

does not affect the respective value of the best response $p^{\sharp}_{\mathrm{g},i}$.

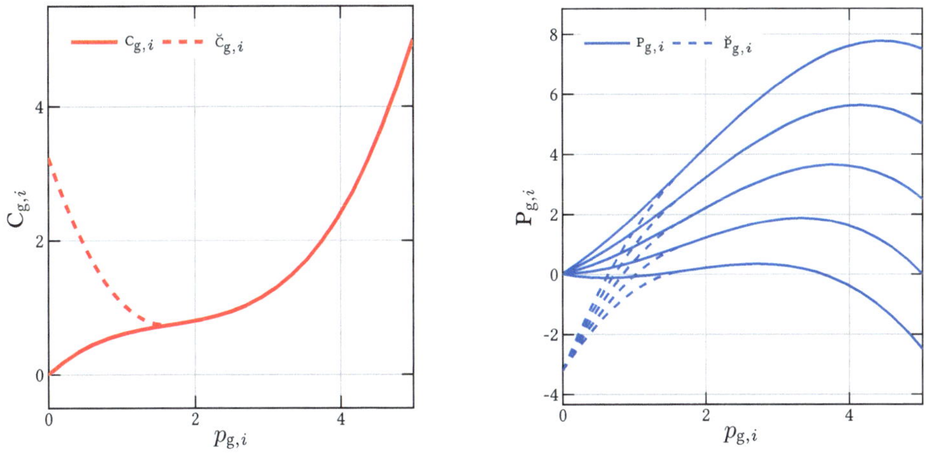

(a) cost function **(b)** profit functions

Figure B.2: Plot of the nonconvex cost and profit functions (solid lines) and their respective "virtual" substitutes resulting from convexification (dashed lines). Notably, the convexified profit functions exhibit the same maxima as their nonconvex counterparts.

B.4 Parameter Values for Example 4

The units of the parameters are given in p.u., except $T'_{\mathrm{d0},i}$ which is given in seconds.

Table B.1: Numerical values of the nodal parameters used in Example 4.

Node	Type	A_i	B_{ii}	Γ_i	$T'_{\mathrm{d0},i}$	$X_{\mathrm{d},i} - X'_{\mathrm{d},i}$
1	$\mathcal{S}$	1.60	−5.5	5.22	6.45	0.0159
2	$\mathcal{S}$	1.22	−5.5	3.98	7.68	0.0213
3	$\mathcal{I}$	1.38	−3.3	0.45	–	–
4	$\mathcal{I}$	1.42	−3.1	0.42	–	–
5	$\mathcal{I}$	1.40	−7.0	0.44	–	–
6	$\mathcal{L}$	1.30	−2.0	–	–	–
7	$\mathcal{L}$	1.30	−2.0	–	–	–

Table B.2: Numerical values of the line parameters used in Example 4.

B_{12}	B_{15}	B_{16}	B_{23}	B_{25}	B_{34}	B_{45}	B_{57}
1.27	1.4	2.0	1.4	2.05	1.1	1.0	2.0

C Supplementary Material to *Real-Time Incentives by Zonal Pricing* (Chapter 4)

C.1 Proof of Theorem 4.25

Proof. Recall the physical plant system (3.121) and define $\boldsymbol{K} := (\mathrm{diag}(\boldsymbol{\kappa}))^{-1}$. Now consider the Lyapunov function candidate

$$\begin{aligned} V(\widetilde{\boldsymbol{x}}) = \widetilde{H}_{\mathrm{p}}(\widetilde{\boldsymbol{x}}_{\mathrm{p}}) + \frac{1}{2}\Big(& (\widetilde{\boldsymbol{p}}_{\mathrm{g}})^{\mathsf{T}} \boldsymbol{K}\boldsymbol{\tau}_{\mathrm{g}}\widetilde{\boldsymbol{p}}_{\mathrm{g}} + \widetilde{\boldsymbol{\mu}}_{\mathrm{g}+}^{\mathsf{T}} \boldsymbol{K}\boldsymbol{\tau}_{\mu_{\mathrm{g}+}}\widetilde{\boldsymbol{\mu}}_{\mathrm{g}+} + \widetilde{\boldsymbol{\mu}}_{\mathrm{g}-}^{\mathsf{T}} \boldsymbol{K}\boldsymbol{\tau}_{\mu_{\mathrm{g}-}}\widetilde{\boldsymbol{\mu}}_{\mathrm{g}-} \\ & + \widetilde{\boldsymbol{\lambda}}^{\mathsf{T}}\boldsymbol{\tau}_{\lambda}\widetilde{\boldsymbol{\lambda}} + \widetilde{\boldsymbol{\nu}}^{\mathsf{T}}\boldsymbol{\tau}_{\nu}\widetilde{\boldsymbol{\nu}} + \widetilde{\boldsymbol{U}}_{\mathrm{f}}^{\mathsf{T}}\boldsymbol{\tau}_{U_{\mathrm{f}}}\widetilde{\boldsymbol{U}}_{\mathrm{f}} + \widetilde{\boldsymbol{\mu}}_{\mathcal{S}+}^{\mathsf{T}}\boldsymbol{\tau}_{\mu_{\mathcal{S}+}}\widetilde{\boldsymbol{\mu}}_{\mathcal{S}+} + \widetilde{\boldsymbol{\mu}}_{\mathcal{S}-}^{\mathsf{T}}\boldsymbol{\tau}_{\mu_{\mathcal{S}-}}\widetilde{\boldsymbol{\mu}}_{\mathcal{S}-} \\ & + \widetilde{\boldsymbol{U}}_{\mathcal{I}}^{\mathsf{T}}\boldsymbol{\tau}_{U_{\mathcal{I}}}\widetilde{\boldsymbol{U}}_{\mathcal{I}} + \widetilde{\boldsymbol{\mu}}_{\mathcal{I}+}^{\mathsf{T}}\boldsymbol{\tau}_{\mu_{\mathcal{I}+}}\widetilde{\boldsymbol{\mu}}_{\mathcal{I}+} + \widetilde{\boldsymbol{\mu}}_{\mathcal{I}-}^{\mathsf{T}}\boldsymbol{\tau}_{\mu_{\mathcal{I}-}}\widetilde{\boldsymbol{\mu}}_{\mathcal{I}-}\Big), \end{aligned} \tag{C.1}$$

where $\widetilde{H}_{\mathrm{p}}(\widetilde{\boldsymbol{x}}_{\mathrm{p}}) = H_{\mathrm{p}}(\boldsymbol{x}_{\mathrm{p}}) - (\widetilde{\boldsymbol{x}}_{\mathrm{p}})^{\mathsf{T}}\nabla H_{\mathrm{p}}(\boldsymbol{x}_{\mathrm{p}}^{\star}) - H_{\mathrm{p}}(\boldsymbol{x}_{\mathrm{p}}^{\star})$ is defined as in the proof of Lemma 3.39. Note that the Hessian of $\widetilde{H}_{\mathrm{p}}(\widetilde{\boldsymbol{x}}_{\mathrm{p}})$ at $\widetilde{\boldsymbol{x}}_{\mathrm{p}} = \mathbb{0}$ is equal to the Hessian of $H_{\mathrm{p}}(\boldsymbol{x}_{\mathrm{p}})$ at $\boldsymbol{x}_{\mathrm{p}} = \boldsymbol{x}_{\mathrm{p}}^{\star}$ and thus positive definite by Assumption 3.33. Furthermore, since $\widetilde{H}_{\mathrm{p}}(\mathbb{0}) = 0$ there exists a neighborhood around $\widetilde{\boldsymbol{x}}_{\mathrm{p}} = \mathbb{0}$ with $\widetilde{H}_{\mathrm{p}}(\widetilde{\boldsymbol{x}}_{\mathrm{p}}) \succ 0$. As all terms in (C.1) except $\widetilde{H}_{\mathrm{p}}(\widetilde{\boldsymbol{x}}_{\mathrm{p}})$ are quadratic forms with positive metrics, $\widetilde{H}_{\mathrm{p}}(\widetilde{\boldsymbol{x}}_{\mathrm{p}}) \succ 0$ also implies that $V(\widetilde{\boldsymbol{x}}) \succ 0$ holds in a (possibly larger) neighborhood around $\widetilde{\boldsymbol{x}} = \mathbb{0}$.

The derivative of $V(\widetilde{\boldsymbol{x}})$ along the trajectories of (3.49), (3.110) equals

$$\begin{aligned} \dot{V}(\widetilde{\boldsymbol{x}}) = {} & \widetilde{\boldsymbol{z}}_{\mathrm{p}}^{\mathsf{T}}\boldsymbol{J}_{\boldsymbol{p}}\widetilde{\boldsymbol{z}}_{\mathrm{p}} - \widetilde{\boldsymbol{z}}_{\mathrm{p}}^{\mathsf{T}}(\boldsymbol{\mathcal{R}}_{\mathrm{p}}(\boldsymbol{x}_{\mathrm{p}}) - \boldsymbol{\mathcal{R}}_{\mathrm{p}}(\boldsymbol{x}_{\mathrm{p}}^{\star})) + \widetilde{\boldsymbol{z}}_{\mathrm{p}}^{\mathsf{T}}\boldsymbol{G}_{\mathrm{p}}\widetilde{\boldsymbol{u}}_{\mathrm{p}} \\ & - \widetilde{\boldsymbol{p}}_{\mathrm{g}}^{\mathsf{T}}\boldsymbol{K}(\nabla \mathrm{C}(\boldsymbol{p}_{\mathrm{g}}) - \nabla \mathrm{C}(\boldsymbol{p}_{\mathrm{g}}^{\star})) - \widetilde{\boldsymbol{p}}_{\mathrm{g}}^{\mathsf{T}}\widetilde{\boldsymbol{\omega}}_{\mathrm{g}} - \widetilde{\boldsymbol{p}}_{\mathrm{g}}^{\mathsf{T}}\boldsymbol{K}\widetilde{\boldsymbol{\mu}}_{\mathrm{g}+} \\ & + \widetilde{\boldsymbol{p}}_{\mathrm{g}}^{\mathsf{T}}\boldsymbol{K}\widetilde{\boldsymbol{\mu}}_{\mathrm{g}-} + \widetilde{\boldsymbol{\mu}}_{\mathrm{g}+}^{\mathsf{T}}\boldsymbol{K}\langle \boldsymbol{p}_{\mathrm{g}} - \overline{\boldsymbol{p}}_{\mathrm{g}}\rangle_{\mu_{\mathrm{g}+}}^{+} + \widetilde{\boldsymbol{\mu}}_{\mathrm{g}-}^{\mathsf{T}}\boldsymbol{K}\langle \underline{\boldsymbol{p}}_{\mathrm{g}} - \boldsymbol{p}_{\mathrm{g}}\rangle_{\mu_{\mathrm{g}-}}^{+} \\ & + \widetilde{\boldsymbol{\lambda}}^{\mathsf{T}}\widetilde{\boldsymbol{\varphi}} - \widetilde{\boldsymbol{U}}_{\mathrm{f}}^{\mathsf{T}}\widetilde{\boldsymbol{U}}_{\mathrm{f}} - \widetilde{\boldsymbol{U}}_{\mathrm{f}}^{\mathsf{T}}\widetilde{\boldsymbol{\mu}}_{\mathcal{S}+} + \widetilde{\boldsymbol{U}}_{\mathrm{f}}^{\mathsf{T}}\widetilde{\boldsymbol{\mu}}_{\mathcal{S}-} \\ & + \widetilde{\boldsymbol{\mu}}_{\mathcal{S}+}\langle \boldsymbol{U}_{\mathrm{f}} - \overline{\boldsymbol{U}}_{\boldsymbol{f}}\rangle_{\mu_{\mathcal{S}+}}^{+} + \widetilde{\boldsymbol{\mu}}_{\mathcal{S}-}\langle \underline{\boldsymbol{U}}_{\boldsymbol{f}} - \boldsymbol{U}_{\mathrm{f}}\rangle_{\mu_{\mathcal{S}-}}^{+} \\ & - \widetilde{\boldsymbol{U}}_{\mathcal{I}}^{\mathsf{T}}\widetilde{\boldsymbol{U}}_{\mathcal{I}} - \widetilde{\boldsymbol{U}}_{\mathcal{I}}^{\mathsf{T}}\widetilde{\boldsymbol{\mu}}_{\mathcal{I}+} + \widetilde{\boldsymbol{U}}_{\mathcal{I}}^{\mathsf{T}}\widetilde{\boldsymbol{\mu}}_{\mathcal{I}-} \\ & + \widetilde{\boldsymbol{\mu}}_{\mathcal{I}-}\langle \boldsymbol{U}_{\mathcal{I}} - \overline{\boldsymbol{U}}_{\mathcal{I}}\rangle_{\mu_{\mathcal{I}-}}^{+} + \widetilde{\boldsymbol{\mu}}_{\mathcal{I}-}\langle \underline{\boldsymbol{U}}_{\mathcal{I}} - \boldsymbol{U}_{\mathcal{I}}\rangle_{\mu_{\mathcal{I}-}}^{+}. \end{aligned} \tag{C.2}$$

Rearrangement of the expressions in (C.2) yields

$$\dot{V}(\boldsymbol{x}) = W_1(\boldsymbol{x}) + W_2(\boldsymbol{x}) + W_3(\boldsymbol{x}) + W_4(\boldsymbol{x}) + W_5(\boldsymbol{x}), \tag{C.3}$$

where

$$W_1(\boldsymbol{x}) = \widetilde{\boldsymbol{z}}_{\mathrm{p}}^{\mathsf{T}} \boldsymbol{J}_{\mathrm{p}} \widetilde{\boldsymbol{z}}_{\mathrm{p}}, \tag{C.4}$$

$$W_2(\boldsymbol{x}) = -\widetilde{\boldsymbol{z}}_{\mathrm{p}}^{\mathsf{T}} (\boldsymbol{\mathcal{R}}_{\mathrm{p}}(\boldsymbol{x}_{\mathrm{p}}) - \boldsymbol{\mathcal{R}}_{\mathrm{p}}(\boldsymbol{x}_{\mathrm{p}}^{\star})) + \widetilde{\boldsymbol{\lambda}}^{\mathsf{T}} \widetilde{\boldsymbol{\varphi}}, \tag{C.5}$$

$$W_3(\boldsymbol{x}) = \widetilde{\boldsymbol{z}}_{\mathrm{p}}^{\mathsf{T}} \boldsymbol{G}_{\mathrm{p}} \widetilde{\boldsymbol{u}}_{\mathrm{p}} - \widetilde{\boldsymbol{p}}_{\mathrm{g}}^{\mathsf{T}} \widetilde{\boldsymbol{\omega}}_{\mathrm{g}}, \tag{C.6}$$

$$W_4(\boldsymbol{x}) = -\widetilde{\boldsymbol{p}}_{\mathrm{g}}^{\mathsf{T}} \boldsymbol{K} (\nabla \mathtt{C}(\boldsymbol{p}_{\mathrm{g}}) - \nabla \mathtt{C}(\boldsymbol{p}_{\mathrm{g}}^{\star})) - \widetilde{\boldsymbol{U}}_{\mathrm{f}}^{\mathsf{T}} \widetilde{\boldsymbol{U}}_{\mathrm{f}} - \widetilde{\boldsymbol{U}}_{\mathcal{I}}^{\mathsf{T}} \widetilde{\boldsymbol{U}}_{\mathcal{I}}, \tag{C.7}$$

$$\begin{aligned} W_5(\boldsymbol{x}) = & -\widetilde{\boldsymbol{p}}_{\mathrm{g}}^{\mathsf{T}} \boldsymbol{K} \widetilde{\boldsymbol{\mu}}_{\mathrm{g}+} + \widetilde{\boldsymbol{\mu}}_{\mathrm{g}+}^{\mathsf{T}} \boldsymbol{K} \langle \boldsymbol{p}_{\mathrm{g}} - \overline{\boldsymbol{p}}_{\mathrm{g}} \rangle_{\boldsymbol{\mu}_{\mathrm{g}+}}^{+} \\ & + \widetilde{\boldsymbol{p}}_{\mathrm{g}}^{\mathsf{T}} \boldsymbol{K} \widetilde{\boldsymbol{\mu}}_{\mathrm{g}-} + \widetilde{\boldsymbol{\mu}}_{\mathrm{g}-}^{\mathsf{T}} \boldsymbol{K} \langle \underline{\boldsymbol{p}}_{\mathrm{g}} - \boldsymbol{p}_{\mathrm{g}} \rangle_{\boldsymbol{\mu}_{\mathrm{g}-}}^{+} \\ & - \widetilde{\boldsymbol{U}}_{\mathrm{f}}^{\mathsf{T}} \widetilde{\boldsymbol{\mu}}_{\mathcal{S}+} + \widetilde{\boldsymbol{\mu}}_{\mathcal{S}+} \langle \boldsymbol{U}_{\mathrm{f}} - \overline{\boldsymbol{U}}_{\mathrm{f}} \rangle_{\boldsymbol{\mu}_{\mathcal{S}+}}^{+} \\ & + \widetilde{\boldsymbol{U}}_{\mathrm{f}}^{\mathsf{T}} \widetilde{\boldsymbol{\mu}}_{\mathcal{S}-} + \widetilde{\boldsymbol{\mu}}_{\mathcal{S}-} \langle \underline{\boldsymbol{U}}_{\mathrm{f}} - \boldsymbol{U}_{\mathrm{f}} \rangle_{\boldsymbol{\mu}_{\mathcal{S}-}}^{+} \\ & - \widetilde{\boldsymbol{U}}_{\mathcal{I}}^{\mathsf{T}} \widetilde{\boldsymbol{\mu}}_{\mathcal{I}+} + \widetilde{\boldsymbol{\mu}}_{\mathcal{I}+} \langle \boldsymbol{U}_{\mathcal{I}} - \overline{\boldsymbol{U}}_{\mathcal{I}} \rangle_{\boldsymbol{\mu}_{\mathcal{I}+}}^{+} \\ & + \widetilde{\boldsymbol{U}}_{\mathcal{I}}^{\mathsf{T}} \widetilde{\boldsymbol{\mu}}_{\mathcal{I}-} + \widetilde{\boldsymbol{\mu}}_{\mathcal{I}-} \langle \underline{\boldsymbol{U}}_{\mathcal{I}} - \boldsymbol{U}_{\mathcal{I}} \rangle_{\boldsymbol{\mu}_{\mathcal{I}-}}^{+}. \end{aligned} \tag{C.8}$$

Due to skew-symmetry of $\boldsymbol{J}_{\mathrm{p}}$ it holds that $W_1(\boldsymbol{x}) = 0$. The term $W_2(\boldsymbol{x})$ is less than or equal to zero whenever the conditions of Lemmas 3.39 and 3.41 are fulfilled. Moreover, with $\widetilde{\boldsymbol{z}}_{\mathrm{p}}^{\mathsf{T}} \boldsymbol{G}_{\mathrm{p}} = \nabla H_{\mathrm{p}}(\boldsymbol{L}) = \boldsymbol{\omega}_{\mathrm{g}} = \widetilde{\boldsymbol{\omega}}_{\mathrm{g}}$, we get $W_3(\boldsymbol{x}) = 0$. Due to (strict) convexity of $\mathtt{C}_{\mathrm{g},i}(p_{\mathrm{g},i})$, the relationship

$$-\frac{1}{\kappa_i}(p_{\mathrm{g},i} - p_{\mathrm{g},i}^{\star})(\nabla \mathtt{C}_{\mathrm{g},i}(p_{\mathrm{g},i}) - \nabla \mathtt{C}_{\mathrm{g},i}(p_{\mathrm{g},i}^{\star})) < 0 \tag{C.9}$$

holds for each $i \in \mathcal{V}_{\mathrm{g}}$. Thus, the first summand in $W_4(\boldsymbol{x})$ is less than or equal to zero. The same holds for the quadratic terms $-(\boldsymbol{U}_{\mathrm{f}} - \boldsymbol{U}_{\mathrm{f}}^{\star})^{\mathsf{T}}(\boldsymbol{U}_{\mathrm{f}} - \boldsymbol{U}_{\mathrm{f}}^{\star})$ and $-(\boldsymbol{U}_{\mathcal{I}} - \boldsymbol{U}_{\mathcal{I}}^{\star})^{\mathsf{T}}(\boldsymbol{U}_{\mathcal{I}} - \boldsymbol{U}_{\mathcal{I}}^{\star})$, which yields $W_4(\boldsymbol{x}) \leq 0$.

Finally, since the inequalities $\widetilde{\boldsymbol{\mu}}^{\mathsf{T}} \langle \boldsymbol{g} \rangle_{\boldsymbol{\mu}}^{+} \leq \widetilde{\boldsymbol{\mu}}^{\mathsf{T}} \boldsymbol{g}$ and $\widetilde{\boldsymbol{\mu}}^{\mathsf{T}} \boldsymbol{g}^{\star} \leq \mathbb{0}$ hold for each convex function $\boldsymbol{g} = \mathrm{col}_i\{g_i\}$ [SDv15, Proposition 3] (see also the proofs of Lemma 3.41 and Theorem 3.44), it can be concluded that

$$(\mu_{x_i+} - \mu_{x_i+}^{\star}) \langle x_i - \overline{x}_i \rangle_{\mu_{x_i+}}^{+} \leq (\mu_{x_i+} - \mu_{x_i+}^{\star})(x_i - x_i^{\star}) \tag{C.10}$$

and

$$(\mu_{x_i-} - \mu_{x_i-}^{\star}) \langle \underline{x}_i - x_i \rangle_{\mu_{x_i-}}^{+} \leq -(\mu_{x_i-} - \mu_{x_i-}^{\star})(x_i - x_i^{\star}). \tag{C.11}$$

Accordingly, each row in (C.8) is less than or equal to zero.

All in all, it follows that $\dot{V} \leq 0$ holds in a neighborhood around $\boldsymbol{x}^{\star}$, thus $\boldsymbol{x}^{\star}$ is stable in the sense of Lyapunov [Kha02, Theorem 4.1]. □

D Supplementary Material to *Optimal Control of Port-Hamiltonian Systems* (Chapter 5)

D.1 Single-Player Case

$H(\boldsymbol{x})$ as Natural Control-Lyapunov Function Candidate

Lemma D.1. *Consider the purely dissipative input-state-output PHS*

$$\boldsymbol{x} = -\boldsymbol{R}(\boldsymbol{x})\frac{\partial H(\boldsymbol{x})}{\partial \boldsymbol{x}} + \boldsymbol{G}(\boldsymbol{x})\boldsymbol{u}, \tag{D.1a}$$

$$\boldsymbol{y} = \boldsymbol{G}^\top(\boldsymbol{x})\frac{\partial H(\boldsymbol{x})}{\partial \boldsymbol{x}} \tag{D.1b}$$

with $\boldsymbol{R}(\boldsymbol{x}) = \boldsymbol{R}^\top(\boldsymbol{x}) \succcurlyeq 0$ and $H(\boldsymbol{x}) \succ 0$. Then (D.1) *is asymptotically stabilizable at the origin by means of a state feedback $\boldsymbol{u}(\boldsymbol{x})$, if and only if $H(\boldsymbol{x})$ is a CLF for* (D.1).

Proof. Let $\boldsymbol{g}(\boldsymbol{x}) \in \mathbb{R}^n$ be an arbitrary but fixed vector. Then for all $\boldsymbol{k}(\boldsymbol{x}) \in \mathbb{R}^n \backslash \{\mathbb{0}\}$ it trivially holds that $\boldsymbol{k}^\top(\boldsymbol{x})\boldsymbol{g}(\boldsymbol{x}) = 0$ is equivalent to $\boldsymbol{g}(\boldsymbol{x}) = \mathbb{0}$. Negating this statement yields

$$\exists\, \boldsymbol{k}(\boldsymbol{x}) \in \mathbb{R}^n : \boldsymbol{k}^\top(\boldsymbol{x})\boldsymbol{g}(\boldsymbol{x}) > 0 \qquad \Longleftrightarrow \qquad \boldsymbol{g}^\top(\boldsymbol{x})\boldsymbol{g}(\boldsymbol{x}) > 0. \tag{D.2}$$

By choosing $\boldsymbol{g}(\boldsymbol{x}) = \boldsymbol{G}^\top(\boldsymbol{x})\nabla H(\boldsymbol{x})$, with $\boldsymbol{G}(\boldsymbol{x})$ and $H(\boldsymbol{x})$ as in (D.1), equation (D.2) reads

$$\exists\, \boldsymbol{k}(\boldsymbol{x}) \in \mathbb{R}^n : \boldsymbol{k}^\top(\boldsymbol{x})\boldsymbol{G}^\top(\boldsymbol{x})\nabla H(\boldsymbol{x}) > 0 \iff (\nabla H(\boldsymbol{x}))^\top \boldsymbol{G}(\boldsymbol{x})\boldsymbol{G}^\top(\boldsymbol{x})\nabla H(\boldsymbol{x}) > 0. \tag{D.3}$$

Alternatively, choosing $\boldsymbol{g}(\boldsymbol{x}) = \sqrt{\boldsymbol{R}(\boldsymbol{x})}\nabla H(\boldsymbol{x})$ with $\boldsymbol{R}(\boldsymbol{x}) = \boldsymbol{R}^\top(\boldsymbol{x}) \succcurlyeq 0$, (D.2) takes the form

$$\exists\, \hat{\boldsymbol{k}}(\boldsymbol{x}) \in \mathbb{R}^n : \hat{\boldsymbol{k}}^\top(\boldsymbol{x})\sqrt{\boldsymbol{R}(\boldsymbol{x})}\nabla H(\boldsymbol{x}) > 0 \iff (\nabla H(\boldsymbol{x}))^\top \boldsymbol{R}(\boldsymbol{x})\nabla H(\boldsymbol{x}) > 0. \tag{D.4}$$

Now we restrict the vector $\boldsymbol{k}(\boldsymbol{x})$ in (D.2) to the form $\boldsymbol{k}(\boldsymbol{x}) = \boldsymbol{G}^\mathsf{T}(\boldsymbol{x})\nabla\widetilde{V}(\boldsymbol{x})$ and $\hat{\boldsymbol{k}}(\boldsymbol{x})$ to $\hat{\boldsymbol{k}}(\boldsymbol{x}) = \sqrt{\boldsymbol{R}(\boldsymbol{x})}\nabla\hat{V}(\boldsymbol{x})$ with $\widetilde{V}(\boldsymbol{x}), \hat{V}(\boldsymbol{x}) \succ 0$, which leads to

$$\overbrace{\exists\, \widetilde{V}(\boldsymbol{x}) \succ 0 : \; (\nabla\widetilde{V}(\boldsymbol{x}))^\mathsf{T}\boldsymbol{G}(\boldsymbol{x})\boldsymbol{G}^\mathsf{T}(\boldsymbol{x})\nabla H(\boldsymbol{x}) > 0}^{\mathtt{X}_1} \implies \overbrace{(\nabla H(\boldsymbol{x}))^\mathsf{T}\boldsymbol{G}\boldsymbol{G}^\mathsf{T}\nabla H(\boldsymbol{x}) > 0}^{\mathtt{X}_2}, \tag{D.5a}$$

$$\overbrace{\exists\, \hat{V}(\boldsymbol{x}) \succ 0 : \; (\nabla\hat{V}(\boldsymbol{x}))^\mathsf{T}\boldsymbol{R}(\boldsymbol{x})\nabla H(\boldsymbol{x}) > 0}^{\mathtt{X}_3} \implies \overbrace{(\nabla H(\boldsymbol{x}))^\mathsf{T}\boldsymbol{R}(\boldsymbol{x})\nabla H(\boldsymbol{x}) > 0}^{\mathtt{X}_4}. \tag{D.5b}$$

Conjunction of the two statements in (D.5) and contraposition yields $\overline{\mathtt{X}_2} \wedge \overline{\mathtt{X}_4} \implies \overline{\mathtt{X}_1} \wedge \overline{\mathtt{X}_3}$, which can be written as

$$(\nabla H(\boldsymbol{x}))^\mathsf{T}\left(\boldsymbol{R}(\boldsymbol{x}) + \boldsymbol{G}(\boldsymbol{x})\boldsymbol{G}^\mathsf{T}(\boldsymbol{x})\right)\nabla H(\boldsymbol{x}) = 0 \qquad \implies \qquad \overline{\mathtt{X}_1} \wedge \overline{\mathtt{X}_3}. \tag{D.6}$$

The left-hand side of (D.6) holds, if and only if $H(\boldsymbol{x})$ is *no* CLF for the PHS (D.1). The right-hand side of (D.6) implies that there also exists no $V(\boldsymbol{x}) \succ 0$ with $(\nabla V(\boldsymbol{x}))^\mathsf{T}(\boldsymbol{R}(\boldsymbol{x}) + \boldsymbol{G}(\boldsymbol{x})\boldsymbol{G}^\mathsf{T}(\boldsymbol{x}))\nabla \boldsymbol{H}(\boldsymbol{x})$, i.e. there exists no CLF for the PHS (D.1). Thus, we conclude that there exists no CLF for (D.1), if $H(\boldsymbol{x})$ is no CLF for (D.1). The inverse statement is trivial: If there exists no CLF for (D.1), then also $H(\boldsymbol{x})$ is no CLF for (D.1). By Artstein's theorem [Art83] it is known that the existence of a CLF is necessary and sufficient for the asymptotic stabilizability of (D.1) at the origin. Hence we conclude that $H(\boldsymbol{x})$ being a CLF is necessary and sufficient for asymptotic stabilizability of (D.1) at the origin □

Openness of $\mathbb{W}^+$

Lemma D.2. *$\mathbb{W}^+$ is an open set, i.e. for each $\boldsymbol{w} \in \mathbb{W}^+$ there exists an $\varepsilon > 0$ such that each $\boldsymbol{w}' \in \mathbb{R}^r$ with $\|\boldsymbol{w} - \boldsymbol{w}'\| < \varepsilon$ lies within $\mathbb{W}^+$.*

Proof. To prove that $\mathbb{W}^+$ is an open set, two auxiliary sets $\mathbb{Y}^+$ and $\mathbb{Y}^+_{\text{cyl}}$ are introduced which can be proven to be open more conveniently. Finally, openness of $\mathbb{W}^+$ is concluded by a canonical projection of $\mathbb{Y}^+_{\text{cyl}}$:

Let $\mathbb{Y}^+ := \{(\boldsymbol{x}, \boldsymbol{w}) \in \mathbb{R}^n \times \mathbb{R}^r \; : \; V(\boldsymbol{x}, \boldsymbol{w}) > 0\}$ be the set of all $(\boldsymbol{x}, \boldsymbol{w})$ with $V(\boldsymbol{x}, \boldsymbol{w}) > 0$. Then $\mathbb{Y}^+$ is the preimage of $\mathbb{R}^+$, i.e. $\mathbb{Y}^+ = V^{-1}(\mathbb{R}^+)$. Each preimage of a continuous function is open whenever the corresponding image is open [Cro05, Theorem 2.9]. Since $V(\boldsymbol{x}, \boldsymbol{w})$ is continuous and $\mathbb{R}^+$ is open, $\mathbb{Y}^+$ must also be open. Furthermore, we note that $\mathbb{Y}^+$ is nonempty due to the fact that $V(\boldsymbol{x}, \mathbb{0}_r) = H(\boldsymbol{x}) > 0$ is fulfilled by definition for all $\boldsymbol{x} \in \mathbb{R}^n \backslash \{\mathbb{0}\}$. Now consider the $\mathbb{Y}^+$*-inner cylinder* (see Fig. D.1) with $\mathbb{Y}^+_{\text{cyl}}(\underline{\boldsymbol{w}}, \overline{\boldsymbol{w}}) := \{(\boldsymbol{x}, \boldsymbol{w}) \in \mathbb{Y}^+ \; : \boldsymbol{w} \in]\underline{\boldsymbol{w}}, \overline{\boldsymbol{w}}[, (\tilde{\boldsymbol{x}}, \boldsymbol{w}) \in \mathbb{Y}^+ \; \forall\, \tilde{\boldsymbol{x}} \in \mathbb{R}^n \backslash \{\mathbb{0}\}\}$. Obviously, $\mathbb{Y}^+_{\text{cyl}}(\underline{\boldsymbol{w}}, \overline{\boldsymbol{w}}) \subseteq \mathbb{Y}^+$, and according to Lemma D.3 (see below), $\mathbb{Y}^+_{\text{cyl}}(\underline{\boldsymbol{w}}, \overline{\boldsymbol{w}})$ is open for each pair $(\underline{\boldsymbol{w}}, \overline{\boldsymbol{w}}) \in \mathbb{R}^r \times \mathbb{R}^r$.

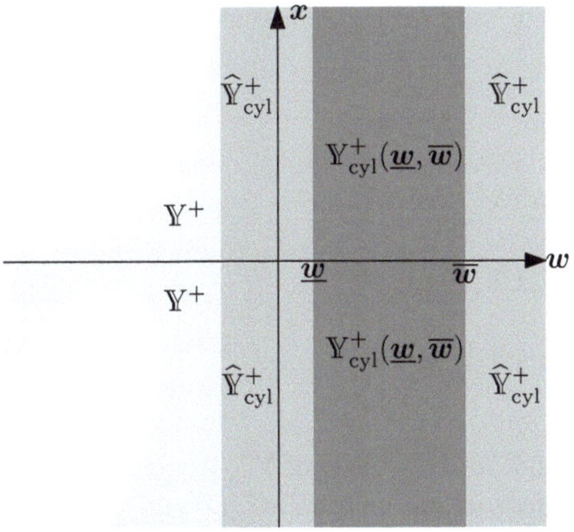

Figure D.1: $\mathbb{Y}^+$ with inner cylinder $\mathbb{Y}^+_{\mathrm{cyl}}(\underline{\boldsymbol{w}}, \overline{\boldsymbol{w}}) \subseteq \mathbb{Y}^+$.

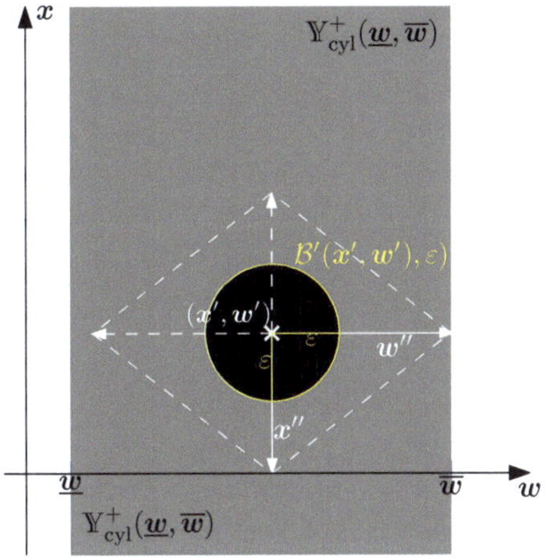

Figure D.2: Each inner cylinder $\mathbb{Y}^+_{\mathrm{cyl}}(\underline{\boldsymbol{w}}, \overline{\boldsymbol{w}})$ is an open set.

With the help of $\mathbb{Y}^+_{\mathrm{cyl}}(\underline{\boldsymbol{w}}, \overline{\boldsymbol{w}})$, we can define the *maximum* $\mathbb{Y}^+$*-inner cylinder*

$$\widehat{\mathbb{Y}}^+_{\mathrm{cyl}} := \bigcup_{\underline{\boldsymbol{w}}, \overline{\boldsymbol{w}} \in \mathbb{R}^r} \mathbb{Y}^+_{\mathrm{cyl}}(\underline{\boldsymbol{w}}, \overline{\boldsymbol{w}}) \tag{D.7}$$

as the union of all possible cylinders $\mathbb{Y}^+_{\mathrm{cyl}}(\underline{\boldsymbol{w}}, \overline{\boldsymbol{w}})$, see Fig. D.1. Since the union of open sets is open [Sin19, Theorem 1.1.9], $\widehat{\mathbb{Y}}^+_{\mathrm{cyl}}$ is open as well. With the canonical projection $\mathrm{proj} : \mathbb{R}^n \times \mathbb{R}^r \to \mathbb{R}^r$, the set $\mathbb{W}^+$ can be interpreted as $\mathbb{W}^+ = \mathrm{proj}(\widehat{\mathbb{Y}}^+_{\mathrm{cyl}})$, i.e. the canonical projection of $\widehat{\mathbb{Y}}^+_{\mathrm{cyl}}$ in $\mathbb{R}^r$. Since projection maps are open maps [Deo18, p. 5], $\mathbb{W}^+$ is an open set. □

Lemma D.3. *The* $\mathbb{Y}^+$*-inner cylinder*

$$\mathbb{Y}^+_{\mathrm{cyl}}(\underline{\boldsymbol{w}}, \overline{\boldsymbol{w}}) := \left\{(\boldsymbol{x}, \boldsymbol{w}) \in \mathbb{Y}^+ : \boldsymbol{w} \in]\underline{\boldsymbol{w}}, \overline{\boldsymbol{w}}[, (\widetilde{\boldsymbol{x}}, \boldsymbol{w}) \in \mathbb{Y}^+ \; \forall \, \widetilde{\boldsymbol{x}} \in \mathbb{R}^n \backslash \{\mathbb{0}\}\right\}$$

with $\underline{\boldsymbol{w}}, \overline{\boldsymbol{w}} \in \mathbb{R}^r$ *is an open set.*

Proof. If $\underline{\boldsymbol{w}} \geq \overline{\boldsymbol{w}}$, then the interval $]\underline{\boldsymbol{w}}, \overline{\boldsymbol{w}}[$ is improper, thus $\mathbb{Y}^+_{\mathrm{cyl}}(\underline{\boldsymbol{w}}, \overline{\boldsymbol{w}}) = \emptyset$. Since empty sets are trivially open, the proof is complete. If $\underline{\boldsymbol{w}} < \overline{\boldsymbol{w}}$, let $(\boldsymbol{x}', \boldsymbol{w}') \in \mathbb{Y}^+_{\mathrm{cyl}}(\underline{\boldsymbol{w}}, \overline{\boldsymbol{w}})$ be an arbitrary point within the cylinder (see Fig. D.2). Now define

$$\boldsymbol{x}'' = \mathrm{col}_i\{\|x'_i\|\}, \qquad i = 1, \dots, n, \tag{D.8}$$

$$\boldsymbol{w}'' = \mathrm{col}_j\{\min\{\overline{w}_j - w'_j, w'_j - \underline{w}_j\}\}, \qquad j = 1, \dots, r, \tag{D.9}$$

where $\boldsymbol{x}''$ denotes the component-wise distances between x'_i and 0, and $\boldsymbol{w}''$ denotes the component-wise distances between w'_j and the lower or upper bounds $\underline{w}_j$ or $\overline{w}_j$, respectively.

From Fig. D.2 it can be seen that by definition of $\boldsymbol{x}''$ and $\boldsymbol{w}''$, it holds that

$$(\boldsymbol{x}' + \frac{1}{2}\boldsymbol{x}'', \boldsymbol{w}') \in \operatorname{int} \mathbb{Y}^+_{\text{cyl}}(\underline{\boldsymbol{w}}, \overline{\boldsymbol{w}}), \tag{D.10}$$

$$(\boldsymbol{x}' - \frac{1}{2}\boldsymbol{x}'', \boldsymbol{w}') \in \operatorname{int} \mathbb{Y}^+_{\text{cyl}}(\underline{\boldsymbol{w}}, \overline{\boldsymbol{w}}), \tag{D.11}$$

$$(\boldsymbol{x}', \boldsymbol{w}' + \frac{1}{2}\boldsymbol{w}'') \in \operatorname{int} \mathbb{Y}^+_{\text{cyl}}(\underline{\boldsymbol{w}}, \overline{\boldsymbol{w}}), \tag{D.12}$$

$$(\boldsymbol{x}', \boldsymbol{w}' - \frac{1}{2}\boldsymbol{w}'') \in \operatorname{int} \mathbb{Y}^+_{\text{cyl}}(\underline{\boldsymbol{w}}, \overline{\boldsymbol{w}}). \tag{D.13}$$

Moreover, it can be seen that both the "upper" and "lower" subset of $\mathbb{Y}^+_{\text{cyl}}(\underline{\boldsymbol{w}}, \overline{\boldsymbol{w}})$ are convex sets. Thus, it is obvious that we can always construct an open ball $\mathcal{B}'((\boldsymbol{x}', \boldsymbol{w}'), \varepsilon)$ around $(\boldsymbol{x}', \boldsymbol{w}')$ with radius

$$\varepsilon = \frac{1}{2} \cdot \left\| \begin{bmatrix} \boldsymbol{x}'' \\ \boldsymbol{w}'' \end{bmatrix} \right\|_\infty \tag{D.14}$$

that lies completely in $\mathbb{Y}^+_{\text{cyl}}(\underline{\boldsymbol{w}}, \overline{\boldsymbol{w}})$. Hence, $\mathbb{Y}^+_{\text{cyl}}(\underline{\boldsymbol{w}}, \overline{\boldsymbol{w}})$ is an open set. □

D.2 Multi-Player Case

Proof of Lemma 5.43

Proof. For linear ISO-PHSs, it holds that

$$\mathcal{X}_{\boldsymbol{G}_i} = \{\boldsymbol{x} \in \mathbb{R}^n : \boldsymbol{G}_i^\top \boldsymbol{Q}\boldsymbol{x} = \mathbb{0}\} = \ker(\boldsymbol{G}_i^\top \boldsymbol{Q}). \tag{D.15}$$

Moreover, in [HJ12, Observation 7.1.6] it is stated that for each positive semidefinite matrix $\boldsymbol{A} \in \mathbb{R}^{n \times n}$, the equality $\boldsymbol{x}^\top \boldsymbol{A}\boldsymbol{x} = 0$ holds, if and only if $\boldsymbol{x} \in \ker(\boldsymbol{A})$. Thus for each $i \in \mathbb{P}$, condition (5.108) is equivalent to

$$\ker(\boldsymbol{G}_i^\top \boldsymbol{Q}) \cap \ker(\boldsymbol{Q}^\top \boldsymbol{R}\boldsymbol{Q}) \cap \left(\bigcap_{\substack{j=1 \\ j \neq i}}^{N} \ker(\Upsilon_j \cdot \boldsymbol{Q}^\top \boldsymbol{G}_j \boldsymbol{S}_j^{-1} \boldsymbol{G}_j^\top \boldsymbol{Q}) \right) = \emptyset. \tag{D.16}$$

Due to the fact that $\Upsilon_j > 0$ and $\ker(\boldsymbol{G}_i^\top \boldsymbol{Q}) \subseteq \ker(\boldsymbol{Q}^\top \boldsymbol{G}_i \boldsymbol{S}_i^{-1} \boldsymbol{G}_i^\top \boldsymbol{Q})$, (D.16) can be written as follows:

$$\begin{aligned}
&\bigcup_{i=1}^{N} \left(\ker(\boldsymbol{G}_i^\top \boldsymbol{Q}) \cap \ker(\boldsymbol{Q}^\top \boldsymbol{R} \boldsymbol{Q}) \cap \left(\bigcap_{\substack{j=1\\ j\neq i}}^{N} \ker(\boldsymbol{Q}^\top \boldsymbol{G}_j \boldsymbol{S}_j^{-1} \boldsymbol{G}_j^\top \boldsymbol{Q}) \right)\right) \\
&= \ker(\boldsymbol{Q}^\top \boldsymbol{R} \boldsymbol{Q}) \cap \bigcup_{i=1}^{N} \left(\ker(\boldsymbol{G}_i^\top \boldsymbol{Q}) \cap \left(\bigcap_{\substack{j=1\\ j\neq i}}^{N} \ker(\boldsymbol{Q}^\top \boldsymbol{G}_j \boldsymbol{S}_j^{-1} \boldsymbol{G}_j^\top \boldsymbol{Q}) \right)\right) \\
&= \ker(\boldsymbol{Q}^\top \boldsymbol{R} \boldsymbol{Q}) \cap \left(\bigcap_{i=1}^{N} \ker(\boldsymbol{G}_i^\top \boldsymbol{Q}) \right) \cap \left(\bigcup_{i=1}^{N} \left(\bigcap_{\substack{j=1\\ j\neq i}}^{N} \ker(\boldsymbol{Q}^\top \boldsymbol{G}_j \boldsymbol{S}_j^{-1} \boldsymbol{G}_j^\top \boldsymbol{Q}) \right)\right) \\
&= \emptyset.
\end{aligned} \tag{D.17}$$

With $\ker(\boldsymbol{Q}\boldsymbol{A}) = \ker \boldsymbol{A}$ for each regular matrix $\boldsymbol{Q}$ together with Lemma D.4 (see below), (D.17) is equivalent to (5.117). □

Lemma D.4. *For each positive definite matrix $\boldsymbol{C} \in \mathbb{R}^N$, it holds that*

$$\ker(\boldsymbol{A}\boldsymbol{C}) \cap \ker(\boldsymbol{B}\boldsymbol{C}) = \emptyset \quad \Longleftrightarrow \quad \ker(\boldsymbol{A}) \cap \ker(\boldsymbol{B}) = \emptyset. \tag{D.18}$$

Proof. The left-hand side of (D.18) can be written as follows

$$\nexists \boldsymbol{x} \in \mathbb{R}^n \backslash \{\mathbb{0}\} : \qquad (\boldsymbol{A}\boldsymbol{C}\boldsymbol{x} = \mathbb{0}) \wedge (\boldsymbol{B}\boldsymbol{C}\boldsymbol{x} = \mathbb{0}). \tag{D.19}$$

With $\boldsymbol{z} := \boldsymbol{C}\boldsymbol{x}$ this is equivalent to

$$\nexists \boldsymbol{z} \in \mathbb{R}^n \backslash \{\mathbb{0}\} : \qquad (\boldsymbol{A}\boldsymbol{z} = \mathbb{0}) \wedge (\boldsymbol{B}\boldsymbol{z} = \mathbb{0}). \tag{D.20}$$

Accordingly, (D.20) states that the kernels of $\boldsymbol{A}$ and $\boldsymbol{B}$ are disjoint. This is equivalent to the right-hand side of (D.18) . □

Proof of Corollary 5.45

Proof. As a consequence of (5.94), it follows that $\boldsymbol{V}(\boldsymbol{x})$ is equivalent to the vector of value functions, if and only if $\Upsilon_i(t) \equiv 1$ for all $i \in \mathbb{P}$. With (5.105), this means that

$$q_i^2(\boldsymbol{x}) + q_i^1(\boldsymbol{x}) + \sum_{\substack{j=1\\ j\neq i}}^{N} q_{ij}^2(\boldsymbol{x}) + q_i^0(\boldsymbol{x}) = 0, \tag{D.21}$$

which is equivalent to

$$\begin{aligned}
&-\frac{1}{2}\left(\frac{\partial V_i}{\partial \boldsymbol{x}}\right)^{\mathsf{T}} \boldsymbol{G}_i(\boldsymbol{x})(\boldsymbol{S}_i(\boldsymbol{x}))^{-1}\boldsymbol{G}_i^{\mathsf{T}}(\boldsymbol{x})\frac{\partial V_i}{\partial \boldsymbol{x}} - \sum_{\substack{j=1\\ j\neq i}}^{N}\left(\frac{\partial V_i}{\partial \boldsymbol{x}}\right)^{\mathsf{T}} \boldsymbol{G}_j(\boldsymbol{x})(\boldsymbol{S}_j(\boldsymbol{x}))^{-1}\boldsymbol{G}_j^{\mathsf{T}}(\boldsymbol{x})\frac{\partial V_j(\boldsymbol{x})}{\partial \boldsymbol{x}}\\
&+\left(\frac{\partial V_i}{\partial \boldsymbol{x}}\right)^{\mathsf{T}}(\boldsymbol{J}(\boldsymbol{x})-\boldsymbol{R}(\boldsymbol{x}))\frac{\partial H}{\partial \boldsymbol{x}} + \frac{1}{2}\ell_i(\boldsymbol{x})\\
&=0. \qquad \text{(D.22)}
\end{aligned}$$

By inserting (5.118), (5.126), and multiplying by -1, we get

$$\begin{aligned}
&\left(\left(\frac{\partial H}{\partial \boldsymbol{x}}\right)^{\mathsf{T}} + \boldsymbol{w}_i^{\mathsf{T}}\frac{\partial \Xi_i}{\partial \boldsymbol{x}}\right)\boldsymbol{K}_i(\boldsymbol{x})\left(\frac{\partial H}{\partial \boldsymbol{x}} + \left(\frac{\partial \Xi_i}{\partial \boldsymbol{x}}\right)^{\mathsf{T}}\boldsymbol{w}_i\right)\\
&+2\sum_{\substack{j=1\\ j\neq i}}^{N}\left(\left(\frac{\partial H}{\partial \boldsymbol{x}}\right)^{\mathsf{T}} + \boldsymbol{w}_i^{\mathsf{T}}\frac{\partial \Xi_i}{\partial \boldsymbol{x}}\right)\boldsymbol{K}_j(\boldsymbol{x})\left(\frac{\partial H}{\partial \boldsymbol{x}} + \left(\frac{\partial \Xi_j}{\partial \boldsymbol{x}}\right)^{\mathsf{T}}\boldsymbol{w}_j\right)\\
&-2\boldsymbol{w}_i^{\mathsf{T}}\frac{\partial \Xi_i}{\partial \boldsymbol{x}}(\boldsymbol{J}(\boldsymbol{x})-\boldsymbol{R}(\boldsymbol{x}))\frac{\partial H}{\partial \boldsymbol{x}} + 2\left(\frac{\partial H}{\partial \boldsymbol{x}}\right)^{\mathsf{T}}\boldsymbol{R}(\boldsymbol{x})\frac{\partial H}{\partial \boldsymbol{x}} - \frac{1}{2}\ell_i(\boldsymbol{x})\\
&=0. \qquad \text{(D.23)}
\end{aligned}$$

Rearranged by $\boldsymbol{w}_i$, this yields

$$\begin{aligned}
&\left(\boldsymbol{w}_i^{\mathsf{T}}\frac{\partial \Xi_i}{\partial \boldsymbol{x}}\right)\boldsymbol{K}_i(\boldsymbol{x})\left(\left(\frac{\partial \Xi_i}{\partial \boldsymbol{x}}\right)^{\mathsf{T}}\boldsymbol{w}_i\right)\\
&+2\left(\left(\frac{\partial H}{\partial \boldsymbol{x}}\right)^{\mathsf{T}}\boldsymbol{K}_i(\boldsymbol{x})\left(\frac{\partial \Xi_i}{\partial \boldsymbol{x}}\right)^{\mathsf{T}}\boldsymbol{w}_i\right) + 2\sum_{\substack{j=1\\ j\neq i}}\left(\left(\frac{\partial H}{\partial \boldsymbol{x}}\right)^{\mathsf{T}} + \boldsymbol{w}_j^{\mathsf{T}}\frac{\partial \Xi_j}{\partial \boldsymbol{x}}\right)\boldsymbol{K}_j^{\mathsf{T}}(\boldsymbol{x})\left(\frac{\partial \Xi_i}{\partial \boldsymbol{x}}\right)^{\mathsf{T}}\boldsymbol{w}_i\\
&+2\left(\frac{\partial H}{\partial \boldsymbol{x}}\right)^{\mathsf{T}}(\boldsymbol{J}(\boldsymbol{x})+\boldsymbol{R}(\boldsymbol{x}))\left(\frac{\partial \Xi_i}{\partial \boldsymbol{x}}\right)^{\mathsf{T}}\boldsymbol{w}_i\\
&+\left(\frac{\partial H}{\partial \boldsymbol{x}}\right)^{\mathsf{T}}\boldsymbol{K}_i(\boldsymbol{x})\frac{\partial H}{\partial \boldsymbol{x}} + 2\sum_{\substack{j=1\\ j\neq i}}^{N}\left(\left(\frac{\partial H}{\partial \boldsymbol{x}}\right)^{\mathsf{T}} + \boldsymbol{w}_j^{\mathsf{T}}\frac{\partial \Xi_j}{\partial \boldsymbol{x}}\right)\boldsymbol{K}_j^{\mathsf{T}}(\boldsymbol{x})\frac{\partial H}{\partial \boldsymbol{x}}\\
&+2\left(\frac{\partial H}{\partial \boldsymbol{x}}\right)^{\mathsf{T}}\boldsymbol{R}(\boldsymbol{x})\frac{\partial H}{\partial \boldsymbol{x}} - \frac{1}{2}\ell_i(\boldsymbol{x})\\
&=0, \qquad \text{(D.24)}
\end{aligned}$$

thus (D.24) can be written in the form (5.121)–(5.126). □

E Supplementary Material to *Case Studies* (Chapter 6)

E.1 Benchmark Model Parameters

Table E.1: Numerical values of the nodal parameters used in Case Studies I and II.

No.	Type	A_i	Γ_i	$X_{d,i} - X'_{d,i}$	$T'_{d0,i}$
1	$\mathcal{L}$	1.61	–	–	–
2	$\mathcal{S}$	1.68	24.73	7.28	0.127
3	$\mathcal{S}$	1.61	20.84	7.41	0.185
4	$\mathcal{I}$	1.65	4.48	–	–
5	$\mathcal{L}$	1.60	–	–	–
6	$\mathcal{S}$	1.54	22.10	6.84	0.147
7	$\mathcal{L}$	1.36	–	–	–
8	$\mathcal{S}$	1.40	23.62	7.63	0.173
9	$\mathcal{S}$	1.38	21.17	7.44	0.168
10	$\mathcal{S}$	1.61	24.61	7.14	0.150
11	$\mathcal{I}$	1.44	4.91	–	–
12	$\mathcal{L}$	1.44	–	–	–
13	$\mathcal{L}$	1.65	–	–	–
14	$\mathcal{I}$	1.50	5.47	–	–
15	$\mathcal{I}$	1.62	5.37	–	–
16	$\mathcal{I}$	1.50	4.55	–	–
17	$\mathcal{I}$	1.33	5.06	–	–
18	$\mathcal{I}$	1.63	4.49	–	–
19	$\mathcal{S}$	1.33	26.97	6.95	0.152
20	$\mathcal{L}$	1.33	–	–	–
21	$\mathcal{S}$	1.29	25.00	6.58	0.150
22	$\mathcal{I}$	1.45	4.56	–	–
23	$\mathcal{L}$	1.28	–	–	–
24	$\mathcal{I}$	1.39	4.55	–	–
25	$\mathcal{I}$	1.31	5.06	–	–
26	$\mathcal{I}$	1.33	4.66	–	–
27	$\mathcal{L}$	1.24	–	–	–
28	$\mathcal{L}$	1.68	–	–	–
29	$\mathcal{S}$	1.35	22.51	6.72	0.186
30	$\mathcal{S}$	1.66	22.54	6.46	0.150

Table E.1: Numerical values of the nodal parameters used in Case Studies I and II (*cont'd*)

No.	Type	A_i	Γ_i	$X_{d,i} - X'_{d,i}$	$T'_{d0,i}$
31	$\mathcal{L}$	1.32	–	–	–
32	$\mathcal{S}$	1.50	25.82	6.64	0.180
33	$\mathcal{I}$	1.69	4.07	–	–
34	$\mathcal{S}$	1.69	25.43	7.04	0.189
35	$\mathcal{L}$	1.43	–	–	–
36	$\mathcal{I}$	1.60	4.44	–	–
37	$\mathcal{S}$	1.67	25.13	7.63	0.149
38	$\mathcal{L}$	1.46	–	–	–
39	$\mathcal{S}$	1.69	26.98	7.58	0.147
40	$\mathcal{S}$	1.29	25.48	7.58	0.144
41	$\mathcal{S}$	1.39	25.31	6.40	0.147
42	$\mathcal{I}$	1.65	4.94	–	–
43	$\mathcal{L}$	1.51	–	–	–
44	$\mathcal{S}$	1.33	23.21	6.65	0.162
45	$\mathcal{I}$	1.30	4.07	–	–
46	$\mathcal{I}$	1.36	4.14	–	–
47	$\mathcal{L}$	1.48	–	–	–
48	$\mathcal{S}$	1.28	22.43	7.28	0.189
49	$\mathcal{I}$	1.61	4.54	–	–
50	$\mathcal{I}$	1.37	4.53	–	–
51	$\mathcal{L}$	1.35	–	–	–
52	$\mathcal{L}$	1.41	–	–	–
53	$\mathcal{I}$	1.58	5.37	–	–
54	$\mathcal{L}$	1.59	–	–	–
55	$\mathcal{S}$	1.40	20.56	7.07	0.123
56	$\mathcal{L}$	1.33	–	–	–
57	$\mathcal{L}$	1.25	–	–	–

References

Public References

[AB95] Aumann, Robert; Brandenburger, Adam: Epistemic Conditions for Nash Equilibrium. In: *Econometrica* 63 (1995), No. 5, p. 1161–1180

[ABAH+20] Al-Badi, Abdullah H.; Ahshan, Razzaqul; Hosseinzadeh, Nasser; Ghorbani, Reza; Hossain, Eklas: Survey of Smart Grid Concepts and Technological Demonstrations Worldwide Emphasizing on the Oman Perspective. In: *Applied System Innovation* 3 (2020), No. 1, p. 5

[ABEPCR18] Avila-Becerril, Sofia; Espinosa-Pérez, Gerardo; Canseco-Rodal, Raul: On the control of power flows in microgrids. In: *IEEE Conference on Decision and Control (CDC)*, IEEE, 2018, p. 3252–3257

[ABMEPG18] Avila-Becerril, Sofía; Montoya, Oscar D.; Espinosa-Pérez, Gerardo; Garcés, Alejandro: Control of a Detailed Model of Microgrids from a Hamiltonian Approach. In: *IFAC-PapersOnLine* 51 (2018), No. 3, p. 187–192

[ACRK19] Alladi, Tejasvi; Chamola, Vinay; Rodrigues, Joel J. P. C.; Kozlov, Sergei A.: Blockchain in Smart Grids: A Review on Different Use Cases. In: *Sensors* 19 (2019), No. 22

[ACRMA19] Aoues, Said; Cardoso-Ribeiro, Flavio L.; Matignon, Denis; Alazard, Daniel: Modeling and Control of a Rotating Flexible Spacecraft: A Port-Hamiltonian Approach. In: *IEEE Transactions on Control Systems Technology* 27 (2019), No. 1, 355–362

[ADJS13] Andreasson, Martin; Dimarogonas, Dimos V.; Johansson, Karl H.; Sandberg, Henrik: Distributed vs. centralized power systems frequency control. In: *European Control Conference (ECC), 2013*, IEEE, 2013, p. 3524–3529

[Ahl18] Ahlheim, Michael: Environmental Economics, the Bioeconomy and the Role of Government. Version: 2018. In: Lewandowski, Iris (ed.): *Bioeconomy*. Cham: Springer International Publishing, 2018, p. 317–329

[AHU58] Arrow, Kenneth J.; Hurwicz, Leonid; Uzawa, Hirofumi: *Studies in Linear and Non-Linear Programming*. Redwood City: Stanford University Press, 1958

[AL13] Abouheaf, Mohammed I.; Lewis, Frank L.: Multi-agent differential graphical games: Nash online adaptive learning solutions. In: *IEEE Conference on Decision and Control (CDC)*, IEEE, 2013, p. 5803–5809

[ALCB15] Alshehri, Khaled; Liu, Ji; Chen, Xudong; Başar, Tamer: A Stackelberg game for multi-period demand response management in the smart grid. In: *IEEE Conference on Decision and Control (CDC)*, IEEE, 2015, p. 5889–5894

[And] Andersson, Göran: *Skript zur Vorlesung Elektrische Energiesysteme: Vorlesungsteil Energieübertragung*, ETH Zürich, Diss.

[ANHMIT20] Abapour, Saeed; Nazari-Heris, Morteza; Mohammadi-Ivatloo, Behnam; Tarafdar Hagh, Mehrdad: Game Theory Approaches for the Solution of Power System Problems: A Comprehensive Review. In: *Archives of Computational Methods in Engineering* 27 (2020), No. 1, p. 81–103

[Ant94] Antipin, Anatoly: Minimization of convex functions on covex sets by means of differential equations. In: *Differential Equations* 30 (1994), No. 9, p. 1365–1375

[AOS+13a] Atzeni, Italo; Ordonez, Luis G.; Scutari, Gesualdo; Palomar, Daniel P.; Fonollosa, Javier R.: Noncooperative and Cooperative Optimization of Distributed Energy Generation and Storage in the Demand-Side of the Smart Grid. In: *IEEE Transactions on Signal Processing* 61 (2013), No. 10, p. 2454–2472

[AOS+13b] Atzeni, Italo; Ordonez, Luis G.; Scutari, Gesualdo; Palomar, Daniel P.; Fonollosa, Javier R.: Demand-Side Management via Distributed Energy Generation and Storage Optimization. In: *IEEE Transactions on Smart Grid* 4 (2013), No. 2, p. 866–876

[AP21] Anaya, Karim L.; Pollitt, Michael G.: How to Procure Flexibility Services within the Electricity Distribution System: Lessons from an International Review of Innovation Projects. In: *Energies* 14 (2021), No. 15, p. 4475

[APM16] Arcak, Murat; Packard, Andrew; Meissen, Chris: *Networks of dissipative systems: Compositional certification of stability, performance, and safety.* Cham: Springer, 2016 (SpringerBriefs in electrical and computer engineering, Control, automation and robotics)

[ARO+18] Ahmad, Fiaz; Rasool, Akhtar; Ozsoy, Emre; Sekar, Raja; Sabanovic, Asif; Elitaş, Meltem: Distribution system state estimation-A step towards smart grid. In: *Renewable and Sustainable Energy Reviews* 81 (2018), p. 2659–2671

[Art83] Artstein, Zvi: Stabilization with relaxed controls. In: *Nonlinear Analysis: Theory, Methods & Applications* 7 (1983), No. 11, p. 1163–1173

[AS74] Alsac, Ongun; Stott, Brian: Optimal Load Flow with Steady-State Security. In: *IEEE Transactions on Power Apparatus and Systems* PAS-93 (1974), No. 3, p. 745–751

[AS17] Altmann, Robert; Schulze, Philipp: A port-Hamiltonian formulation of the Navier–Stokes equations for reactive flows. In: *Systems & Control Letters* 100 (2017), p. 51–55

[ASB01] ALBERTO, Luís F.C.; SILVA, Flávio H.; BRETAS, Newton G.: Direct methods for transient stability analysis in power systems: state of art and future perspectives. In: *IEEE Porto Power Tech Proceedings (Cat. No.01EX502)*, IEEE, 2001, p. 6

[ASF+20] AZAR, Ahmad T.; SERRANO, Fernando E.; FLORES, Marco A.; VAIDYANATHAN, Sundarapandian; ZHU, Quanmin: Adaptive neural-fuzzy and backstepping controller for port-Hamiltonian systems. In: *International Journal of Computer Applications in Technology* 62 (2020), No. 1, p. 1–12

[ATM+20] ABLEITNER, Liliane; TIEFENBECK, Verena; MEEUW, Arne; WÖRNER, Anselma; FLEISCH, Elgar; WORTMANN, Felix: User behavior in a real-world peer-to-peer electricity market. In: *Applied Energy* 270 (2020), p. 115061

[Aum76] AUMANN, Robert J.: Agreeing to Disagree. In: *The Annals of Statistics* 4 (1976), No. 6, p. 1236–1239

[AvJ17] ADIBI, Mahya; VAN DER WOUDE, Jacob W.; JELTSEMA, Dimitri: A port-Hamiltonian approach to secondary voltage control of microgrids. In: *IEEE PES Innovative Smart Grid Technologies Conference Europe (ISGT-Europe)*, IEEE, 2017, p. 1–6

[BA03] BRETAS, Newton G.; ALBERTO, Luís F.C.: Lyapunov function for power systems with transfer conductances: extension of the invariance principle. In: *IEEE Transactions on Power Systems* 18 (2003), No. 2, p. 769–777

[BAA+21] BAASHAR, Yahia; ALKAWSI, Gamal; ALKAHTANI, Ammar A.; HASHIM, Wahidah; RAZALI, Rina A.; TIONG, Sieh K.: Toward Blockchain Technology in the Energy Environment. In: *Sustainability* 13 (2021), No. 16, p. 9008

[Bak21] BAKER, Kyri: Solutions of DC OPF are Never AC Feasible. In: *ACM International Conference on Future Energy Systems*, ACM, 2021, p. 264–268

[Bap14] BAPAT, Ravindra B.: *Graphs and Matrices*. London: Springer, 2014

[Bau16] BAUSO, Dario: *Advances in design and control*. Vol. 30: *Game theory with engineering applications*. Philadelphia: Society for Industrial and Applied Mathematics, 2016

[BBM+14] BRUCKNER, Thomas; BASHMAKOV, Igor A.; MULUGETTA, Yacob; CHUM, Helena; DE LA VEGA NAVARRO, ANGEL; EDMONDS, James A.; FAAIJ, Andre; FUNGTAMMASAN, Bundit; GARG, Amit; HERTWICH, Edgar; HONNERY, Damon R.; INFIELD, David; KAINUMA, Mikiko; KHENNAS, Smail; KIM, Suduk; NIMIR, Hassan B.; RIAHI, Keywan; STRACHAN, Neil; WISER, Ryan; ZHANG, Xiliang: Energy Systems. Version: 2014. In: EDENHOFER, Ottmar (ed.); PICHS-MADRUGA, Ramón (ed.); SOKONA, Youba (ed.); FARAHANI, Ellie (ed.); KADNER, Susanne (ed.); SEYBOTH, Kristin (ed.); ADLER, Anna (ed.); BAUM, Ina (ed.); BRUNNER, Steffen (ed.); EICKEMEIER, Patrick (ed.); KRIEMANN, Benjamin (ed.); SAVOLAINEN, Jussi (ed.); SCHLÖMER, Steffen (ed.); VON STECHOW, Christoph (ed.); ZWICKEL, Timm (ed.); MINX, Jan C. (ed.): *Climate Change 2014: Mitigation of Climate Change.*

Contribution of Working Group III to the Fifth Assessment Report of the Intergovernmental Panel on Climate Change. Cambridge: Cambridge University Press, 2014, p. 511–598

[BBPS09] Borrelli, Francesco; Baotic, Mato; Pekar, Jaroslav; Stewart, Greg: On the complexity of explicit MPC laws. In: *European Control Conference (ECC)*, IEEE, 2009, p. 2408–2413

[BD15] Bolognani, Saverio; Dörfler, Florian: Fast power system analysis via implicit linearization of the power flow manifold. In: *Annual Allerton Conference on Communication, Control, and Computing (Allerton)*, IEEE, 2015, p. 402–409

[BD19] Bernstein, Andrey; Dall'Anese, Emiliano: Real-Time Feedback-Based Optimization of Distribution Grids: A Unified Approach. In: *IEEE Transactions on Control of Network Systems* 6 (2019), No. 3, p. 1197–1209

[BDE12] Brunner, Florian D.; Dürr, Hans-Bernd; Ebenbauer, Christian: Feedback design for multi-agent systems: A saddle point approach. In: *IEEE Conference on Decision and Control (CDC)*, IEEE, 2012, p. 3783–3789

[BDST17] Bergna-Diaz, Gilbert; Sanchez, Santiago; Tedeschi, Elisabetta: Port-Hamiltonian modelling of Modular Multilevel Converters with fixed equilibrium point. In: *International Conference on Ecological Vehicles and Renewable Energies*, 2017, p. 1–12

[BGM+14] Brunekreeft, Gert; Goto, Mika; Meyer, Roland; Maruyama, Masahiro; Hattori, Toru: *Bremen Energy Working Papers*. Vol. 16: *Unbundling of electricity transmission system operators in Germany: An experience report*. 2014

[BH07] Beck, Hans-Peter; Hesse, Ralf: Virtual synchronous machine. In: *International Conference on Electrical Power Quality and Utilisation*, IEEE, 2007

[BHJ16] Bertsch, Joachim; Hagspiel, Simeon; Just, Lisa: Congestion management in power systems. In: *Journal of Regulatory Economics* 50 (2016), No. 3, p. 290–327

[BHK78] Breton, A.; Haurie, A.; Kalocsai, R.: Efficient management of interconnected power systems: A game-theoretic approach. In: *Automatica* 14 (1978), No. 5, p. 443–452

[Bic04] Bicchieri, Cristina: Rationality and Game Theory. Version: 2004. In: Mele, Alfred R. (ed.); Rawling, Piers (ed.): *The Oxford Handbook of Rationality*. Oxford University Press, 2004, p. 182–205

[BJ16] Bian, Tao; Jiang, Zhong-Ping: Value iteration, adaptive dynamic programming, and optimal control of nonlinear systems. In: *IEEE Conference on Decision and Control (CDC)*, IEEE, 2016, p. 3375–3380

[BL11] Battaglini, Antonella; Lilliestam, Johan; Heinrich Böll Foundation (ed.): *On Transmission Grid Governance*. Berlin, 2011 (ERENE Issue Paper)

[BLMH+21] Belgioioso, Giuseppe; Liao-McPherson, Dominic; Hudoba de Badyn, Mathias; Bolognani, Saverio; Lygeros, John; Dörfler, Florian: *Sampled-Data Online Feedback Equilibrium Seeking: Stability and Tracking*. http://arxiv.org/pdf/2103.13988v2. Version: 2021

[BM76] Bondy, John A.; Murthy, Uppaluri S. R.: *Graph theory with applications*. New York: North Holland, 1976

[BM17] Bager, Simon; Mundaca, Luis: Making 'Smart Meters' smarter? Insights from a behavioural economics pilot field experiment in Copenhagen, Denmark. In: *Energy Research & Social Science* 28 (2017), p. 68–76

[BMC+20] Behabtu, Henok A.; Messagie, Maarten; Coosemans, Thierry; Berecibar, Maitane; Anlay Fante, Kinde; Kebede, Abraham A.; van Mierlo, Joeri: A Review of Energy Storage Technologies' Application Potentials in Renewable Energy Sources Grid Integration. In: *Sustainability* 12 (2020), No. 24, p. 10511

[BO99] Başar, Tamer; Olsder, Geert J.: *Classics in Applied Mathematics*. Vol. 23: *Dynamic noncooperative game theory*. 2nd edition. Philadelphia: SIAM, 1999

[Bor93] Bork, Robert H.: *The antitrust paradox: A policy at war with itself*. New York: Free Press, 1993

[Bos14] Bosch, Siegfried: *Lineare Algebra*. Berlin, Heidelberg: Springer, 2014

[BPP+08] Barklund, E.; Pogaku, Nagaraju; Prodanovic, Milan; Hernandez-Aramburo, Carlos A.; Green, Tim C.: Energy Management in Autonomous Microgrid Using Stability-Constrained Droop Control of Inverters. In: *IEEE Transactions on Power Electronics* 23 (2008), No. 5, p. 2346–2352

[Bre11] Bressan, Alberto: Noncooperative Differential Games. In: *Milan Journal of Mathematics* 79 (2011), No. 2, p. 357–427

[BS13] Bevrani, Hassan; Shokoohi, Shoresh: An Intelligent Droop Control for Simultaneous Voltage and Frequency Regulation in Islanded Microgrids. In: *IEEE Transactions on Smart Grid* 4 (2013), No. 3, p. 1505–1513

[BT17] Baldivieso Monasterios, Pablo R.; Trodden, Paul: Low-Complexity Distributed Predictive Automatic Generation Control With Guaranteed Properties. In: *IEEE Transactions on Smart Grid* 8 (2017), No. 6, p. 3045–3054

[Bun20] Bundesnetzagentur: *Market manipulation in the wholesale energy market in connection with imbalances*. https://www.bundesnetzagentur.de/SharedDocs/Downloads/EN/BNetzA/PressSection/PressReleases/2020/202009027_Marketmanipulation.pdf;jsessionid=5E3AA156517305CF462FDA42F423382D?__blob=publicationFile&v=2. Version: 07.09.2020

[Bun21] Bundesverband der Energie- und Wasserwirtschaft e.V.: *Strompreisanalyse Juni 2021: Haushalte und Industrie*. https://www.bdew.de/media/documents/BDEW-Strompreisanalyse_no_halbjaehrlich_Ba_online_10062021.pdf. Version: 2021

[BV00] Bergen, Arthur R.; Vittal, Vijay: *Power systems analysis.* 2nd edition. Upper Saddle River, NJ: Prentice Hall, 2000

[BV15] Boyd, Stephen P.; Vandenberghe, Lieven: *Convex optimization.* 18th print edition. Cambridge: Cambridge Univ. Press, 2015

[Cab17] Cabral, Luis M. B.: *Introduction to industrial organization.* 2nd edition. Cambridge, London: The MIT Press, 2017

[CB04] Curtis, Jess W.; Beard, Randal W.: Satisficing: A New Approach to Constructive Nonlinear Control. In: *IEEE Transactions on Automatic Control* 49 (2004), No. 7, p. 1090–1102

[CBK17] Chakraborty, Pratyush; Baeyens, Enrique; Khargonekar, Pramod P.: Distributed control of flexible demand using proportional allocation mechanism in a smart grid: Game theoretic interaction and price of anarchy. In: *Sustainable Energy, Grids and Networks* 12 (2017), p. 30–39

[CBP+19] Chakraborty, Pratyush; Baeyens, Enrique; Poolla, Bala K.; Khargonekar, Pramod P.; Varaiya, Pravin: Sharing Storage in a Smart Grid: A Coalitional Game Approach. In: *IEEE Transactions on Smart Grid* 10 (2019), No. 4, p. 4379–4390

[CC18] Cavraro, Guido; Carli, Ruggero: Local and Distributed Voltage Control Algorithms in Distribution Networks. In: *IEEE Transactions on Power Systems* 33 (2018), No. 2, p. 1420–1430

[CC20] Cherukuri, Ashish; Cortés, Jorge: Iterative Bidding in Electricity Markets: Rationality and Robustness. In: *IEEE Transactions on Network Science and Engineering* 7 (2020), No. 3, p. 1265–1281

[CCG02] Coelho, Ernane A.; Cortizo, Porfirio C.; Garcia, Pedro F.: Small-signal stability for parallel-connected inverters in stand-alone AC supply systems. In: *IEEE Transactions on Industry Applications* 38 (2002), No. 2, p. 533–542

[CDA93] Chandorkar, Mukul C.; Divan, Deepakraj M.; Adapa, Rambabu: Control of parallel connected inverters in standalone AC supply systems. In: *IEEE Transactions on Industry Applications* 29 (1993), No. 1, p. 136–143

[CGB+19] Cupelli, Marco; Gurumurthy, Sriram K.; Bhanderi, Siddharth; Cupelli, Lisette; Monti, Antonello: Power Sharing Control in Microgrids - an Approach Guaranteeing Large Signal Stability. In: *2019 IEEE Conference on Power Electronics and Renewable Energy (CPERE)*, IEEE, 2019, p. 379–384

[CGHM13] Castaños, Fernando; Gromov, Dmitry; Hayward, Vincent; Michalska, Hannah: Implicit and explicit representations of continuous-time port-Hamiltonian systems. In: *Systems & Control Letters* 62 (2013), No. 4, p. 324–330

[CGM19] Coelho, Paulo; Gomes, Mário; Moreira, Carlos: Smart Metering Technology. Version: 2019. In: Zambroni de Souza, Antonio Carlos (ed.); Castilla, Miguel (ed.): *Microgrids Design and Implementation.* Cham: Springer International Publishing, 2019, p. 97–137

[CHMG14] CLUDIUS, Johanna; HERMANN, Hauke; MATTHES, Felix C.; GRAICHEN, Verena: The merit order effect of wind and photovoltaic electricity generation in Germany 2008–2016: Estimation and distributional implications. In: *Energy Economics* 44 (2014), p. 302–313

[CK19] CHIS, Adriana; KOIVUNEN, Visa: Coalitional Game-Based Cost Optimization of Energy Portfolio in Smart Grid Communities. In: *IEEE Transactions on Smart Grid* 10 (2019), No. 2, p. 1960–1970

[CLI+17] CVIJIC, Sanja; LANG, Jeffrey; ILIC, Marija; BABAEI, Saman; STEFOPOULOS, George: Reliable adaptive optimization demonstration using big data. In: *IEEE Power & Energy Society General Meeting*, IEEE, 2017, p. 1–5

[CLLV14] CHEN, He; LI, Yonghui; LOUIE, Raymond H. Y.; VUCETIC, Branka: Autonomous Demand Side Management Based on Energy Consumption Scheduling and Instantaneous Load Billing: An Aggregative Game Approach. In: *IEEE Transactions on Smart Grid* 5 (2014), No. 4, p. 1744–1754

[CMC16] CHERUKURI, Ashish; MALLADA, Enrique; CORTÉS, Jorge: Asymptotic convergence of constrained primal–dual dynamics. In: *Systems & Control Letters* 87 (2016), p. 10–15

[Con14] CONNECT ENERGY ECONOMICS GMBH: *Leitstudie Strommarkt - Arbeitspaket Optimierung des Strommarktdesigns [Pilot study electricity market: Working package optimization of electricity market design]: Final report for Federal Ministry for Economic Affairs and Energy.* 2014

[Con19] CONSENTEC: Untersuchung zur Beschaffung von Redispatch [Procurement of Redispatch]: Project (055/17) on behalf of the Federal Ministry for Economic Affairs and Energy. (2019). `https://www.bmwi.de/Redaktion/DE/Publikationen/Studien/untersuchung-zur-beschaffung-von-redispatch.pdf`

[Cou21] COUNCIL OF THE EUROPEAN UNION: Regulation (EU) 2021/1119 establishing the framework for achieving climate neutrality ('European Climate Law'). In: *Official Journal of the European Union* (2021), No. L 243, p. 1–17

[Cro05] CROSSLEY, Martin D.: *Essential Topology.* London: Springer, 2005 (Springer undergraduate mathematics series)

[CS13] CALABRIA, Mauro; SCHUMACHER, Walter: Impact of inverter clustering on the small-signal stability of a grid. In: PAGE, Bernd (ed.); FLEISCHER, Andreas G. (ed.); GÖBEL, Johannes (ed.); WOHLGEMUTH, Volker (ed.): *Proceedings of the 27th Conference on Environmental Informatics - Informatics for Environmental Protection, Sustainable Development and Risk Management.* Aachen: Shaker Verlag, 2013

[CSD+20] CHERUKURI, Ashish; STEGINK, Tjerk W.; DE PERSIS, Claudio; VAN DER SCHAFT, Arjan J.; CORTÉS, Jorge: *Frequency-driven market mechanisms for optimal dispatch in power networks.* `http://carmenere.ucsd.edu/jorge/publications/data/2020_StChPeScCo-auto.pdf`. Version: 2020

[CSML13] Christofides, Panagiotis D.; Scattolini, Riccardo; Muñoz de la Peña, David; Liu, Jinfeng: Distributed model predictive control: A tutorial review and future research directions. In: *Computers & Chemical Engineering* 51 (2013), p. 21–41

[CSP+19] Chamorro, Harold R.; Sanchez, Andres C.; Pantoja, Andres; Zelinka, Ivan; Gonzalez-Longatt, Francisco; Sood, Vijay K.: A network control system for hydro plants to counteract the non-synchronous generation integration. In: *International Journal of Electrical Power & Energy Systems* 105 (2019), No. 1, p. 404–419

[CSPB19] Colombino, Marcello; Simpson-Porco, John W.; Bernstein, Andrey: Towards robustness guarantees for feedback-based optimization. In: *IEEE Conference on Decision and Control (CDC)*, IEEE, 2019, p. 6207–6214

[CT14] Caliskan, Sina Y.; Tabuada, Paulo: Compositional Transient Stability Analysis of Multimachine Power Networks. In: *IEEE Transactions on Control of Network Systems* 1 (2014), No. 1, p. 4–14

[CTFS18] Cucuzzella, Michele; Trip, Sebastian; Ferrara, Antonella; Scherpen, Jacquelien M. A.: Cooperative Voltage Control in AC Microgrids. In: *IEEE Conference on Decision and Control (CDC)*, IEEE, 2018, p. 6723–6728

[CVV17] Coletta, Guido; Vaccaro, Alfredo; Villacci, Domenico: A review of the enabling methodologies for PMUs-based dynamic thermal rating of power transmission lines. In: *Electric Power Systems Research* 152 (2017), p. 257–270

[CY17] Chen, Lijun; You, Seungil: Reverse and Forward Engineering of Frequency Control in Power Networks. In: *IEEE Transactions on Automatic Control* 62 (2017), No. 9, p. 4631–4638

[CY19] Cheng, Lefeng; Yu, Tao: Game-Theoretic Approaches Applied to Transactions in the Open and Ever-Growing Electricity Markets From the Perspective of Power Demand Response: An Overview. In: *IEEE Access* 7 (2019), p. 25727–25762

[CYWL15] Chen, Haoyong; Ye, Rong; Wang, Xiaodong; Lu, Runge: Cooperative Control of Power System Load and Frequency by Using Differential Games. In: *IEEE Transactions on Control Systems Technology* 23 (2015), No. 3, p. 882–897

[CZ18] Chen, Juntao; Zhu, Quanyan: A Stackelberg Game Approach for Two-Level Distributed Energy Management in Smart Grids. In: *IEEE Transactions on Smart Grid* 9 (2018), No. 6, p. 6554–6565

[Das95] Dastidar, Krishnendu G.: On the existence of pure strategy Bertrand equilibrium. In: *Economic Theory* 5 (1995), No. 1, p. 19–32

[DBK+18] Das, Choton K.; Bass, Octavian; Kothapalli, Ganesh; Mahmoud, Thair S.; Habibi, Daryoush: Overview of energy storage systems in distribution networks: Placement, sizing, operation, and power quality. In: *Renewable and Sustainable Energy Reviews* 91 (2018), p. 1205–1230

[DBMS09] Duindam, Vincent (ed.); Bruyninckx, Herman (ed.); Macchelli, Alessandro (ed.); Stramigioli, Stefano (ed.): *Modeling and Control of Complex Physical Systems: The Port-Hamiltonian Approach.* Berlin, Heidelberg: Springer, 2009

[DBSPG19] Dörfler, Florian; Bolognani, Saverio; Simpson-Porco, John W.; Grammatico, Sergio: Distributed Control and Optimization for Autonomous Power Grids. In: *European Control Conference (ECC),* IEEE, 2019, p. 2436–2453

[DCB13] Dörfler, Florian; Chertkov, Michael; Bullo, Francesco: Synchronization in complex oscillator networks and smart grids. In: *Proceedings of the National Academy of Sciences of the United States of America* 110 (2013), No. 6, p. 2005–2010

[dD21] d'Aspremont, Claude; Dos Santos Ferreira, Rodolphe: *The economics of competition, collusion and in-between.* Cham: Palgrave Macmillan, 2021

[DdGP10] DeLellis, Pietro; diBernardo, Mario; Garofalo, Franco; Porfiri, Maurizio: Evolution of Complex Networks via Edge Snapping. In: *IEEE Transactions on Circuits and Systems I: Regular Papers* 57 (2010), No. 8, p. 2132–2143

[DE18] Dissanayake, Anushka M.; Ekneligoda, Nishantha C.: Game theoretic transient control of parallel connected inverters in islanded microgrids. In: *IEEE Power & Energy Society Innovative Smart Grid Technologies Conference (ISGT),* IEEE, 2018, p. 1–5

[DEA$^+$17] Daki, Houda; El Hannani, Asmaa; Aqqal, Abdelhak; Haidine, Abdelfattah; Dahbi, Aziz. Big Data management in smart grid: concepts, requirements and implementation. In: *Journal of Big Data* 4 (2017), No. 1

[Deb59] Debreu, Gerard: *Monograph / Cowles Foundation for Research in Economics at Yale University.* Vol. 17: *Theory of value: An axiomatic analysis of economic equilibrium.* New York: Wiley, 1959

[Deo18] Deo, Satya: *Algebraic Topology : A Primer.* 2nd edition. Singapore: Springer, 2018 (Texts and Readings in Mathematics)

[Deu14] Deutscher Bundestag: Gesetz zur grundlegenden Reform des Erneuerbare-Energien-Gesetzes und zur Änderung weiterer Bestimmungen des Energiewirtschaftsrechts. In: *Bundesgesetzblatt I* Vol. 33. 2014, p. 1066–1132

[Deu19] Deutscher Bundestag: Bundes-Klimaschutzgesetz (KSG). In: *Bundesgesetzblatt I* Vol. 48. 2019, p. 2513–2521

[Deu20] Deutscher Bundestag: Gesetz zur Änderung des Erneuerbare-Energien-Gesetzes und weiterer energierechtlicher Vorschriften. In: *Bundesgesetzblatt I* Vol. 65. 2020, p. 3138–3205

[DG17] Dörfler, Florian; Grammatico, Sergio: Gather-and-broadcast frequency control in power systems. In: *Automatica* 79 (2017), p. 296–305

[DKH19] De Viveiro, Gustavo; Kurdziel, Marie-Jeanne; Hagemann, Markus: *Transition towards a decarbonised electricity sector – A framework of analysis for power system transformation*, NewClimate Institute, Re-expertise, 2019

[DM16] De Persis, Claudio; Monshizadeh, Nima: A modular design of incremental Lyapunov functions for microgrid control with power sharing. In: *European Control Conference (ECC)*, IEEE, 2016, p. 1501–1506

[DM19] De Persis, Claudio; Monshizadeh, Nima: A Feedback Control Algorithm to Steer Networks to a Cournot–Nash Equilibrium. In: *IEEE Transactions on Control of Network Systems* 6 (2019), No. 4, p. 1486–1497

[DMSD16] De Persis, Claudio; Monshizadeh, Nima; Schiffer, Johannes; Dörfler, Florian: A Lyapunov approach to control of microgrids with a network-preserved differential-algebraic model. In: *IEEE Conference on Decision and Control (CDC)*, IEEE, 2016, p. 2595–2600

[DMSP17] Dvijotham, Krishnamurthy; Mallada, Enrique; Simpson-Porco, John W.: High-Voltage Solution in Radial Power Networks: Existence, Properties, and Equivalent Algorithms. In: *IEEE Control Systems Letters* 1 (2017), No. 2, p. 322–327

[Dou88] Douglass, Dale A.: Weather-dependent versus static thermal line ratings (power overhead lines). In: *IEEE Transactions on Power Delivery* 3 (1988), No. 2, p. 742–753

[dQQA19] do Prado, Josue; Qiao, Wei; Qu, Liyan; Agüero, Julio: The Next-Generation Retail Electricity Market in the Context of Distributed Energy Resources: Vision and Integrating Framework. In: *Energies* 12 (2019), No. 3, p. 491

[DRVA10] Del Carpio Huayllas, Tesoro E.; Ramos, Dorel S.; Vasquez-Arnez, Ricardo L.: Microgrid systems: Current status and challenges. In: *IEEE/PES Transmission and Distribution Conference and Exposition: Latin America (T&D-LA)*, IEEE, 2010, p. 7–12

[DSE+21] Dkhili, Nouha; Salas, David; Eynard, Julien; Thil, Stéphane; Grieu, Stéphane: Innovative Application of Model-Based Predictive Control for Low-Voltage Power Distribution Grids with Significant Distributed Generation. In: *Energies* 14 (2021), No. 6, p. 1773

[DSF15] D'Arco, Salvatore; Suul, Jon A.; Fosso, Olav B.: A Virtual Synchronous Machine implementation for distributed control of power converters in Smart-Grids. In: *Electric Power Systems Research* 122 (2015), No. 11, p. 180–197

[DSPB16] Dörfler, Florian; Simpson-Porco, John W.; Bullo, Francesco: Breaking the Hierarchy: Distributed Control and Economic Optimality in Microgrids. In: *IEEE Transactions on Control of Network Systems* 3 (2016), No. 3, p. 241–253

[DT68] Dommel, Hermann; Tinney, William: Optimal Power Flow Solutions. In: *IEEE Transactions on Power Apparatus and Systems* PAS-87 (1968), No. 10, p. 1866–1876

[Dv03] DIELMANN, Klaus-Peter; VAN DER VELDEN, Alwin: Virtual power plants (VPP) - a new perspective for energy generation? In: *International Scientific and Practical Conference of Students, Post-graduates Modern Techniques and Technologies*, IEEE, 2003, p. 18–20

[ED18] EMMERICH, Michael T. M.; DEUTZ, André H.: A tutorial on multiobjective optimization: fundamentals and evolutionary methods. In: *Natural computing* 17 (2018), No. 3, p. 585–609

[EE19] ENTSO-E: *Report on Deterministic Frequency Deviations.* https://consultations.entsoe.eu/system-development/deterministic_frequency_deviations_report/user_uploads/report_deterministic_frequency_deviations_final-draft-for-consultation.pdf. Version: 2019

[EE21] ENTSO-E: *Costs of Congestion Management.* https://transparency.entsoe.eu/congestion-management/r2/costs/show. Version: 2021

[EES+19] EL RAHI, Georges; ETESAMI, Seyed R.; SAAD, Walid; MANDAYAM, Narayan B.; POOR, Harold V.: Managing Price Uncertainty in Prosumer-Centric Energy Trading: A Prospect-Theoretic Stackelberg Game Approach. In: *IEEE Transactions on Smart Grid* 10 (2019), No. 1, p. 702–713

[EMG+21] ELA, Erik; MILLS, Andrew; GIMON, Eric; HOGAN, Mike; BOUCHEZ, Nicole; GIACOMONI, Anthony; NG, Hok; GONZALEZ, Jim; DESOCIO, Mike: Electricity Market of the Future: Potential North American Designs Without Fuel Costs. In: *IEEE Power and Energy Magazine* 19 (2021), No. 1, p. 41–52

[Eng06] ENGWERDA, Jacob C.: *LQ Dynamic Optimization and Differential Games.* Hoboken: John Wiley & Sons, Ltd, 2006

[ES13] ENGWERDA, Jacob C.; SALMAH: Necessary and Sufficient Conditions for Feedback Nash Equilibria for the Affine-Quadratic Differential Game. In: *Journal of Optimization Theory and Applications* 157 (2013), No. 2, p. 552–563

[Eur20] EUROPEAN COMMISSION: *National Energy and Climate Plans: Member State contributions to the EU's 2030 climate ambition.* https://ec.europa.eu/commission/presscorner/detail/en/fs_20_1611. Version: 2020

[FAP18] FAN, Songli; AI, Qian; PIAO, Longjian: Bargaining-based cooperative energy trading for distribution company and demand response. In: *Applied Energy* 226 (2018), No. 14, p. 469–482

[FCL13] FARIVAR, Masoud; CHEN, Lijun; LOW, Steven: Equilibrium and dynamics of local voltage control in distribution systems. In: *IEEE Conference on Decision and Control (CDC)*, IEEE, 2013, p. 4329–4334

[FCSP+20] FARROKHABADI, Mostafa; CAÑIZARES, Claudio A.; SIMPSON-PORCO, John W.; NASR, Ehsan; FAN, Lingling; MENDOZA-ARAYA, Patricio A.; TONKOSKI, Reinaldo; TAMRAKAR, Ujjwol; HATZIARGYRIOU, Nikos D.; LAGOS, Dimitris; WIES, Richard W.; PAOLONE, Mario; LISERRE, Marco; MEEGAHAPOLA, Lasantha;

KABALAN, Mahmoud; HAJIMIRAGHA, Amir H.; PERALTA, Dario; ELIZONDO, Marcelo A.; SCHNEIDER, Kevin P.; TUFFNER, Francis K.; REILLY, Jim: Microgrid Stability Definitions, Analysis, and Examples. In: *IEEE Transactions on Power Systems* 35 (2020), No. 1, p. 13–29

[Fek21] FEKETE, Patrick: *Redispatch in Deutschland: Auswertung der Transparenzdaten April 2013 bis einschließlich Dezember 2020.* 2021

[FH16] FALAIZE, Antoine; HÉLIE, Thomas: Passive Guaranteed Simulation of Analog Audio Circuits: A Port-Hamiltonian Approach. In: *Applied Sciences* 6 (2016), No. 10, p. 273

[FHS03] FUJIMOTO, Kenji; HORIUCHI, Tetsu; SUGIE, Toshiharu: Optimal control of Hamiltonian systems with input constraints via iterative learning. In: *IEEE Conference on Decision and Control (CDC)*, IEEE, 2003, p. 4387–4392

[FK08a] FREEMAN, Randy A.; KOKOTOVIĆ, Petar V.: *Robust nonlinear control design: State-space and Lyapunov techniques.* Reprint of the 1996 edition. Boston: Birkhäuser, 2008 (Modern Birkhäuser classics)

[FK08b] FUJIMOTO, Kenji; KOYAMA, Ikuo: Iterative Feedback Tuning for Hamiltonian Systems. In: *IFAC Proceedings Volumes* 41 (2008), No. 2, p. 15678–15683

[FMC17] FELE, Filiberto; MAESTRE, Jose M.; CAMACHO, Eduardo F.: Coalitional Control: Cooperative Game Theory and Control. In: *IEEE Control Systems* 37 (2017), No. 1, p. 53–69

[FP10] FEIJER, Diego; PAGANINI, Fernando: Stability of primal–dual gradient dynamics and applications to network optimization. In: *Automatica* 46 (2010), No. 12, p. 1974–1981

[FR16] FRANK, Stephen; REBENNACK, Steffen: An introduction to optimal power flow: Theory, formulation, and examples. In: *IIE Transactions* 48 (2016), No. 12, p. 1172–1197

[FS12] FARINA, Marcello; SCATTOLINI, Riccardo: Distributed predictive control: A non-cooperative algorithm with neighbor-to-neighbor communication for linear systems. In: *Automatica* 48 (2012), No. 6, p. 1088–1096

[FS21] FREUND, Svenne; SCHMITZ, Gerhard: Implementation of model predictive control in a large-sized, low-energy office building. In: *Building and Environment* 197 (2021), p. 107830

[FSK+20] FIGGENER, Jan; STENZEL, Peter; KAIRIES, Kai-Philipp; LINSSEN, Jochen; HABERSCHUSZ, David; WESSELS, Oliver; ANGENENDT, Georg; ROBINIUS, Martin; STOLTEN, Detlef; SAUER, Dirk U.: The development of stationary battery storage systems in Germany – A market review. In: *Journal of Energy Storage* 29 (2020), p. 101153

[Fuj03] FUJIMOTO, Kenji: Optimal control of Hamiltonian systems via iterative learning. In: *SICE 2003 Annual Conference*, IEEE, 2003, p. 2617–2622

[FYL06] Freeman, Randy A.; Yang, Peng; Lynch, Kevin M.: Stability and Convergence Properties of Dynamic Average Consensus Estimators. In: *IEEE Conference on Decision and Control (CDC)*, IEEE, 2006, p. 338–343

[FZO+13] Fiaz, Shaik; Zonetti, Daniele; Ortega, Romeo; Scherpen, Jacquelien M.; van der Schaft, Arjan J.: A port-Hamiltonian approach to power network modeling and analysis. In: *European Journal of Control* 19 (2013), No. 6, p. 477–485

[GA20] Ghosh, Arnob; Aggarwal, Vaneet: Penalty Based Control Mechanism for Strategic Prosumers in a Distribution Network. In: *Energies* 13 (2020), No. 2, p. 452

[GCLL13] Guerrero, Josep M.; Chandorkar, Mukul; Lee, Tzung-Lin; Loh, Poh C.: Advanced Control Architectures for Intelligent Microgrids—Part I: Decentralized and Hierarchical Control. In: *IEEE Transactions on Industrial Electronics* 60 (2013), No. 4, p. 1254–1262

[GCS09] Gerelli, Oscar; Carloni, Raffaella; Stramigioli, Stefano: Port-Based Modeling and Optimal Control for a new Very Versatile Energy Efficient Actuator. In: *IFAC Proceedings Volumes* 42 (2009), No. 16, p. 493–498

[GK65] Golub, Gene; Kahan, William: Calculating the Singular Values and Pseudo-Inverse of a Matrix. In: *Journal of the Society of Industrial and Applied Mathematics: Series B, Numerical Analysis* 2 (1965), No. 2, p. 205–224

[GMG+07] Guerrero, Josep M.; Matas, Jos; Garcia de Vicuna, Luis; Castilla, Miguel; Miret, Jaume: Decentralized Control for Parallel Operation of Distributed Generation Inverters Using Resistive Output Impedance. In: *IEEE Transactions on Industrial Electronics* 54 (2007), No. 2, p. 994–1004

[Goe17] Goebel, Rafal: Stability and robustness for saddle-point dynamics through monotone mappings. In: *Systems & Control Letters* 108 (2017), p. 16–22

[Gov16] Government of Canada: *Pan-Canadian Framework on Clean Growth and Climate Change: Canada's plan to address climate change.* Environment and Climate Change Canada, 2016

[GS17] Greenlaw, Steven A.; Shapiro, David: *Principles of Microeconomics 2e.* Rice University, 2017

[GSAB05] Guedes, Renato B.; Silva, Flávio H.; Alberto, Luís F.C.; Bretas, Newton G.: Large disturbance voltage stability assessment using extended lyapunov function and considering voltage dependent active loads. In: *IEEE Power Engineering Society General Meeting*, IEEE, 2005, p. 682–689

[GSC17] Groothuis, Stefan S.; Stramigioli, Stefano; Carloni, Raffaella: Modeling Robotic Manipulators Powered by Variable Stiffness Actuators: A Graph-Theoretic and Port-Hamiltonian Formalism. In: *IEEE Transactions on Robotics* 33 (2017), No. 4, p. 807–818

[Gür18] Gür, Turgut M.: Review of electrical energy storage technologies, materials and systems: challenges and prospects for large-scale grid storage. In: *Energy & Environmental Science* 11 (2018), No. 10, p. 2696–2767

[HBD21] Hauswirth, Adrian; Bolognani, Saverio; Dörfler, Florian: Projected Dynamical Systems on Irregular, Non-Euclidean Domains for Nonlinear Optimization. In: *SIAM Journal on Control and Optimization* 59 (2021), No. 1, p. 635–668

[HBHD16] Hauswirth, Adrian; Bolognani, Saverio; Hug, Gabriela; Dörfler, Florian: Projected gradient descent on Riemannian manifolds with applications to online power system optimization. In: *Annual Allerton Conference on Communication, Control, and Computing (Allerton)*, IEEE, 2016, p. 225–232

[HBHD21a] Hauswirth, Adrian; Bolognani, Saverio; Hug, Gabriela; Dörfler, Florian: *Optimization Algorithms as Robust Feedback Controllers.* `http://arxiv.org/pdf/2103.11329v1`. Version: 2021

[HBHD21b] Hauswirth, Adrian; Bolognani, Saverio; Hug, Gabriela; Dörfler, Florian: Timescale Separation in Autonomous Optimization. In: *IEEE Transactions on Automatic Control* 66 (2021), No. 2, p. 611–624

[HDA21] Hashemipour, Naser; Del Crespo Granado, Pedro; Aghaei, Jamshid: Dynamic allocation of peer-to-peer clusters in virtual local electricity markets: A marketplace for EV flexibility. In: *Energy* 236 (2021), p. 121428

[HG91] Huneault, Maurice; Galiana, Francisco D.: A survey of the optimal power flow literature. In: *IEEE Transactions on Power Systems* 6 (1991), No. 2, p. 762–770

[HGK18] Hammer, Bernhard; Gong, Kuangye; Konigorski, Ulrich: Modeling and control of inverter-based microgrids. In: *IFAC-PapersOnLine* 51 (2018), No. 2, p. 19–24

[HJ12] Horn, Roger A.; Johnson, Charles R.: *Matrix Analysis.* Cambridge: Cambridge University Press, 2012

[HKK$^+$05] Hutchison, David; Kanade, Takeo; Kittler, Josef; Kleinberg, Jon M.; Mattern, Friedemann; Mitchell, John C.; Naor, Moni; Nierstrasz, Oscar; Pandu Rangan, C.; Steffen, Bernhard; Sudan, Madhu; Terzopoulos, Demetri; Tygar, Dough; Vardi, Moshe Y.; Weikum, Gerhard; Brandes, Ulrik; Erlebach, Thomas: *Network Analysis.* Vol. 3418. Berlin, Heidelberg: Springer, 2005

[HKZ12] Haurie, Alain; Krawczyk, Jacek B.; Zaccour, Georges: *Games and Dynamic Games.* Vol. 1. World Scientific, 2012

[HMM19] Han, Liyang; Morstyn, Thomas; McCulloch, Malcolm: Incentivizing Prosumer Coalitions With Energy Management Using Cooperative Game Theory. In: *IEEE Transactions on Power Systems* 34 (2019), No. 1, p. 303–313

[HMM+20] Hauschild, Sarah-Alexa; Marheineke, Nicole; Mehrmann, Volker; Mohring, Jan; Badlyan, Arbi M.; Rein, Markus; Schmidt, Martin: Port-Hamiltonian Modeling of District Heating Networks. In: Reis, Timo (ed.); Grundel, Sara (ed.); Schöps, Sebastian (ed.): *Progress in Differential-Algebraic Equations II*. Cham: Springer International Publishing, 2020, p. 333–355

[HMMD13] Hooshmand, Ali; Mohammadpour, Javad; Malki, Heidar; Daneshi, Hossein: Power system dynamic scheduling with high penetration of renewable sources. In: *American Control Conference (ACC)*, IEEE, 2013, p. 5827–5832

[HMR+20] Hatziargyriou, Nikos D.; Milanovic, Jovica V.; Rahmann, Claudia; Ajjarapu, Venkataramana; Cañizares, Claudio A.; Erlich, Istvan; Hill, David; Hiskens, Ian A.; Kamwa, Innocent; Pal, Bikash; Pourbeik, Pouyan; Sanchez-Gasca, Juan J.; Stankovic, Aleksandar M.; van Cutsem, Thierry; Vittal, Vijay; Vournas, Costas: Definition and Classification of Power System Stability Revisited & Extended. In: *IEEE Transactions on Power Systems* 36 (2020), No. 4, p. 3271–3281

[HOBD20] Hauswirth, Adrian; Ortmann, Lukas; Bolognani, Saverio; Dörfler, Florian: Limit Behavior and the Role of Augmentation in Projected Saddle Flows for Convex Optimization. In: *IFAC-PapersOnLine* 53 (2020), No. 2, p. 5511–5517

[Hog92] Hogan, William W.: Contract networks for electric power transmission. In: *Journal of Regulatory Economics* 4 (1992), No. 3, p. 211–242

[HRR20] Holguin, Juan P.; Rodriguez, David C.; Ramos, Gustavo: Reverse Power Flow (RPF) Detection and Impact on Protection Coordination of Distribution Systems. In: *IEEE Transactions on Industry Applications* 56 (2020), No. 3, p. 2393–2401

[HSMT19] Hirth, Lion; Schlecht, Ingmar; Maurer, Christoph; Tersteegen, Bernd: Cost- or market-based? Future redispatch procurement in Germany: Conclusions from the project "Beschaffung von Redispatch". (2019). https://www.bmwi.de/Redaktion/EN/Publikationen/Studien/future-redispatch-procurement-in-germany.html

[HYC+18] Huang, Wei-Tzer; Yao, Kai-Chao; Chen, Ming-Ku; Wang, Feng-Ying; Zhu, Cang-Hui; Chang, Yung-Ruei; Lee, Yih-Der; Ho, Yuan-Hsiang: Derivation and Application of a New Transmission Loss Formula for Power System Economic Dispatch. In: *Energies* 11 (2018), No. 2, p. 417

[HZB+17] Hauswirth, Adrian; Zanardi, Alessandro; Bolognani, Saverio; Dörfler, Florian; Hug, Gabriela: Online optimization in closed loop on the power flow manifold. In: *IEEE Manchester PowerTech*, IEEE, 2017, p. 1–6

[HZZ20] He, Xiangbai (ed.); Zhang, Hao (ed.); Zahar, Alexander (ed.): *Climate change law in China in global context*. Abingdon, New York: Routledge, 2020 (Routledge advances in climate change research)

[IEE21] IEEE: *57 Bus Test System*. https://icseg.iti.illinois.edu/ieee-57-bus-system/. Version: 2021

[Int20] International Energy Acency: *India 2020 Energy Policy Review*. Paris: OECD Publishing, 2020 (IEA Energy Policy Reviews)

[JAD16] Jouini, Taouba; Arghir, Catalin; Dörfler, Florian: Grid-Friendly Matching of Synchronous Machines by Tapping into the DC Storage. In: *IFAC-PapersOnLine* 49 (2016), No. 22, p. 192–197

[JB20] Jansen, Luca L.; Buzna, Luboš: Evaluation of Modeling Differences of Nodal vs. Zonal Pricing Based Electricity Markets: Optimization Models and Network Representation. In: *International Conference on Emerging eLearning Technologies and Applications (ICETA)*, IEEE, 2020, p. 236–242

[JJ13] Jiang, Zhong-Ping; Jiang, Yu: Robust adaptive dynamic programming for linear and nonlinear systems: An overview. In: *European Journal of Control* 19 (2013), No. 5, p. 417–425

[JLv09] Jokic, Andrej; Lazar, Mircea; van den Bosch, Paul: On Constrained Steady-State Regulation: Dynamic KKT Controllers. In: *IEEE Transactions on Automatic Control* 54 (2009), No. 9, p. 2250–2254

[JOv17] Jeltsema, Dimitri; Ortega, Romeo; van der Schaft, Arjan J.: *Energy-Based Modeling and Control of Physical Systems*. `http://www.math.rug.nl/arjan/DownloadLectures/hangzhou.pdf`. Version: 2017 (Hangzhou Workshop, April 6 - 7)

[JW15] Just, Sebastian; Weber, Christoph: Strategic behavior in the German balancing energy mechanism: incentives, evidence, costs and solutions. In: *Journal of Regulatory Economics* 48 (2015), No. 2, p. 218–243

[JW16] Ju, Chengquan; Wang, Peng: Optimal power flow with worst-case scenarios considering uncertainties of loads and renewables. In: *International Conference on Probabilistic Methods Applied to Power Systems (PMAPS)*, IEEE, 2016, p. 1–7

[JY03] Jia, Naixiong; Yokoyama, Ryuichi: Profit allocation of independent power producers based on cooperative Game theory. In: *International Journal of Electrical Power & Energy Systems* 25 (2003), No. 8, p. 633–641

[KA01] Kokotović, Petar; Arcak, Murat: Constructive nonlinear control: a historical perspective. In: *Automatica* 37 (2001), No. 5, p. 637–662

[KBP85] Kwatny, Harry G.; Bahar, Leon Y.; Pasrija, Arun K.: Energy-like Lyapunov functions for power system stability analysis. In: *IEEE Transactions on Circuits and Systems* 32 (1985), No. 11, p. 1140–1149

[KD18] Kappagantu, Ramakrishna; Daniel, S. A.: Challenges and issues of smart grid implementation: A case of Indian scenario. In: *Journal of Electrical Systems and Information Technology* 5 (2018), No. 3, p. 453–467

[KH19] Koch, Christopher; Hirth, Lion: Short-term electricity trading for system balancing: An empirical analysis of the role of intraday trading in balancing Germany's electricity system. In: *Renewable and Sustainable Energy Reviews* 113 (2019), p. 109275

[Kha02] KHALIL, Hassan K.: *Nonlinear Systems: Third Edition.* Upper Saddle River, New Jersey: Prentice Hall, 2002

[KJK+10] KIM, Jong-Yul; JEON, Jin-Hong; KIM, Seul-Ki; CHO, Changhee; PARK, June H.; KIM, Hak-Man; NAM, Kee-Young: Cooperative Control Strategy of Energy Storage System and Microsources for Stabilizing the Microgrid during Islanded Operation. In: *IEEE Transactions on Power Electronics* 25 (2010), No. 12, p. 3037–3048

[KKŠ20] KUHAR, Urban; KOSEC, Gregor; ŠVIGELJ, Aleš: *Observability of Power-Distribution Systems.* Cham: Springer International Publishing, 2020

[KM17] KREUZ, Sebastian; MÜSGENS, Felix: The German Energiewende and its roll-out of renewable energies: An economic perspective. In: *Frontiers in Energy* 11 (2017), No. 2, p. 126–134

[KO93] KEENAN, Donald C.; O'BRIEN, Mike J.: Competition, collusion, and chaos. In: *Journal of Economic Dynamics and Control* 17 (1993), No. 3, p. 327–353

[KO20] KULOVESI, Kati; OBERTHÜR, Sebastian: Assessing the EU's 2030 Climate and Energy Policy Framework: Incremental change toward radical transformation? In: *Review of European, Comparative & International Environmental Law* 29 (2020), p. 151–166

[Kos56] KOSE, Tairoku: Solutions of Saddle Value Problems by Differential Equations. In: *Econometrica* 24 (1956), No. 1, p. 59–70

[KPP+17] KARTHIKEYAN, Nainar; POKHREL, Basanta R.; PILLAI, Jayakrishnan R.; BAK-JENSEN, Birgitte; FREDERIKSEN, Kenn H. B.: Demand response in low voltage distribution networks with high PV penetration. In: *International Universities Power Engineering Conference (UPEC)*, IEEE, 2017, p. 1–6

[KPS15] KANNAN, G.; PADMA SUBRAMANIAN, D.; SIVA SUBRAMANIAN, S.: Reactive Power Pricing Using Group Search Optimization in Deregulated Electricity Market. In: KAMALAKANNAN, C. (ed.); SURESH, L. P. (ed.); DASH, Subhransu S. (ed.); PANIGRAHI, Bijaya K. (ed.): *Power Electronics and Renewable Energy Systems.* New Delhi: Springer India, 2015, p. 305–312

[KRK+18] KHALILI, Ramtin; RABIEYAN, Hadi; KHODADADI, Abolfazl; ZAKER, Behrooz; KARRARI, Mehdi; KARRARI, Shahab: Mathematical Modelling and Parameter Estimation of an Industrial Steam Turbine-Generator Based on Operational Data. In: *IFAC-PapersOnLine* 51 (2018), No. 2, p. 214–219

[Kug01] KUGI, Andreas: *Lecture Notes in Control and Information Sciences.* Vol. 260: *Nonlinear control based on physical models: Electrical, mechanical and hydraulic systems.* London: Springer, 2001

[Küh01] KÜHN, Kai-Uwe: Fighting collusion by regulating communication between firms. In: *Economic Policy* 16 (2001), No. 32, p. 168–204

[Kun94] KUNDUR, Prabha S.: *Power system stability and control*. Indian edition. Mc Graw Hill Education (India) Private Limited, 1994 (The EPRI power system engineering series)

[KvC+19] KIA, Solmaz S.; VAN SCOY, Bryan; CORTÉS, Jorge; FREEMAN, Randy A.; LYNCH, Kevin M.; MARTINEZ, Sonia: Tutorial on Dynamic Average Consensus: The Problem, Its Applications, and the Algorithms. In: *IEEE Control Systems* 39 (2019), No. 3, p. 40–72

[KZGB16] KEKATOS, Vassilis; ZHANG, Liang; GIANNAKIS, Georgios B.; BALDICK, Ross: Voltage Regulation Algorithms for Multiphase Power Distribution Grids. In: *IEEE Transactions on Power Systems* 31 (2016), No. 5, p. 3913–3923

[LAPG17] LATIF, Aadil; AHMAD, Ishtiaq; PALENSKY, Peter; GAWLIK, Wolfgang: Zone based optimal reactive power dispatch in smart distribution network using distributed generation. In: *Workshop on Modeling and Simulation of Cyber-Physical Energy Systems (MSCPES)*, IEEE, 2017, p. 1–6

[Las02] LASSETER, Robert H.: MicroGrids. In: *IEEE Power Engineering Society Winter Meeting. Conference Proceedings (Cat. No.02CH37309)*, IEEE, 2002, p. 305–308

[LC12] LEDGERWOOD, Shaun D.; CARPENTER, Paul R.: A Framework for the Analysis of Market Manipulation. In: *Review of Law & Economics* 8 (2012), No. 1

[LCZL14] LI, Na; CHEN, Lijun; ZHAO, Changhong; LOW, Steven H.: Connecting automatic generation control and economic dispatch from an optimization view. In: *American Control Conference (ACC)*, IEEE, 2014

[LGCZ15] LEE, Joohyung; GUO, Jun; CHOI, Jun K.; ZUKERMAN, Moshe: Distributed Energy Trading in Microgrids: A Game-Theoretic Model and Its Equilibrium Analysis. In: *IEEE Transactions on Industrial Electronics* 62 (2015), No. 6, p. 3524–3533

[LH16] LO PRETE, Chiara; HOBBS, Benjamin F.: A cooperative game theoretic analysis of incentives for microgrids in regulated electricity markets. In: *Applied Energy* 169 (2016), No. 3, p. 524–541

[LHK19] LEHMANN, Nico; HUBER, Julian; KIESLING, Andreas: Flexibility in the context of a cellular system model. In: *International Conference on the European Energy Market (EEM)*, IEEE, 2019, p. 1–6

[LK09] LI, Yun W.; KAO, Ching-Nan: An Accurate Power Control Strategy for Power-Electronics-Interfaced Distributed Generation Units Operating in a Low-Voltage Multibus Microgrid. In: *IEEE Transactions on Power Electronics* 24 (2009), No. 12, p. 2977–2988

[LKF15] LUCIA, Sergio; KÖGEL, Markus; FINDEISEN, Rolf: Contract-based Predictive Control of Distributed Systems with Plug and Play Capabilities. In: *IFAC-PapersOnLine* 48 (2015), No. 23, p. 205–211

[LLL+19] Li, Zhi; Liu, Yanzhu; Liu, Di; Ding, Xueying; Zhang, Mimi; Liu, Yuanyuan: A Differential Game Model of Energy Demand Side Management for Micro Grid. In: *IEEE International Conference on Energy Internet (ICEI)*, IEEE, 2019, p. 351–355

[LLL20] Liu, Dongmei; Liu, Liu; Lu, Yufeng: LQ-Optimal Control of Boundary Control Systems. In: *Iranian Journal of Science and Technology, Transactions of Electrical Engineering* 44 (2020), No. 1, 403–412

[LLW14] Liu, Derong; Li, Hongliang; Wang, Ding: Online Synchronous Approximate Optimal Learning Algorithm for Multi-Player Non-Zero-Sum Games With Unknown Dynamics. In: *IEEE Transactions on Systems, Man, and Cybernetics: Systems* 44 (2014), No. 8, p. 1015–1027

[Low14] Low, Steven H.: Convex Relaxation of Optimal Power Flow—Part I: Formulations and Equivalence. In: *IEEE Transactions on Control of Network Systems* 1 (2014), No. 1, p. 15–27

[LPWL07] Lowe, Philip; Pucinskaite, Ingrida; Webster, William; Lindberg, Patrick: Effective unbundling of energy transmission networks: lessons from the Energy Sector Inquiry. In: *Competition Policy Newsletter* (2007), No. 1, p. 23–34

[LQD14] Li, Na; Qu, Guannan; Dahleh, Munther: Real-time decentralized voltage control in distribution networks. In: *Annual Allerton Conference on Communication, Control, and Computing (Allerton)*, IEEE, 2014, p. 582–588

[LS15] Leon, Andres E.; Solsona, Jorge A.: Sub-Synchronous Interaction Damping Control for DFIG Wind Turbines. In: *IEEE Transactions on Power Systems* 30 (2015), No. 1, p. 419–428

[LSK05] Laaksonen, Hannu; Saari, Pekka; Komulainen, Risto: Voltage and frequency control of inverter based weak LV network microgrid. In: *International Conference on Future Power Systems*, IEEE, 2005, p. 1–6

[LSM+20] Louati, Haithem; Scheuermann, Tobias; Maschke, Bernhard; Zanota, Marie-Line; Vicente, Jerome; Kotyczka, Paul; Pitault, Isabelle: Network-Based Modeling of Transport Phenomena in Solid and Fluid Phases of Open-Cell Foams: Construction of Graphs. In: *Advanced Engineering Materials* 22 (2020), No. 5, p. 1901468

[LSZ18] Liu, Hao J.; Shi, Wei; Zhu, Hao: Distributed Voltage Control in Distribution Networks: Online and Robust Implementations. In: *IEEE Transactions on Smart Grid* 9 (2018), No. 6, p. 6106–6117

[LVS12] Lewis, Frank L.; Vrabie, Draguna L.; Syrmos, Vassilis L.: *Optimal control*. 3rd edition. Hoboken: Wiley, 2012

[LW18] Lamoline, François; Winkin, Joseph J.: On LQG control of stochastic port-Hamiltonian systems on infinite-dimensional spaces. In: *International Symposium on Mathematical Theory of Networks and Systems*, 2018, p. 197–203

[LZC16] Li, Na; Zhao, Changhong; Chen, Lijun: Connecting Automatic Generation Control and Economic Dispatch From an Optimization View. In: *IEEE Transactions on Control of Network Systems* 3 (2016), No. 3, p. 254–264

[Mag47] Magnusson, Philip C.: The Transient-Energy Method of Calculating Stability. In: *Transactions of the American Institute of Electrical Engineers* 66 (1947), No. 1, p. 747–755

[MBB12] Machowski, Jan; Bialek, Janusz W.; Bumby, James R.: *Power system dynamics: Stability and control.* 2nd edition, reprinted with corrections. Chichester: Wiley, 2012

[MBR+17] Martini, Luciano; Brunner, Helfried; Rodriguez, Emilio; Caerts, Chris; Strasser, Thomas I.; Burt, Graeme M.: Grid of the future and the need for a decentralised control architecture: the web-of-cells concept. In: *CIRED - Open Access Proceedings Journal* 2017 (2017), No. 1, p. 1162–1166

[MC11] Mousavi, Omid A.; Cherkaoui, Rachid: *Literature Survey on Fundamental Issues of Voltage and Reactive Power Control*, Ecole Polytechnique Fédérale de Lausanne, Literature Survey, 2011

[MCA95] Mullins, S. H.; Charlesworth, W. W.; Anderson, D. C.: A New Method for Solving Mixed Sets of Equality and Inequality Constraints. In: *Journal of Mechanical Design* 117 (1995), No. 2A, p. 322–328

[MD77] Martin, R. W.; Dillon, T. S.: Solutions of the problem of stochastic optimal control of hydro-thermal power systems. In: *IFAC Proceedings Volumes* 10 (1977), No. 16, p. 257–264

[MDS+18] Monshizadeh, Pooya; De Persis, Claudio; Stegink, Tjerk W.; Monshizadeh, Nima; van der Schaft, Arjan J.: Stability and frequency regulation of inverters with capacitive inertia. In: *IEEE Conference on Decision and Control (CDC)*, IEEE, 2018, p. 5696–5701

[Meh85] Mehlmann, Alexander: *Mathematical systems in economics.* Vol. 95: *Differentialspiele - die Analyse dynamischer Konfliktsituationen.* Königstein (Taunus): Hain, 1985

[MEH+17] Matthes, Felix; Emele, Lukas; Hermann, Hauke; Loreck, Charlotte; Peter, Frank; Ziegenhagen, Inka; Cook, Vanessa: *Germany's Electric Future: Coal phase-out 2035.* Berlin: Umweltstiftung WWF - Deutschland, 2017

[MFMC+08] Marquis-Favre, Wilfrid; Mouhib, Omar; Chereji, Bogdan; Thomasset, Daniel; Pousin, Jérôme; Picq, Martine: Bond graph formulation of an optimal control problem for linear time invariant systems. In: *Journal of the Franklin Institute* 345 (2008), No. 4, 349–373

[MGMH14] Martín Martínez, Sergio; Gómez Lázaro, Emilio; Molina Garcia, Angel; Honrubia Escribano, Andrés: Impact of wind power curtailments on the Spanish Power System operation. In: *IEEE PES General Meeting | Conference & Exposition*, IEEE, 2014, p. 1–5

[MGR+18] Mengelkamp, Esther; Gärttner, Johannes; Rock, Kerstin; Kessler, Scott; Orsini, Lawrence; Weinhardt, Christof: Designing microgrid energy markets. In: *Applied Energy* 210 (2018), p. 870–880

[MJMF+09] Mouhib, Omar; Jardin, Audrey; Marquis-Favre, Wilfrid; Bideaux, Eric; Thomasset, Daniel: Optimal control problem in bond graph formalism. In: *Simulation Modelling Practice and Theory* 17 (2009), No. 1, p. 240–256

[MKR+20] Mahmud, Khizir; Khan, Behram; Ravishankar, Jayashri; Ahmadi, Abdollah; Siano, Pierluigi: An internet of energy framework with distributed energy resources, prosumers and small-scale virtual power plants: An overview. In: *Renewable and Sustainable Energy Reviews* 127 (2020), p. 109840

[MLS14] Modares, Hamidreza; Lewis, Frank L.; Sistani, Mohammad-Bagher N.: Online solution of nonquadratic two-player zero-sum games arising in the H∞ control of constrained input systems. In: *International Journal of Adaptive Control and Signal Processing* 28 (2014), No. 3-5, p. 232–254

[MMM20] Moncecchi, Matteo; Meneghello, Stefano; Merlo, Marco: A Game Theoretic Approach for Energy Sharing in the Italian Renewable Energy Communities. In: *Applied Sciences* 10 (2020), No. 22, p. 8166

[MMMT12] Marano, Alejandro; Maza Ortega, José M.; Martínez Ramos, José L.; Trebolle, David: Voltage control of active distribution networks by means of dispersed generation. In: *CIRED workshop*, IET, 2012, p. 248

[MMO17] Mondal, Ayan; Misra, Sudip; Obaidat, Mohammad S.: Distributed Home Energy Management System With Storage in Smart Grid Using Game Theory. In: *IEEE Systems Journal* 11 (2017), No. 3, p. 1857–1866

[MMR+21] Marchand, Sophie; Monsalve, Cristian; Reimann, Thorsten; Heckmann, Wolfram; Ungerland, Jakob; Lauer, Hagen; Ruhe, Stephan; Krauss, Christoph: Microgrid Systems: Towards a Technical Performance Assessment Frame. In: *Energies* 14 (2021), No. 8, p. 2161

[MMS07] Macchelli, Alessandro; Melchiorri, Claudio; Stramigioli, Stefano: Port-Based Modeling of a Flexible Link. In: *IEEE Transactions on Robotics* 23 (2007), No. 4, p. 650–660

[MMS09] Macchelli, Alessandro; Melchiorri, Claudio; Stramigioli, Stefano: Port-Based Modeling and Simulation of Mechanical Systems With Rigid and Flexible Links. In: *IEEE Transactions on Robotics* 25 (2009), No. 5, p. 1016–1029

[MNC15] Monaco, Salvatore; Normand-Cyrot, Dorothée: On optimality of passivity based controllers in discrete time. In: *Systems & Control Letters* 75 (2015), p. 117–123

[MNRQ17] Mojica-Nava, Eduardo; Rivera, Sergio; Quijano, Nicanor: Game-theoretic dispatch control in microgrids considering network losses and renewable distributed energy resources integration. In: *IET Generation, Transmission & Distribution* 11 (2017), No. 6, p. 1583–1590

[MNSS18] Mazouchi, Majid; Naghibi-Sistani, Mohammad B.; Sani, Seyed Kamal H.: A novel distributed optimal adaptive control algorithm for nonlinear multi-agent differential graphical games. In: *IEEE/CAA Journal of Automatica Sinica* 5 (2018), No. 1, p. 331–341

[Moo06] Moore, James C.: *General Equilibrium and Welfare Economics: An Introduction.* Berlin, Heidelberg: Springer Berlin Heidelberg, 2006

[MQ18] Mouche, Pierre von; Quartieri, Federico: Cournot equilibrium uniqueness via demi-concavity. In: *Optimization* 67 (2018), No. 4, p. 441–455

[MR11] Müller, Christian; Rehkopf, Andreas: Optimale Betriebsführung eines virtuellen Kraftwerks auf Basis von gasbetriebenen Mikro-Blockheizkraftwerken. In: *at - Automatisierungstechnik* 59 (2011), No. 3, p. 180–186

[MRF11] Möller, Christoph; Rachev, Svetlozar T.; Fabozzi, Frank J.: Balancing energy strategies in electricity portfolio management. In: *Energy Economics* 33 (2011), No. 1, p. 2–11

[MRHD17] MacDougall, Pamela; Ran, Bob; Huitema, George B.; Deconinck, Geert: Multi-goal optimization of competing aggregators using a web-of-cells approach. In: *IEEE PES Innovative Smart Grid Technologies Conference Europe (ISGT-Europe)*, IEEE, 2017, p. 1–6

[MS21] Maiwald, Jens; Schütte, Tino: Decentralised Electricity Markets and Proactive Customer Behaviour. In: *Energies* 14 (2021), No. 3, p. 781

[MSM17] Mahmoudian Esfahani, Mohammad; Sheikh, Ahmed; Mohammed, Osama: Adaptive real-time congestion management in smart power systems using a real-time hybrid optimization algorithm. In: *Electric Power Systems Research* 150 (2017), No. 2, p. 118–128

[MWH+18] Ma, Tengfei; Wu, Junyong; Hao, Liangliang; Yan, Huaguang; Li, Dezhi: A Real-Time Pricing Scheme for Energy Management in Integrated Energy Systems: A Stackelberg Game Approach. In: *Energies* 11 (2018), No. 10, p. 2858

[MWY+19] Ma, Kai; Wang, Congshan; Yang, Jie; Hua, Changchun; Guan, Xinping: Pricing Mechanism With Noncooperative Game and Revenue Sharing Contract in Electricity Market. In: *IEEE Transactions on Cybernetics* 49 (2019), No. 1, p. 97–106

[MY16] Mahmoudian Esfahani, Mohammad; Yousefi, Gholam R.: Real Time Congestion Management in Power Systems Considering Quasi-Dynamic Thermal Rating and Congestion Clearing Time. In: *IEEE Transactions on Industrial Informatics* 12 (2016), No. 2, p. 745–754

[MZL14] Mallada, Enrique; Zhao, Changhong; Low, Steven H.: Optimal load-side control for frequency regulation in smart grids. In: *Annual Allerton Conference on Communication, Control, and Computing (Allerton)*, IEEE, 2014, p. 731–738

[MZL17] Mallada, Enrique; Zhao, Changhong; Low, Steven H.: Optimal Load-Side Control for Frequency Regulation in Smart Grids. In: *IEEE Transactions on Automatic Control* 62 (2017), No. 12, p. 6294–6309

[Nar84] Narasimhamurthi, Natarajan: On the existence of energy function for power systems with transmission losses. In: *IEEE Transactions on Circuits and Systems* 31 (1984), No. 2, p. 199–203

[NBM+20] Navon, Aviad; Ben Yosef, Gefen; Machlev, Ram; Shapira, Shmuel; Roy Chowdhury, Nilanjan; Belikov, Juri; Orda, Ariel; Levron, Yoash: Applications of Game Theory to Design and Operation of Modern Power Systems: A Comprehensive Review. In: *Energies* 13 (2020), No. 15, p. 3982

[NCB+18] Nagarajan, Adarsh; Coddington, Michael; Brown, David; Hassan, Sheikh; Franciosa, Leonardo; Sison-Lebrilla, Elaine: *Studies on the Effects of High Renewable Penetrations on Driving Point Impedance and Voltage Regulator Performance*, National Renewable Energy Laboratory, Technical Report NREL/TP-5D00-70517, 2018

[NFM+11] Narendra, Krish; Fedirchuk, David; Midence, René; Zhang, Nan; Mulawarman, Adi; Mysore, Pratap; Sood, Vijay: New microprocessor based relay to monitor and protect power systems against sub-harmonics. In: *IEEE Electrical Power and Energy Conference*, IEEE, 2011, p. 438–443

[NFT19] Nahata, Pulkit; Ferrari-Trecate, Giancarlo: Passivity-based Voltage and Frequency Stabilization in AC microgrids. In: *European Control Conference (ECC)*, IEEE, 2019, p. 1890–1895

[NKR13] Nguyen, Phuong H.; Kling, Wil L.; Ribeiro, Paulo F.: A Game Theory Strategy to Integrate Distributed Agent-Based Functions in Smart Grids. In: *IEEE Transactions on Smart Grid* 4 (2013), No. 1, p. 568–576

[NL85] Nowaihi, Ali al; Levine, Paul L.: The stability of the cournot oligopoly model: A reassessment. In: *Journal of Economic Theory* 35 (1985), No. 2, p. 307–321

[NLJB14] Nageshrao, Subramanya P.; Lopes, Gabriel A.; Jeltsema, Dimitri; Babuška, Robert: Passivity-based reinforcement learning control of a 2-DOF manipulator arm. In: *Mechatronics* 24 (2014), No. 8, p. 1001–1007

[NLJB16] Nageshrao, Subramanya P.; Lopes, Gabriel A.; Jeltsema, Dimitri; Babuška, Robert: Port-Hamiltonian Systems in Adaptive and Learning Control: A Survey. In: *IEEE Transactions on Automatic Control* 61 (2016), No. 5, p. 1223–1238

[Nor17] Norwegian Ministry of Climate and Environment: *Norway's Climate Strategy for 2030: a transformational approach within a European cooperation framework: (Meld. St. 41).* https://www.regjeringen.no/contentassets/7d3c209f821248da8d4727713ab9619c/en-gb/pdfs/stm201620170041000engpdfs.pdf. Version: 2017

[NOS+15] Namerikawa, Toru; Okubo, Norio; Sato, Ryutaro; Okawa, Yoshihiro; Ono, Masahiro: Real-Time Pricing Mechanism for Electricity Market With Built-In Incentive for Participation. In: *IEEE Transactions on Smart Grid* 6 (2015), No. 6, p. 2714–2724

[NP96] Nevistić, Vesna; Primbs, James A.: *Constrained Nonlinear Optimal Control: A Converse HJB Approach.* https://authors.library.caltech.edu/28127/1/CDS96-021.pdf. Version: 1996 (Technical Memorandum No. CIT-CDS 96-021)

[NSH15] Nguyen, Hung K.; Song, Ju B.; Han, Zhu: Distributed Demand Side Management with Energy Storage in Smart Grid. In: *IEEE Transactions on Parallel and Distributed Systems* 26 (2015), No. 12, p. 3346–3357

[NW06] Nocedal, Jorge; Wright, Stephen J.: *Numerical Optimization.* 2nd edition. Springer Science+Business Media LLC, 2006 (Springer Series in Operations Research and Financial Engineering)

[NY21] Naval, Natalia; Yusta, Jose M.: Virtual power plant models and electricity markets - A review. In: *Renewable and Sustainable Energy Reviews* 149 (2021), p. 111393

[OCN+17] Oh, Seaseung; Chae, Suyong; Neely, Jason; Baek, Jongbok; Cook, Marvin: Efficient Model Predictive Control Strategies for Resource Management in an Islanded Microgrid. In: *Energies* 10 (2017), No. 7, p. 1008

[OFM+20] Okura, Yuki; Fujimoto, Kenji; Maruta, Ichiro; Saito, Akio; Ikeda, Hidetoshi: Bayesian Inference for Path Following Control of Port-Hamiltonian Systems with Training Trajectory Data. In: *SICE Journal of Control, Measurement, and System Integration* 13 (2020), No. 2, p. 40–46

[OGA+05] Ortega, Romeo; Galaz, Martha; Astolfi, Alessandro; Sun, Yuanzhang; Shen, Tielong: Transient stabilization of multimachine power systems with nontrivial transfer conductances. In: *IEEE Transactions on Automatic Control* 50 (2005), No. 1, p. 60–75

[OGC04] Ortega, Romeo; García-Canseco, Eloísa: Interconnection and Damping Assignment Passivity-Based Control: A Survey. In: *European Journal of Control* 10 (2004), No. 5, p. 432–450

[OHC+20] Ortmann, Lukas; Hauswirth, Adrian; Caduff, Ivo; Dörfler, Florian; Bolognani, Saverio: Experimental validation of feedback optimization in power distribution grids. In: *Electric Power Systems Research* 189 (2020), p. 106782

[OMSE+14] Olivares, Daniel E.; Mehrizi-Sani, Ali; Etemadi, Amir H.; Cañizares, Claudio A.; Iravani, Reza; Kazerani, Mehrdad; Hajimiragha, Amir H.; Gomis-Bellmunt, Oriol; Saeedifard, Maryam; Palma-Behnke, Rodrigo; Jimenez-Estevez, Guillermo A.; Hatziargyriou, Nikos D.: Trends in Microgrid Control. In: *IEEE Transactions on Smart Grid* 5 (2014), No. 4, p. 1905–1919

[OPA+11] Ortega, Romeo; Praly, Laurent; Astolfi, Alessandro; Lee, Junggi; Nam, Kwanghee: Estimation of Rotor Position and Speed of Permanent Magnet Synchronous Motors With Guaranteed Stability. In: *IEEE Transactions on Control Systems Technology* 19 (2011), No. 3, p. 601–614

[OSFM07] Olfati-Saber, Reza; Fax, J. A.; Murray, Richard M.: Consensus and Cooperation in Networked Multi-Agent Systems. In: *Proceedings of the IEEE* 95 (2007), No. 1, p. 215–233

[OvCA08] Ortega, Romeo; van der Schaft, Arjan J.; Castaños, Fernando; Astolfi, Alessandro: Control by Interconnection and Standard Passivity-Based Control of Port-Hamiltonian Systems. In: *IEEE Transactions on Automatic Control* 53 (2008), No. 11, p. 2527–2542

[OvME02] Ortega, Romeo; van der Schaft, Arjan J.; Maschke, Bernhard; Escobar, Gerardo: Interconnection and damping assignment passivity-based control of port-controlled Hamiltonian systems. In: *Automatica* 38 (2002), No. 4, p. 585–596

[OvMM01] Ortega, Romeo; van der Schaft, Arjan J.; Mareels, Iven M. Y.; Maschke, Bernhard: Putting energy back in control. In: *IEEE Control Systems* 21 (2001), No. 2, p. 18–33

[OW15] Overbye, Thomas J.; Weber, James: Smart Grid Wide-Area Transmission System Visualization. In: *Engineering* 1 (2015), No. 4, p. 466–474

[PBD20] Picallo, Miguel; Bolognani, Saverio; Dörfler, Florian: Closing the loop: Dynamic state estimation and feedback optimization of power grids. In: *Electric Power Systems Research* 189 (2020), p. 106753

[PCH+20] Pfeifer, Martin; Caspart, Sven; Hampel, Silja; Muller, Charles; Krebs, Stefan; Hohmann, Sören: Explicit port-Hamiltonian formulation of multi-bond graphs for an automated model generation. In: *Automatica* 120 (2020), p. 109121

[PCK+21] Park, Sanguk; Cho, Keonhee; Kim, Seunghwan; Yoon, Guwon; Choi, Myeong-In; Park, Sangmin; Park, Sehyun: Distributed Energy IoT-Based Real-Time Virtual Energy Prosumer Business Model for Distributed Power Resource. In: *Sensors* 21 (2021), No. 13

[PEH18] Pierre, Brian J.; Elkhatib, Mohamed E.; Hoke, Andy: PV Inverter Fault Response Including Momentary Cessation, Frequency-Watt, and Virtual Inertia. In: *World Conference on Photovoltaic Energy Conversion (WCPEC)*, IEEE, 2018, p. 3660–3665

[PFA+19] Prettico, Giuseppe; Flammini, Marco G.; Andreadou, Nikoleta; Vitiello, Sila; Fulli, Gianluca; Masera, Marcelo: *Distribution System Operators observatory 2018: Overview of the electricity distribution system in Europe.* Luxembourg: Publications Office of the European Union, 2019

[PLB+16] Park, Sangdon; Lee, Joohyung; Bae, Sohee; Hwang, Ganguk; Choi, Jun K.: Contribution-Based Energy-Trading Mechanism in Microgrids for Future Smart Grid: A Game Theoretic Approach. In: *IEEE Transactions on Industrial Electronics* 63 (2016), No. 7, p. 4255–4265

[PLd20] Poplavskaya, Ksenia; Lago, Jesus; de Vries, Laurens: Effect of market design on strategic bidding behavior: Model-based analysis of European electricity balancing markets. In: *Applied Energy* 270 (2020), No. 6, p. 115130

[PMC79] Papavassilopoulos, George P.; Medanic, Juraj V.; Cruz, Jose B.: On the existence of Nash strategies and solutions to coupled riccati equations in linear-quadratic games. In: *Journal of Optimization Theory and Applications* 28 (1979), No. 1, p. 49–76

[Pol09a] Polyak, Roman A.: On the local quadratic convergence of the primal–dual augmented Lagrangian method. In: *Optimization Methods and Software* 24 (2009), No. 3, p. 369–379

[Pol09b] Polyak, Roman A.: Regularized Newton method for unconstrained convex optimization. In: *Mathematical Programming, Series B* 120 (2009), No. 1, p. 125–145

[PRG14] Parisio, Alessandra; Rikos, Evangelos; Glielmo, Luigi: A Model Predictive Control Approach to Microgrid Operation Optimization. In: *IEEE Transactions on Control Systems Technology* 22 (2014), No. 5, p. 1813–1827

[PRK+17] Pratt, Annabelle; Ruth, Mark; Krishnamurthy, Dheepak; Sparn, Bethany; Lunacek, Monte; Jones, Wesley; Mittal, Saurabh; Wu, Hongyu; Marks, Jesse: Hardware-in-the-loop simulation of a distribution system with air conditioners under model predictive control. In: *IEEE Power & Energy Society General Meeting*, IEEE, 2017, p. 1–5

[PZCL11] Pei, Wenhui; Zhang, Chenghui; Cui, Naxin; Li, Ke: Port-controlled Hamiltonian optimal control of induction motor system for electric vehicles. In: *Proceedings of the 30th Chinese Control Conference* (2011), p. 6229–6234

[PZS+18] Pfeifer, Martin; Zimmerlin, Martin; Schwarzendorfer, Martin; Sauter, Patrick S.; Krebs, Stefan; Leibfried, Thomas; Hohmann, Sören: Weighted Least Squares State Estimation for Coupled Power and Gas Distribution Networks. In: *International Universities Power Engineering Conference (UPEC)*, IEEE, 2018, p. 1–6

[RCPP20] Razzanelli, Matteo; Crisostomi, Emanuele; Pallottino, Lucia; Pannocchia, Gabriele: Distributed model predictive control for energy management in a network of microgrids using the dual decomposition method. In: *Optimal Control Applications and Methods* 41 (2020), No. 1, p. 25–41

[RCT17] Rikos, Evangelos; Cabiati, Mattia; Tornelli, Carlo: Adaptive frequency containment and balance restoration controls in a distribution network. In: *IEEE PES Innovative Smart Grid Technologies Conference Europe (ISGT-Europe)*, IEEE, 2017, p. 1–6

[Ren07] REN, Wei: Multi-vehicle consensus with a time-varying reference state. In: *Systems & Control Letters* 56 (2007), No. 7-8, p. 474–483

[Ris21] RISSE, Oliver: *Grid Congestion as a Challenge for the Electricity System: Options for a Future Market Design: Position paper.* Munich: acatech – National Academy of Science and Engineering, 2021 (Series on Science-based Policy Advice)

[RLBR12] ROCABERT, Joan; LUNA, Alvaro; BLAABJERG, Frede; RODRÍGUEZ, Pedro: Control of Power Converters in AC Microgrids. In: *IEEE Transactions on Power Electronics* 27 (2012), No. 11, p. 4734–4749

[RLP+15] ROGELJ, Joeri; LUDERER, Gunnar; PIETZCKER, Robert C.; KRIEGLER, Elmar; SCHAEFFER, Michiel; KREY, Volker; RIAHI, Keywan: Energy system transformations for limiting end-of-century warming to below 1.5 °C. In: *Nature Climate Change* 5 (2015), No. 6, p. 519–527

[RN20] RAI, Alan; NUNN, Oliver: On the impact of increasing penetration of variable renewables on electricity spot price extremes in Australia. In: *Economic analysis and policy* 67 (2020), p. 67–86

[RS19] RICHTER, Bent; STAUDT, Philipp: Perspectives on data availability and market approaches to congestion management. In: *it - Information Technology* 61 (2019), No. 2-3, p. 73–85

[RSFT15] RIVERSO, Stefano; SARZO, Fabio; FERRARI-TRECATE, Giancarlo: Plug-and-Play Voltage and Frequency Control of Islanded Microgrids With Meshed Topology. In: *IEEE Transactions on Smart Grid* 6 (2015), No. 3, p. 1176–1184

[RT06] REINISCH, Walter; TEZUKA, Tetsuo: Market Power and Trading Strategies on the Electricity Market: A Market Design View. In: *IEEE Transactions on Power Systems* 21 (2006), No. 3, p. 1180–1190

[RYI+16] ROSCOE, Andrew J.; YU, Mengran; IERNA, Richard; ZHU, Jiebei; DYŚKO, Adam; URDAL, Helge; BOOTH, Campbell: A VSM (virtual synchronous machine) convertor control model suitable for RMS studies for resolving system operator/owner challenges. In: *Wind Integration Workshop* (2016), p. 1–8

[RYNC14] ROSHANY-YAMCHI, Samira; NEGENBORN, Rudy R.; CORNELIO, Avi A.: Nash-Based Distributed MPC for Multi-Rate Systems. Version: 2014. In: MAESTRE, José M. (ed.); NEGENBORN, Rudy R. (ed.): *Distributed Model Predictive Control Made Easy* Vol. 69. Dordrecht: Springer Netherlands, 2014, p. 341–353

[Sac01] SACKMANN, Martin: *Modifizierte Optimale Regelung: Nichtlinearer Entwurf unter Verwendung der Hyperstabilitätstheorie*, Technische Universität Karlsruhe, Dissertation, 2001

[SALB05] SILVA, Flávio H.; ALBERTO, Luís F.C.; LONDON, Joao B.; BRETAS, Newton G.: Smooth perturbation on a classical energy function for lossy power system stability analysis. In: *IEEE Transactions on Circuits and Systems I: Regular Papers* 52 (2005), No. 1, p. 222–229

[Sau11] Sauer, Peter W.: Time-scale features and their applications in electric power system dynamic modeling and analysis. In: *American Control Conference (ACC)*, IEEE, 2011, p. 4155–4159

[SBCA19] Shiltz, Dylan J.; Baros, Stefanos; Cvetkovic, Milos; Annaswamy, Anuradha M.: Integration of Automatic Generation Control and Demand Response via a Dynamic Regulation Market Mechanism. In: *IEEE Transactions on Control Systems Technology* 27 (2019), No. 2, p. 631–646

[SBNL15] Sprangers, Olivier; Babuška, Robert; Nageshrao, Subramanya P.; Lopes, Gabriel A. D.: Reinforcement learning for port-hamiltonian systems. In: *IEEE Transactions on Cybernetics* 45 (2015), No. 5, p. 1003–1013

[SCA16] Shiltz, Dylan J.; Cvetkovic, Milos; Annaswamy, Anuradha M.: An Integrated Dynamic Market Mechanism for Real-Time Markets and Frequency Regulation. In: *IEEE Transactions on Sustainable Energy* 7 (2016), No. 2, 875–885

[SCD+19] Stegink, Tjerk W.; Cherukuri, Ashish; De Persis, Claudio; van der Schaft, Arjan J.; Cortés, Jorge: Hybrid Interconnection of Iterative Bidding and Power Network Dynamics for Frequency Regulation and Optimal Dispatch. In: *IEEE Transactions on Control of Network Systems* 6 (2019), No. 2, p. 572–585

[Sch14] Schaber, Katrin: *Integration of Variable Renewable Energies in the European power system: a model-based analysis of transmission grid extensions and energy sector coupling*, Technische Universität München, Dissertation, 2014

[Sch15] Schiffer, Johannes: *Stability and power sharing in microgrids*, Technische Universität Berlin, Dissertation, 2015

[Sch21] Schmidt, Tobias S.: Spurring low-carbon electrosynthesis through energy and innovation policy. In: *iScience* 24 (2021), No. 2, p. 102045

[SCTB88] Schweppe, Fred C.; Caramanis, Michael C.; Tabors, Richard D.; Bohn, Roger E.: *Spot Pricing of Electricity*. Boston: Springer, 1988 (The Kluwer International Series in Engineering and Computer Science, Power Electronics & Power Systems)

[SDF17] Schiffer, Johannes; Dörfler, Florian; Fridman, Emilia: Robustness of distributed averaging control in power systems: Time delays & dynamic communication topology. In: *Automatica* 80 (2017), p. 261–271

[SDJD15] Sinha, Mohit; Dörfler, Florian; Johnson, Brian B.; Dhople, Sairaj V.: Virtual Oscillator Control subsumes droop control. In: *American Control Conference (ACC)*, IEEE, 2015, p. 2353–2358

[SDS88] Santos, A.; Deckmann, Sigmar M.; Soares, Secundino: A dual augmented Lagrangian approach for optimal power flow. In: *IEEE Transactions on Power Systems* 3 (1988), No. 3, p. 1020–1025

[SDv15] Stegink, Tjerk W.; De Persis, Claudio; van der Schaft, Arjan J.: Port-Hamiltonian Formulation of the Gradient Method Applied to Smart Grids. In: *IFAC-PapersOnLine* 48 (2015), No. 13, p. 13–18

[SDv16] Stegink, Tjerk W.; De Persis, Claudio; van der Schaft, Arjan J.: Optimal power dispatch in networks of high-dimensional models of synchronous machines. In: *IEEE Conference on Decision and Control (CDC)*, IEEE, 2016, p. 4110–4115

[SDv17a] Stegink, Tjerk W.; De Persis, Claudio; van der Schaft, Arjan J.: Stabilization of Structure-Preserving Power Networks with Market Dynamics. In: *IFAC-PapersOnLine* 50 (2017), p. 6737–6742

[SDv17b] Stegink, Tjerk W.; De Persis, Claudio; van der Schaft, Arjan J.: A Unifying Energy-Based Approach to Stability of Power Grids With Market Dynamics. In: *IEEE Transactions on Automatic Control* 62 (2017), No. 6, p. 2612–2622

[SDv19] Stegink, Tjerk W.; De Persis, Claudio; van der Schaft, Arjan J.: An energy-based analysis of reduced-order models of (networked) synchronous machines. In: *Mathematical and Computer Modelling of Dynamical Systems* 25 (2019), No. 1, p. 1–39

[SESJ13] Sokoler, Leo E.; Edlund, Kristian; Standardi, Laura; Jorgensen, John B.: A decomposition algorithm for optimal control of distributed energy system. In: *IEEE PES ISGT Europe 2013*, IEEE, 2013, p. 1–5

[SFC+18] Serale, Gianluca; Fiorentini, Massimo; Capozzoli, Alfonso; Bernardini, Daniele; Bemporad, Alberto: Model Predictive Control (MPC) for Enhancing Building and HVAC System Energy Efficiency: Problem Formulation, Applications and Opportunities. In: *Energies* 11 (2018), No. 3, p. 631

[SFO15] Schiffer, Johannes; Fridman, Emilia; Ortega, Romeo: Stability of a class of delayed port-Hamiltonian systems with application to droop-controlled microgrids. In: *IEEE Conference on Decision and Control (CDC)*, IEEE, 2015, p. 6391–6396

[SGAM18] Stadler, Paul; Girardin, Luc; Ashouri, Araz; Maréchal, François: Contribution of Model Predictive Control in the Integration of Renewable Energy Sources within the Built Environment. In: *Frontiers in Energy Research* 6 (2018)

[SGV14] Shafiee, Qobad; Guerrero, Josep M.; Vasquez, Juan C.: Distributed Secondary Control for Islanded Microgrids—A Novel Approach. In: *IEEE Transactions on Power Electronics* 29 (2014), No. 2, p. 1018–1031

[SH69] Starr, A. W.; Ho, Y. C.: Nonzero-sum differential games. In: *Journal of Optimization Theory and Applications* 3 (1969), No. 3, p. 184–206

[SHH19] Sarfati, Mahir; Hesamzadeh, Mohammad R.; Holmberg, Pär: Production efficiency of nodal and zonal pricing in imperfectly competitive electricity markets. In: *Energy Strategy Reviews* 24 (2019), No. 1, p. 193–206

[SHP11] Saad, Walid; Han, Zhu; Poor, Harold V.: Coalitional Game Theory for Cooperative Micro-Grid Distribution Networks. In: *IEEE International Conference on Communications Workshops (ICC)*, IEEE, 2011, p. 1–5

[SHPB12] Saad, Walid; Han, Zhu; Poor, Harold V.; Başar, Tamer: Game-Theoretic Methods for the Smart Grid: An Overview of Microgrid Systems, Demand-Side Management, and Smart Grid Communications. In: *IEEE Signal Processing Magazine* 29 (2012), No. 5, p. 86–105

[Sin19] Singh, Tej B.: *Introduction to Topology*. Singapore: Springer, 2019 (Springer eBooks)

[SJK97] Sepulchre, Rodolphe; Janković, Mrdjan; Kokotović, Petar V.: *Constructive Nonlinear Control*. London: Springer, 1997

[SK00] Sackmann, Martin S.; Krebs, Volker G.: Modified Optimal Control: Global Asymptotic Stabilization of Nonlinear Systems. In: *IFAC Proceedings Volumes* 33 (2000), No. 13, p. 199–204

[SKSJ21] Song, Jae G.; Kang, Eung S.; Shin, Hyeon W.; Jang, Ju W.: A Smart Contract-Based P2P Energy Trading System with Dynamic Pricing on Ethereum Blockchain. In: *Sensors* 21 (2021), No. 6

[SLL02] Song, Haili; Liu, Chen-Ching; Lawarree, Jacques: Nash equilibrium bidding strategies in a bilateral electricity market. In: *IEEE Transactions on Power Systems* 17 (2002), No. 1, p. 73–79

[SMKH19] Strehle, Felix; Malan, Albertus J.; Krebs, Stefan; Hohmann, Sören: A Port-Hamiltonian Approach to Plug-and-Play Voltage and Frequency Control in Islanded Inverter-Based AC Microgrids. In: *IEEE Conference on Decision and Control (CDC)*, IEEE, 2019, p. 4648–4655

[SMKH20] Strehle, Felix; Malan, Albertus J.; Krebs, Stefan; Hohmann, Sören: Passivity Conditions for Plug-and-Play Operation of Nonlinear Static AC Loads. In: *IFAC-PapersOnLine* 53 (2020), No. 2, p. 12237–12243

[SNM+21] Strehle, Felix; Nahata, Pulkit; Malan, Albertus J.; Hohmann, Sören; Ferrari-Trecate, Giancarlo: A Unified Passivity-Based Framework for Control of Modular Islanded AC Microgrids. In: *IEEE Transactions on Control Systems Technology* (2021), p. 1–17

[SOA+14] Schiffer, Johannes; Ortega, Romeo; Astolfi, Alessandro; Raisch, Jörg; Sezi, Tevfik: Conditions for stability of droop-controlled inverter-based microgrids. In: *Automatica* 50 (2014), No. 10, p. 2457–2469

[Son89] Sontag, Eduardo D.: A 'universal' construction of Artstein's theorem on nonlinear stabilization. In: *Systems & Control Letters* 13 (1989), No. 2, p. 117–123

[SP18] Staffell, Iain; Pfenninger, Stefan: The increasing impact of weather on electricity supply and demand. In: *Energy* 145 (2018), p. 65–78

[SP19] Simpson-Porco, John W.: Equilibrium-Independent Dissipativity With Quadratic Supply Rates. In: *IEEE Transactions on Automatic Control* 64 (2019), No. 4, p. 1440–1455

[SPDB13] Simpson-Porco, John W.; Dörfler, Florian; Bullo, Francesco: Synchronization and power sharing for droop-controlled inverters in islanded microgrids. In: *Automatica* 49 (2013), No. 9, p. 2603–2611

[SPDB17] Simpson-Porco, John W.; Dörfler, Florian; Bullo, Francesco: Voltage Stabilization in Microgrids via Quadratic Droop Control. In: *IEEE Transactions on Automatic Control* 62 (2017), No. 3, p. 1239–1253

[SPPMD16] Simpson-Porco, John W.; Poolla, Bala K.; Monshizadeh, N.; Dörfler, Florian: Quadratic performance of primal-dual methods with application to secondary frequency control of power systems. In: *IEEE Conference on Decision and Control (CDC)*, 2016, p. 1840–1845

[SPSD+15] Simpson-Porco, John W.; Shafiee, Qobad; Dörfler, Florian; Vasquez, Juan C.; Guerrero, Josep M.; Bullo, Francesco: Secondary Frequency and Voltage Control of Islanded Microgrids via Distributed Averaging. In: *IEEE Transactions on Industrial Electronics* 62 (2015), No. 11, p. 7025–7038

[SRRD15] Smith, Jeff; Rylander, Matthew; Rogers, Lindsey; Dugan, Roger: It's All in the Plans: Maximizing the Benefits and Minimizing the Impacts of DERs in an Integrated Grid. In: *IEEE Power and Energy Magazine* 13 (2015), No. 2, p. 20–29

[SSC+19] Sauter, Patrick S.; Solanki, Bharatkumar V.; Cañizares, Claudio A.; Bhattacharya, Kankar; Hohmann, Sören: Electric Thermal Storage System Impact on Northern Communities' Microgrids. In: *IEEE Transactions on Smart Grid* 10 (2019), No. 1, p. 852–863

[Sta19] Staudt, Philipp: *Transmission Congestion Management in Electricity Grids - Designing Markets and Mechanisms*, Karlsruhe Institute of Technology, Dissertation, 2019

[Sto97] Stoft, Steven: Transmission pricing zones: simple or complex? In: *The Electricity Journal* 10 (1997), No. 1, p. 24–31

[STY+18] Sun, Bohao; Tang, Yong; Ye, Lin; Chen, Chaoyu; Zhang, Cihang; Zhong, Wuzhi: A Frequency Control Strategy Considering Large Scale Wind Power Cluster Integration Based on Distributed Model Predictive Control. In: *Energies* 11 (2018), No. 6, p. 1600

[Tay15] Taylor, Joshua A.: *Convex optimization of power systems*. Cambridge: Cambridge University Press, 2015

[TBD16] Trip, Sebastian; Bürger, Mathias; De Persis, Claudio: An internal model approach to (optimal) frequency regulation in power grids with time-varying voltages. In: *Automatica* 64 (2016), p. 240–253

[TBW15] Trepper, Katrin; Bucksteeg, Michael; Weber, Christoph: Market splitting in Germany – New evidence from a three-stage numerical model of Europe. In: *Energy Policy* 87 (2015), No. 1, p. 199–215

[TCL12] Tan, Chee W.; Cai, Desmond W. H.; Lou, Xin: DC optimal power flow: Uniqueness and algorithms. In: *IEEE International Conference on Smart Grid Communications (SmartGridComm)*, IEEE, 2012, p. 641–646

[TFRFT16] Tucci, Michele; Floriduz, Alessandro; Riverso, Stefano; Ferrari-Trecate, Giancarlo: Plug-and-play control of AC islanded microgrids with general topology. In: *European Control Conference (ECC)*, IEEE, 2016, p. 1493–1500

[TFT17] Tucci, Michele; Ferrari-Trecate, Giancarlo: Voltage and frequency control in AC islanded microgrids: a scalable, line-independent design algorithm. In: *IFAC-PapersOnLine* 50 (2017), No. 1, p. 13922–13927

[THL19] Tang, Zhiyuan; Hill, David J.; Liu, Tao: Fast Distributed Reactive Power Control for Voltage Regulation in Distribution Networks. In: *IEEE Transactions on Power Systems* 34 (2019), No. 1, p. 802–805

[TJ13] Tang, Wenyuan; Jain, Rahul: Game-theoretic analysis of the nodal pricing mechanism for electricity markets. In: *IEEE Conference on Decision and Control (CDC)*, IEEE, 2013, p. 562–567

[TJUM97] Tuladhar, Anil; Jin, Hua; Unger, Tom; Mauch, Konrad: Parallel operation of single phase inverter modules with no control interconnections. In: *Applied Power Electronics Conference (APEC)*, IEEE, 1997, p. 94–100

[TM15] Tackx, Koen; Meeus, Leonardo: *Outlook on the European DSO Landscape 2020: The Trends that Will Change the Name of the Game*, Vlerick Business School, White Paper, 2015. `https://www.elder.org.tr/Content/yayinlar/Energy-Outlook-DSO-2020.pdf`

[TMM+10] Tan, Ying; Moase, William H.; Manzie, Chris; Nešić, Dragan; Mareels, Iven M. Y.: Extremum seeking from 1922 to 2010. In: *Chinese Control Conference*. IEEE, 2010, p. 14–26

[TS12] Ton, Dan T.; Smith, Merrill A.: The U.S. Department of Energy's Microgrid Initiative. In: *The Electricity Journal* 25 (2012), No. 8, p. 84–94

[Tv16] Tielens, Pieter; van Hertem, Dirk: The relevance of inertia in power systems. In: *Renewable and Sustainable Energy Reviews* 55 (2016), No. 4, p. 999–1009

[TZS+14] Tushar, Wayes; Zhang, Jian A.; Smith, David B.; Poor, Harold V.; Thiebaux, Sylvie: Prioritizing Consumers in Smart Grid: A Game Theoretic Approach. In: *IEEE Transactions on Smart Grid* 5 (2014), No. 3, p. 1429–1438

[van17] van der Schaft, Arjan J.: *L2-Gain and Passivity Techniques in Nonlinear Control*. Cham: Springer International Publishing, 2017

[Var10] Varian, Hal R.: *Intermediate microeconomics: A modern approach*. 8th edition. New York: Norton, 2010

[VBGQ+20] Velasquez, Miguel A.; Barreiro-Gomez, Julian; Quijano, Nicanor; Cadena, Angela I.; Shahidehpour, Mohammad: Intra-Hour Microgrid Economic Dispatch Based on Model Predictive Control. In: *IEEE Transactions on Smart Grid* 11 (2020), No. 3, p. 1968–1979

[vDD14] van den Bergh, Kenneth; Delarue, Erik; D'haeseleer, William: *DC power flow in unit commitment models*, KU Leuven, TME Working Paper- Energy and Environment, 2014

[VG16] Virulkar, Vasudeo B.; Gotmare, Gajanan V.: Sub-synchronous resonance in series compensated wind farm: A review. In: *Renewable and Sustainable Energy Reviews* 55 (2016), No. 6, p. 1010–1029

[VHRW08] Venkat, Aswin N.; Hiskens, Ian A.; Rawlings, James B.; Wright, Stephen J.: Distributed MPC Strategies With Application to Power System Automatic Generation Control. In: *IEEE Transactions on Control Systems Technology* 16 (2008), No. 6, p. 1192–1206

[vJ14] van der Schaft, Arjan J.; Jeltsema, Dimitri: Port-Hamiltonian Systems Theory: An Introductory Overview. In: *FnT in Systems and Control (Foundations and Trends in Systems and Control)* 1 (2014), No. 2, p. 173–378

[VL10] Vamvoudakis, Kyriakos G.; Lewis, Frank L.: Online Actor-Critic Algorithm to Solve the Continuous-Time Infinite Horizon Optimal Control Problem. In: *Automatica* 46 (2010), No. 5, p. 878–888

[VL11] Vamvoudakis, Kyriakos G.; Lewis, Frank L.: Multi-player non-zero-sum games: Online adaptive learning solution of coupled Hamilton–Jacobi equations. In: *Automatica* 47 (2011), No. 8, p. 1556–1569

[VMH16] Vamvoudakis, Kyriakos G.; Miranda, Marcio F.; Hespanha, Joao P.: Asymptotically Stable Adaptive-Optimal Control Algorithm With Saturating Actuators and Relaxed Persistence of Excitation. In: *IEEE Transactions on Neural Networks and Learning Systems* 27 (2016), No. 11, p. 2386–2398

[VT15] Vu, Thanh L.; Turitsyn, Konstantin: Synchronization stability of lossy and uncertain power grids. In: *American Control Conference (ACC)*, IEEE, 2015, p. 5056–5061

[vV08] van Cutsem, Thierry; Vournas, Costas: *Voltage Stability of Electric Power Systems*. Berlin: Springer, 2008 (Power Electronics and Power Systems)

[Wan15] Wang, Fengyu: *Improving Deterministic Reserve Requirements for Security Constrained Unit Commitment and Scheduling Problems in Power Systems*, Arizona State University, Dissertation, 2015

[WBL+14] Wang, Xiongfei; Blaabjerg, Frede; Liserre, Marco; Chen, Zhe; He, Jinwei; Li, Yunwei: An Active Damper for Stabilizing Power-Electronics-Based AC Systems. In: *IEEE Transactions on Power Electronics* 29 (2014), No. 7, p. 3318–3329

[WCW+15] Wang, Zhaoyu; Chen, Bokan; Wang, Jianhui; Begovic, Miroslav M.; Chen, Chen: Coordinated Energy Management of Networked Microgrids in Distribution Systems. In: *IEEE Transactions on Smart Grid* 6 (2015), No. 1, p. 45–53

[Wei06] Weibull, Jörgen W.: Price competition and convex costs. In: *SSE/EFI Working Paper Series in Economics and Finance* 622 (2006)

[Wei17] Weibelzahl, Martin: Nodal, zonal, or uniform electricity pricing: how to deal with network congestion. In: *Frontiers in Energy* 11 (2017), No. 2, p. 210–232

[WGCS19] Wang, Gang; Giannakis, Georgios B.; Chen, Jie; Sun, Jian: Distribution system state estimation: an overview of recent developments. In: *Frontiers of Information Technology & Electronic Engineering* 20 (2019), No. 1, p. 4–17

[WH18] Wang, Hao; Huang, Jianwei: Incentivizing Energy Trading for Interconnected Microgrids. In: *IEEE Transactions on Smart Grid* 9 (2018), No. 4, p. 2647–2657

[WHK18] Wu, Chenye; Hug, Gabriela; Kar, Soummya: Smart Inverter for Voltage Regulation: Physical and Market Implementation. In: *IEEE Transactions on Power Systems* 33 (2018), No. 6, p. 6181–6192

[WHLM18] Wu, Yongxin; Hamroun, Boussad; Le Gorrec, Yann; Maschke, Bernhard: Reduced order LQG control design for port Hamiltonian systems. In: *Automatica* 95 (2018), p. 86–92

[WK09] Weaver, Wayne W.; Krein, Philip T.: Game-Theoretic Control of Small-Scale Power Systems. In: *IEEE Transactions on Power Delivery* 24 (2009), No. 3, p. 1560–1567

[WL20] Wang, Jianhui; Lu, Xiaonan: Sustainable and Resilient Distribution Systems With Networked Microgrids [Point of View]. In: *Proceedings of the IEEE* 108 (2020), No. 2, p. 238–241

[WLM15] Wei, Wei; Liu, Feng; Mei, Shengwei: Nash Bargain and Complementarity Approach Based Environmental/Economic Dispatch. In: *IEEE Transactions on Power Systems* 30 (2015), No. 3, p. 1548–1549

[WLP+19] Wang, Zhaojian; Liu, Feng; Pang, John Z. F.; Low, Steven H.; Mei, Shengwei: Distributed Optimal Frequency Control Considering a Nonlinear Network-Preserving Model. In: *IEEE Transactions on Power Systems* 34 (2019), No. 1, p. 76–86

[WSE99] Weeren, Arie J. T. M.; Schumacher, Johannes M.; Engwerda, Jacob C.: Asymptotic Analysis of Linear Feedback Nash Equilibria in Nonzero-Sum Linear-Quadratic Differential Games. In: *Journal of Optimization Theory and Applications* 101 (1999), No. 3, p. 693–722

[WSI18] Wu, Xiangyu; Shen, Chen; Iravani, Reza: A Distributed, Cooperative Frequency and Voltage Control for Microgrids. In: *IEEE Transactions on Smart Grid* 9 (2018), No. 4, p. 2764–2776

[WWM16] Wei, Wei; Wang, Jianhui; Mei, Shengwei: Convexification of the Nash Bargaining Based Environmental-Economic Dispatch. In: *IEEE Transactions on Power Systems* 31 (2016), No. 6, p. 5208–5209

[WY13] Wu, Liang; Yang, Jin M.: Load frequency control of area power system with multi-source power generation units based on differential games tracking control. In: *IEEE PES Asia-Pacific Power and Energy Engineering Conference (APPEEC)*, IEEE, 2013, p. 1–6

[WYL21] Wen, Guanghui; Yu, Xinghuo; Liu, Zhiwei: Recent progress on the study of distributed economic dispatch in smart grid: an overview. In: *Frontiers of Information Technology & Electronic Engineering* 22 (2021), No. 1, p. 25–39

[XXM19] Xing, Xiaowen; Xie, Lili; Meng, Hongmin: Cooperative energy management optimization based on distributed MPC in grid-connected microgrids community. In: *International Journal of Electrical Power & Energy Systems* 107 (2019), No. 12, p. 186–199

[XZK+16] Xu, Qianyao; Zhang, Ning; Kang, Chongqing; Xia, Qing; He, Dawei; Liu, Chun; Huang, Yuehui; Cheng, Lu; Bai, Jianhua: A Game Theoretical Pricing Mechanism for Multi-Area Spinning Reserve Trading Considering Wind Power Uncertainty. In: *IEEE Transactions on Power Systems* 31 (2016), No. 2, p. 1084–1095

[XZN13] Xu, Jun; Zou, Yuanyuan; Niu, Yugang: Distributed Predictive Control for Energy Management of Multi-Microgrids Systems. In: *IFAC Proceedings Volumes* 46 (2013), No. 13, p. 551–556

[YH15] Yu, Mengmeng; Hong, Seung H.: A Real-Time Demand-Response Algorithm for Smart Grids: A Stackelberg Game Approach. In: *IEEE Transactions on Smart Grid* 7 (2015), No. 2, p. 879–888

[YLGW19] Yang, Benqiang; Liu, Shuli; Gaterell, Mark; Wang, Yang: Smart metering and systems for low-energy households: challenges, issues and benefits. In: *Advances in Building Energy Research* 13 (2019), No. 1, p. 80–100

[YTL12] Yu, Nanpeng; Tesfatsion, Leigh; Liu, Chen-Ching: Financial Bilateral Contract Negotiation in Wholesale Electricity Markets Using Nash Bargaining Theory. In: *IEEE Transactions on Power Systems* 27 (2012), No. 1, p. 251–267

[YTN13] Yang, Peng; Tang, Gongguo; Nehorai, Arye: A game-theoretic approach for optimal time-of-use electricity pricing. In: *IEEE Transactions on Power Systems* 28 (2013), No. 2, p. 884–892

[ZC06] Zhang, Yankui; Chen, Chen: A Novel Power Injection Model of IPFC for Power Flow Analysis Inclusive of Practical Constraints. In: *IEEE Transactions on Power Systems* 21 (2006), No. 4, p. 1550–1556

[ZC16] Zhou, Xinyang; Chen, Lijun: An incremental local algorithm for better voltage control in distribution networks. In: *IEEE Conference on Decision and Control (CDC)*, IEEE, 2016, p. 2396–2402

[ZCL13] Zhang, Huaguang; Cui, Lili; Luo, Yanhong: Near-Optimal Control for Nonzero-Sum Differential Games of Continuous-Time Nonlinear Systems Using Single-Network ADP. In: *IEEE Transactions on Cybernetics* 43 (2013), No. 1, p. 206–216

[ZDC20] Zhong, Yaofeng D.; Dey, Biswadip; Chakraborty, Amit: Dissipative SymODEN: Encoding Hamiltonian Dynamics with Dissipation and Control into Deep Learning. In: *ICLR Workshop on Integration of Deep Neural Models and Differential Equations (DeepDiffEq)*, 2020, p. 1–6

[ZHB12] Zhu, Quanyan; Han, Zhu; Başar, Tamer: A differential game approach to distributed demand side management in smart grid. In: *IEEE International Conference on Communications (ICC)*, IEEE, 2012, p. 3345–3350

[ZLP15] Zhang, Xuan; Li, Na; Papachristodoulou, Antonis: Achieving real-time economic dispatch in power networks via a saddle point design approach. In: *IEEE Power & Energy Society General Meeting*, IEEE, 2015, p. 1–5

[ZMLB18] Zhao, Changhong; Mallada, Enrique; Low, Steven H.; Bialek, Janusz: Distributed plug-and-play optimal generator and load control for power system frequency regulation. In: *International Journal of Electrical Power & Energy Systems* 101 (2018), p. 1–12

[ZMST11] Zimmerman, Ray D.; Murillo-Sanchez, Carlos E.; Thomas, Robert J.: MATPOWER: Steady-State Operations, Planning, and Analysis Tools for Power Systems Research and Education. In: *IEEE Transactions on Power Systems* 26 (2011), No. 1, p. 12–19

[ZOJS15] Zhang, Meng; Ortega, Romeo; Jeltsema, Dimitri; Su, Hongye: Further deleterious effects of the dissipation obstacle in control-by-interconnection of port-Hamiltonian systems. In: *Automatica* 61 (2015), p. 227–231

[ZP15] Zhang, Xuan; Papachristodoulou, Antonis: A real-time control framework for smart power networks: Design methodology and stability. In: *Automatica* 58 (2015), p. 43–50

[ZPE17] Zweifel, Peter; Praktiknjo, Aaron; Erdmann, Georg: *Energy Economics*. Berlin, Heidelberg: Springer, 2017

[ZPM13] Zugno, Marco; Pinson, Pierre; Madsen, Henrik: Impact of Wind Power Generation on European Cross-Border Power Flows. In: *IEEE Transactions on Power Systems* 28 (2013), No. 4, p. 3566–3575

[ZS10] Zamora, Ramon; Srivastava, Anurag K.: Controls for microgrids with storage: Review, challenges, and research needs. In: *Renewable and Sustainable Energy Reviews* 14 (2010), No. 7, p. 2009–2018

[ZTL12] Zhao, Changhong; Topcu, Ufuk; Low, Steven H.: Swing dynamics as primal-dual algorithm for optimal load control. In: *IEEE International Conference on Smart Grid Communications (SmartGridComm)*, IEEE, 2012, p. 570–575

[ZW10] ZHANG, Yue-Jun; WEI, Yi-Ming: An overview of current research on EU ETS: Evidence from its operating mechanism and economic effect. In: *Applied Energy* 87 (2010), No. 6, p. 1804–1814

Own Publications and Conference Proceedings

[EKH15] ECKERT, Marius; KÖLSCH, Lukas; HOHMANN, Sören: Fractional algebraic identification of the distribution of relaxation times of battery cells. In: *IEEE International Conference on Decision and Control (CDC)*. Osaka, Japan: IEEE, 2015, p. 2101–2108

[KBKH19] KÖLSCH, Lukas; BHATT, Kirtan; KREBS, Stefan; HOHMANN, Sören: Steady-State Optimal Frequency Control for Lossy Power Grids with Distributed Communication. In: *IEEE International Conference on Electrical, Control and Instrumentation Engineering (ICECIE)*. Kuala Lumpur, Malaysia: IEEE, 2019, p. 1–8

[KDB+20] KÖLSCH, Lukas; DUPUIS, Manuel; BHATT, Kirtan; KREBS, Stefan; HOHMANN, Sören: Distributed Frequency Regulation for Heterogeneous Microgrids via Steady State Optimal Control. In: *IEEE Green Technologies Conference (GreenTech)*. Oklahoma City, OK, USA: IEEE, 2020, p. 92–99

[KJSMH22] KÖLSCH, Lukas; JANÉ SONEIRA, Pol; MALAN, Albertus J.; HOHMANN, Sören: Learning Feedback Nash Strategies for Nonlinear Port-Hamiltonian Systems. In: *International Journal of Control* (2022), p. 1–14

[KJSSH21] KÖLSCH, Lukas; JANÉ SONEIRA, Pol; STREHLE, Felix; HOHMANN, Sören: Optimal Control of Port-Hamiltonian Systems: A Continuous-Time Learning Approach. In: *Automatica* 130 (2021), 109725

[KWKH20] KÖLSCH, Lukas; WIENINGER, Katharina; KREBS, Stefan; HOHMANN, Sören: Distributed Frequency and Voltage Control for AC Microgrids based on Primal-Dual Gradient Dynamics. In: *IFAC-PapersOnLine* 53 (2020), No. 2, p. 12229–12236

[KZV+22] KÖLSCH, Lukas; ZELLMANN, Lena; VYAS, Rishabh; PFEIFER, Martin; HOHMANN, Sören: Optimal Distributed Frequency and Voltage Control for Zonal Electricity Markets. In: *IEEE Transactions on Power Systems* 37 (2022), No. 4, p. 2666–2678

[PKH21] PFEIFER, Martin; KÖLSCH, Lukas; HOHMANN, Sören: Automated Design of An Input-State-Output Observer for Linear Port-Hamiltonian Systems with Application to the State Estimation in Unbalanced Distribution Systems. In: *SIAM Conference on Computational Science and Engineering*. Houston, TX, USA: SIAM, 2021

[SPK+18] STREHLE, Felix; PFEIFER, Martin; KÖLSCH, Lukas; DEGÜNTHER, Charlotte; RUF, Johannes; ANDRESEN, Lisa; HOHMANN, Sören: Towards Port-Hamiltonian Modeling of Multi-Carrier Energy Systems: A Case Study for a Coupled Electricity and Gas Distribution System. In: *IFAC-PapersOnLine* 51 (2018), No. 2, p. 463–468

Supervised Theses

[Bha19] Bhatt, Kirtan: *Implementierung einer verteilten optimierungsbasierten Frequenzregelung für heterogene Energieverteilnetze*, Karlsruhe Institute of Technology (KIT), Bachelor's Thesis, 2019

[Dup18] Dupuis, Manuel: *Regelungstechnische Analyse und Modellierung von Systemdienstleistungen für sektorgekoppelte Energiesysteme*, Karlsruhe Institute of Technology (KIT), Bachelor's Thesis, 2018

[JS19] Jané Soneira, Pol: *Optimal Control and Differential Games for Port-Hamiltonian Systems*, Karlsruhe Institute of Technology (KIT), Master's Thesis, 2019

[Ley18] Leyser, Markus: *Spieltheoretische Modellierung von zellularisierten Energienetzen*, Karlsruhe Institute of Technology (KIT), Bachelor's Thesis, 2018

[Nur18] Nursalim, Limanan: *Completely Positive Optimization for Multi-Carrier Energy Distribution Grids*, Karlsruhe Institute of Technology (KIT), Bachelor's Thesis, 2018

[Pfl18] Pflüger, Julian: *Structured Port-Hamiltonian Modeling and Observer Synthesis for Linear Dynamic Systems with Application to Electrical Distribution Networks*, Karlsruhe Institute of Technology (KIT), Master's Thesis, 2018

[Rau21] Rausche, Lukas M.: *Distributed Optimization and Control of Power Systems using Zonal Pricing*, Karlsruhe Institute of Technology (KIT), Master's Thesis, 2021

[Sch18] Schmoll, Markus: *Passivitätsbasierte Regelung für Energienetze auf Basis Port-Hamiltonscher Systeme*, Karlsruhe Institute of Technology (KIT), Master's Thesis, 2018

[Str17] Strehle, Felix: *Modeling of Multi-Carrier Energy Distribution Systems with Port-Hamiltonian Systems*, Karlsruhe Institute of Technology (KIT), Master's Thesis, 2017

[Vie19] Vieth, Jonathan: *Entwurf eines modellprädiktiven Reglers für das sektorengekoppelte Verteilnetz der Stadtwerke Kiel*, Karlsruhe Institute of Technology (KIT), Bachelor's Thesis, 2019

[Vya20] Vyas, Rishabh: *Implementation and Comparative Study of Frequency and Voltage Controllers for AC Microgrids*, Karlsruhe Institute of Technology (KIT), Bachelor's Thesis, 2020

[Wie19] Wieninger, Katharina: *Optimierungsbasierte Frequenz- und Spannungsregelung in zellularen Energienetzen*, Karlsruhe Institute of Technology (KIT), Master's Thesis, 2019

[Yu20] Yu, Tianqi: *Simulation and Validation of Market-based Controllers for Cellular Energy Networks*, Karlsruhe Institute of Technology (KIT), Master's Thesis, 2020

[Zel20] Zellmann, Lena: *Simulation von optimierungsbasierten Regelungsstrategien für zellulare Energienetze*, Karlsruhe Institute of Technology (KIT), Bachelor's Thesis, 2020

Karlsruher Beiträge zur Regelungs- und Steuerungstechnik (ISSN 2511-6312)
Institut für Regelungs- und Steuerungssysteme

Band 01 Diehm, Gunter
Identifikation des menschlichen Bewegungsverhaltens auf der Basis von Primitiven.
ISBN 978-3-7315-0608-9

Band 02 Flad, Michael
Kooperative Regelungskonzepte auf Basis der Spieltheorie und deren Anwendung auf Fahrerassistenzsysteme.
ISBN 978-3-7315-0610-2

Band 03 Eckert, Marius
Modellbasierte Identifikation fraktionaler Systeme und ihre Anwendung auf die Lithium-Ionen-Zelle.
ISBN 978-3-7315-0690-4

Band 04 Krebs, Stefan
Intervallbeobachter für lineare parametervariante Systeme und deren Anwendung auf die Asynchronmaschine.
ISBN 978-3-7315-0857-1

Band 05 Kaspar, Stephan
Fahrdynamikuntersuchungen eines Elektrofahrzeugs mit Einzelrad-Hinterradantrieb.
ISBN 978-3-7315-0916-5

Band 06 Sauter, Patrick S.
Modellierung und zentrale prädiktive Regelung von multimodalen Energieverteilnetzen.
ISBN 978-3-7315-0963-9

Band 07 Kupper, Martin
Verteilte Zustandsschätzung fraktionaler Systeme und ihre Anwendung auf Lithium-Ionen-Batteriesysteme.
ISBN 978-3-7315-0971-4

Band 08 Merkert, Lennart
Optimal Scheduling of Combined Heat and Power Generation Considering Heating Grid Dynamics.
ISBN 978-3-7315-1056-7

Die Bände sind unter www.ksp.kit.edu als PDF frei verfügbar oder als Druckausgabe bestellbar.

Band 09 Ludwig, Julian
Automatisierte kooperative Transition einer Regelungsaufgabe zwischen Mensch und Maschine am Beispiel des hochautomatisierten Fahrens.
ISBN 978-3-7315-1069-7

Band 10 Inga Charaja, Juan Jairo
Inverse Dynamic Game Methods for Identification of Cooperative System Behavior.
ISBN 978-3-7315-1080-2

Band 11 Schnurr, Christoph Xaver
Ein Verfahren zur lexikographischen modellprädiktiven Regelung mit der Anwendung auf eine permanenterregte Synchronmaschine.
ISBN 978-3-7315-1095-6

Band 12 Schwab, Stefan
Guaranteed Verification of Dynamic Systems.
ISBN 978-3-7315-0965-3

Band 13 Pfeifer, Martin
Automated Model Generation and Observer Design for Interconnected Systems: A Port-Hamiltonian Approach.
ISBN 978-3-7315-1135-9

Band 14 König, Alexander
Absicherung hochautomatisierten Fahrens durch passiven virtuellen Dauerlauftest.
ISBN 978-3-7315-1141-0

Band 15 Stark, Oliver
Parameter- und Ordnungsidentifikation von fraktionalen Systemen mit einer Anwendung auf eine Lithium-Ionen-Batteriezelle.
ISBN 978-3-7315-1187-8

Band 16 Köpf, Florian
Adaptive Dynamic Programming: Solltrajektorienfolgeregelung und Konvergenzbedingungen.
ISBN 978-3-7315-1193-9

Band 17 Kölsch, Lukas
Dynamic Incentives for Optimal Control of Competitive Power Systems.
ISBN 978-3-7315-1209-7
